The Cl

Why, despite all we know about the cau[illegible] little effective action been taken to cut greenhouse gas emissions, and what can we do to change that? This book explains the mechanisms and impacts of the climate crisis, traces the history and reasons behind the lack of serious effort to combat it, describes some people's ongoing skepticism and how to shift it, and motivates an urgent program of action. It argues that the pathway to stopping dangerous global heating will require a much larger mobilization of advocacy and activism to impel decision-makers to abandon fossil fuels and transition to renewable energy and electrification embedded in a political and social framework guided by justice principles. It is an excellent resource for students and researchers on the climate crisis, the need for a renewable energy transition, and the current blocks to progress.

ADAM R. ARON is a climate activist and professor of psychology at the University of California, San Diego. His research and teaching focus on the social science of collective action on the climate crisis.

'The Climate Crisis is exceptional for covering the wide range of complex issues involved in climate change. Aron's description of the physical climate system is one of the clearest I've read. He also informs us about impacts, and about the political, economic and psychological obstacles to addressing the crisis. In considering solutions, he draws on both the research literature and the recent efforts of social movements to create change. The Climate Crisis lays the groundwork for thinking carefully about this crucial problem and for taking action.'

Professor Thomas Dietz, Michigan State University; author of Decisions for Sustainability: Facts and Values

'The Climate Crisis does a great job of connecting the climate science ... with the evolving scholarship on how we approach - and could change our approach - to systemic risks and dread of climate change, which is arguably the greatest challenge we face as a species.'

Professor Daniel Kammen, University of California, Berkeley; and Former Science Envoy, Department of State in the Obama-Biden Administration

'This is a tremendous book on the climate crisis, what it consists of, and why we are finding it so difficult to tackle. The insights on social and psychological dynamics are designed to make it much more than just information: this is a how-to manual on how you can become an effective climate activist and advocate. A perfect book for our difficult times. I have already recommended it to everyone I've met since reading it.'

Professor Julia K. Steinberger, University of Lausanne

'This book is a tour de force for anyone interested in understanding climate change and how to overcome barriers to action. With unparalleled clarity, Aron explains the history of the climate change debate, the complex psychology of why we have failed to act and what cutting-edge social and behavioral science research has to say about how to engender the large-scale societal change needed to manage the most existential crisis of our time. Aron's new book offers essential reading for anyone interested in the future of our planet.'

Professor Sander van der Linden, University of Cambridge and former Editor-in-Chief of the Journal of Environmental Psychology.

The Climate Crisis

Science, Impacts, Policy, Psychology, Justice, Social Movements

ADAM R. ARON
University of California, San Diego

Shaftesbury Road, Cambridge CB2 8EA, United Kingdom

One Liberty Plaza, 20th Floor, New York, NY 10006, USA

477 Williamstown Road, Port Melbourne, VIC 3207, Australia

314–321, 3rd Floor, Plot 3, Splendor Forum, Jasola District Centre, New Delhi – 110025, India

103 Penang Road, #05–06/07, Visioncrest Commercial, Singapore 238467

Cambridge University Press is part of Cambridge University Press & Assessment, a department of the University of Cambridge.

We share the University's mission to contribute to society through the pursuit of education, learning and research at the highest international levels of excellence.

www.cambridge.org
Information on this title: www.cambridge.org/9781108833806

DOI: 10.1017/9781108982566

First published 2023

A catalogue record for this publication is available from the British Library.

Library of Congress Cataloging-in-Publication Data
Names: Aron, Adam, 1972– author.
Title: The climate crisis : science, impacts, policy, psychology, sociology, and justice / Adam Aron, University of California, San Diego.
Description: New York, NY : Cambridge University Press, 2023. | Includes bibliographical references and index.
Identifiers: LCCN 2022031380 (print) | LCCN 2022031381 (ebook) | ISBN 9781108833806 (hardback) | ISBN 9781108987158 (paperback) | ISBN 9781108982566 (epub)
Subjects: LCSH: Climatic changes–History. | Climate change mitigation. | Climatic changes–Government policy. | Environmentalism.
Classification: LCC QC903 .A768 2023 (print) | LCC QC903 (ebook) | DDC 333.72–dc23/eng/20211119
LC record available at https://lccn.loc.gov/2022031380
LC ebook record available at https://lccn.loc.gov/2022031381

ISBN 978-1-108-83380-6 Hardback
ISBN 978-1-108-98715-8 Paperback

To Nicole, Luca, Leo, and Hannah.

Contents

Glossary — *page* ix
Acknowledgments — xvii

Introduction — 1

PART I BACKGROUND: HISTORY, SCIENCE, AND CAPITALISM

1 **The History of Human-Caused Global Heating** — 7
2 **Climate Science** — 33
3 **Climate Impacts** — 61
4 **Capitalism and the Climate Crisis** — 84

PART II FROM SKEPTICISM TO BELIEF

5 **Skepticism, Misinformation, and Motivated Cognition** — 111
6 **Science Communication: Countering Skepticism and Delivering Information Clearly** — 140
7 **Elevating Risk Perceptions about Global Heating** — 160

PART III FROM BELIEF TO ACTION

8 **Principles for Just and Effective Action** — 189
9 **A Technical and Social Framework to Guide Climate Action** — 221

10 Building and Taking Collective Action 253

Conclusion 289

Notes 294
Bibliography 332
Index 350

Glossary

Absolute advantage: for instance, a country produces a greater quantity of goods or services with the same quantity of inputs as another country per unit time, and it can do this for many or all goods. (Compare to comparative advantage.)

Additionality requirement: requires showing that emissions reductions would not have occurred without offsets payments, which is notoriously hard to establish (see Clean Development Mechanism / carbon offsets).

Amplifier of climate change: also known as a positive feedback – for instance, increased global heating reduces the ice-albedo effect (reflectance of sunlight by ice and snow), which leads to more absorption of heat in the oceans which increases global heating.

Astroturfing: an approach which masks the actual sponsors of a message to make it seem as if it is coming from grassroots organizations, when in fact there is no grassroots at all.

Attitude roots: the underlying fears, ideologies, worldviews, and identity needs that sustain and motivate surface attitudes (see surface attitudes).

Bioenergy, biofuel, biomass: bioenergy, of which biofuel is one type, is derived from biomass, which refers to any material derived from living or recently living material.

Bioenergy with carbon capture and sequestration (BECCS): biomass is grown and then burned to generate electricity, and the resulting CO_2 is captured and permanently sequestered underground.

Biospheric values: viewing one's and others' actions in terms of the advantages and disadvantages they pose to nature.

Bounded uncertainty: where uncertainty is expressed as a point estimate with a numeric range around it.

Cap-and-trade: the attempt to cap the emissions of corporations by allowing them to trade in pollution credits.

Carbon budget: the maximum amount of cumulative anthropogenic CO_2 emissions that would result in limiting global warming to a specific level with a specific probability, while also accounting for other anthropogenic climate forcers.

Carbon colonialism: when wealthy countries export or outsource their climate and energy crisis to low- and middle-income countries deliberately or otherwise.

Carbon cycle: in which CO_2 is absorbed into the oceans, taken up into plants through photosynthesis, and then released again through the respiration of animals and the decaying of dead plant and animal matter.

Carbon fee and dividend: when a carbon price requires a measure that rebates some portion of that price to affected individuals, such as a credit for low-income citizens.

Carbon leakage: when attempts to cut emissions in one place results in carbon-intensive production leaking out somewhere else.

Carbon neutrality: typically a plan to maintain business-as-usual emissions while attempting to "neutralize" those with carbon offsets.

Carbon sink: the absorption of atmospheric CO_2 by land, plants, and oceans.

Clean Development Mechanism (carbon offsets): allows industrialized countries to invest in putative emissions reduction projects, typically in the developing world, and count them toward their own targets.

Climate fatalism (also defeatism): a form of strong skepticism that leads people to feel that taking action against the climate threat is futile.

Climate forcing: a perturbation of the earth's energy balance, which is measured in watts per square meter (W/m^2) and can be either natural or anthropogenic and either positive or negative.

Climate justice: addresses the ethical aspects of climate change and has many manifestations, including in trade, international agreements, and the application of technical approaches.

Climate sensitivity: how sensitive the climate is to increases in CO_2, typically expressed as the amount of heating corresponding to a doubling of CO_2.

Co-benefits: a communications approach to try to gain support for policies, such as those related to climate change, without directly challenging people's underlying attitudes and beliefs, for example by talking about more jobs.

Cognitive biases: how a set of inbuilt mental shortcuts sometimes distorts our thinking.

Collective efficacy: an individual group member's belief about what the group can do (compare to self-efficacy).

Collective identity: one's sense of membership of a social group.

Committed warming: future warming from greenhouse gases that have already been emitted.

Common but differentiated responsibilities: acknowledges that individual countries have different capabilities for combating climate change; recognizes that, for instance, rich countries should be able to cut emissions sooner with less dire consequences for their populace than less developed countries.

Community-based organizing: a slow approach to building an organization through creating structures, often framed around self-interest more than ideology (contrast with mass mobilization).

Comparative advantage: for instance, a country produces a greater quantity of goods of one kind than another country, but not of another kind: comparative advantage in one good implies a comparative disadvantage in another (compare to absolute advantage, meaning the ability to produce more of all or most goods).

Confirmation bias: the tendency to cherry-pick available evidence to support one's existing knowledge, attitudes, and beliefs.

Consensus gap: the difference between the actual scientific consensus and people's belief in it.

Conspiracy thinking: the belief that powerful vested interests are capable of conspiring with one another to withhold scientific breakthroughs, fabricate events, or withhold evidence.

Culture of uncare: individuals and groups protect entitlement and privilege, and refuse to accept the moral implications of their own self-focus.

Debunk: countering specific misinformation by pulling apart its mistaken premises and conclusions.

Decouple: when the increase of environmental impacts of emissions is smaller than the growth of gross domestic product in a given period (see gross domestic product).

Degrowth: see post-growth

Derisking: instead of divesting from a fund (such as fossil stocks and bonds), which is a moral position, derisking leads to the, perhaps temporary, sale of the funds on the basis that they are risky.

Direct air capture of CO_2: electric-powered devices that filter the gas out of the air so that it can be permanently sequestered underground.

Earth system models: run on supercomputers that represent the weather as a grid of three-dimensional cubes, each representing one small part of the planet

Eco-anxiety: a chronic fear of environmental doom.

Embedded liberalism: in which market processes and corporate activities were embedded in a web of social and political constraints in the form of strong regulations, including laws that limited pollution (corresponds to the immediate post-World War II period in the USA).

Embodied emissions: the emissions arising from goods, services, and commodities that are imported and then consumed by industry, commerce, the household, and the individual.

Emissions scenarios: see representative concentration pathways (RCPs) or shared socioeconomic pathways.

Environmental justice: working to include those individuals who are disproportionately impacted by pollution in the decision-making process, and working to restore their rights.

Epistemic skepticism: skepticism about knowledge that global heating is happening, is human-caused, and is having and will have bad impacts.

Exchange value: the part of the value of an object that relates to inputs such as surplus labor time (compare with use value).

Externalized costs: costs that are generated by producers but carried by society as a whole, for example when a business treats the atmosphere as a dumping ground for the CO_2 that is emitted when making products without paying the cost.

Extractivism: the high-intensity, export-oriented extraction of common ecological goods rooted in colonialism, and which creates large economic profits for the powerful few in the short term, but minimal benefits for the communities from whence the resources are taken.

Extreme event attribution: the science of how extreme events are assigned to climate change.

Faith in institutions: where one fits on a spectrum ranging from cynicism about institutional or political change to overreliance on the system to make change.

False equivalence: where the broadcast media present, for instance, climate change stories as a disagreement between two "equally" valid points of view, when in fact solid science is on one side and opinion on the other.

Frames: the shared meanings and cultural understandings that bind together people in a movement and create resonances among larger parts of the public.

Front groups: organizations that purport to represent one agenda while serving another, for example an organization with an innocuous or patriotic-sounding name that is related to the fossil fuel industry.

Gateway Belief Model: the idea that providing information about the consensus among climate scientists not only increases belief in the level of consensus, but increases beliefs in global heating, human causation, worry about global heating, and support for action.

Global Heating 1-2-3: (1) Fact: CO_2 makes planets hotter than they would otherwise be. (2) Fact: Human activity, especially the burning of fossil fuels, releases CO_2, and adds more of this heat-trapping gas to the atmosphere. (3) Logical conclusion: We must therefore expect that rising CO_2 levels will heat our planet.

Global South: refers broadly to the mostly low-income and often politically marginalized regions of Latin America, Asia, Africa, and Oceania that are also sometimes referred to as the Third World or Periphery to denote regions outside of Europe, North America, and Japan

Global warming potential: the heat absorbed by a greenhouse gas, as a multiple of the heat that would be absorbed by the same mass of CO_2.

Goldilocks zone: the zone that Earth is in – neither too hot nor too cold to support human life as we know it.

Great Acceleration: occurred after World War II as the result of a dramatic, continuous, and roughly simultaneous surge in growth across a large range of measures of human activity, such as gross domestic product, transportation, fossil fuel burning, and the use of industrial fertilizers.

Green hydrogen: using renewable electricity to create hydrogen by electrolysis.

Greenhouse gas effect: energy from the Sun goes through the atmosphere and warms the planet's surface and the atmosphere prevents some of the heat from returning directly to space, resulting in a warmer planet.

Greenwashing: in which business-as-usual practices are represented as forms of climate action.

Gross domestic product: the 'value added' creation by goods and services in a particular country over a specific period of time.

Growth imperative: same thing as the expansionary compulsion, in which businesses must constantly expand and grow in order to compete in the marketplace and meet the expectations of their owners and investors.

Homo economicus: according to Adam Smith, a model of human behavior in which a person has infinite capacity to make rational decisions; according to Kate Raworth, a portrait of humans that has four features: solitary, calculating, competitive, and insatiable (compare to Homo socialis).

Homo socialis: a different model of human behavior, as being trusting, reciprocating, and cooperative (compare to Homo economicus).

Hyperbolic discounting: people's tendency to have a stronger preference for immediate payoffs than for later payoffs even when it will ultimately benefit them less.

Ice-albedo effect: how ice and snow reflect solar radiation.

In-group messaging: a communication strategy that takes advantage of social identity links by having members of a political group talk to others in that group about a topic such as climate change.

Irreducible uncertainty: uncertainty is expressed verbally (not with numeric point estimate and range, see bounded uncertainty).

Jiu jitsu persuasion: a model of persuasion that emphasizes creating change by aligning with (rather than competing with) someone's attitude roots (see attitude roots).

Just transition: the policies that wind down industries that harm the environment must also find ways to safeguard livelihoods.

Life-cycle assessment: examines the multiple types of environmental impacts of a product over its entire lifetime, such as the emissions generated through manufacture and transport, the emissions saved through operation (e.g., of a wind turbine), and the emissions associated with decommissioning as well as saved through recycling.

Mass mobilization: utilizes the power of disruption by quickly drawing together a lot of people and leaving the establishment scrambling (contrast with community building).

Master frames: frames that are shared across social movements.

Mechanistic explanation: explaining a phenomenon such as the greenhouse gas effect in terms of its constituent causal parts.

Meta-analysis: a statistical analysis that combines the results of many individual studies.

Milankovitch cycles: describes the overall effect of changes in the Earth's movement on its climate over thousands of years. These cycles are driven by changes in the Earth's position relative to the Sun, for example the amount of eccentricity, that affect the absorption of sunlight.

Momentum-based organizing: attempts to combine both the short-term explosive potential of mass mobilization (to bring in more people and gain attention) with the benefits of a strong leadership and administrative structure to keep those participants engaged (community building).

Motivated cognition: when our thinking is steered towards particular conclusions by our goals and needs.

Nationally Determined Contribution: the theoretical emissions reductions a country commits to under the Paris Accord of the UN process.

Negative emissions: the opposite of greenhouse emissions, this requires removing CO_2 from the atmosphere and sequestering it underground (for example, see direct air capture of CO_2 and bioenergy with carbon capture and sequestration, BECCS).

Neoliberalism: a new version of classical liberalism's free-market perspective, which prioritizes market principles over social and political concerns and works against the earlier state interventionist approaches of embedded liberalism.

Net zero: the idea that the emissions one generates can be balanced by negative emissions, such as pulling CO_2 out of the atmosphere using geoengineering approaches.

Participatory efficacy: a group member turns up because they believe that their participation is essential to the group's effectiveness.

Permanence: (when referring to carbon offsets) – for instance, that carbon captured in forests through offsets payments will stay in those forests.

Personality: the relatively enduring patterns of feelings and behaviors that distinguish people from one another.

Planned obsolescence: one way to achieve artificial scarcity – build devices so they break early and have to replaced.

Political economy: rather than a narrow focus on economics, this field considers economics embedded in wider political and social contexts.

Post-growth: a movement that argues climate change should prompt a rethink about economic growth, and proposes a set of practices to encourage human thriving while reducing consumption.

Prebunking: when science communication is used to help counter skepticism regarding climate change by correcting specific misinformation.

R-value: a measure of the strength of a correlation between two variables (a value of 1 is a perfect correlation, and 0 is no correlation at all).

Radical hope: hope this is directed toward a future goodness that transcends the current ability to understand what it is.

Rebound effects: when increases in, for example, energy efficiency lead to economic savings which are then used for more consumption which increases emissions.

Remunicipalization: changing private to public ownership of assets or companies, such as bringing a privatized electric utility under public city control.

Representative concentration pathways (RCPs): these express a different degree of climate forcing by the year 2100 as measured by W/m^2: 2.6, 4.5, 6, and 8.5. Each of these scenarios calculates the likely atmospheric concentration of CO_2 by the end of this century by inserting the emissions levels into a carbon cycle model that estimates how much CO_2 will be absorbed by the oceans and land reservoirs and how much will stay in the atmosphere

Response skepticism: refers to doubt about whether and how one responds to climate change at a personal, group, national, and international level.

Self-efficacy: an individual's belief that they can accomplish whatever they want to; that they can use their skills to perform a behavior that will lead to a desired outcome.

Shared socioeconomic pathways: five different ways that climate change might evolve expressed in terms of societal, demographic, and economic changes over the next century (contrast to RCPs).

Shifting baselines syndrome: when a gradual change in how people accept a changing situation makes them think it's normal.

Social identity needs: the human tendency to draw meaning and self-definition from belonging to social groups.

Social norms: the shared standards that guide the behavior of members within a group.

Solastalgia: the pain that occurs from the loss of a comforting place when people's connection to the land is disrupted in the face of large-scale mining or other destruction.

Stagflation: when countries experience a surge in both unemployment and inflation

Steady-state economy: an economy of a stable or slightly changing size.

Stock-and-flow problem: (analogy to explain levels of CO_2 in the atmosphere) – the level of water in a bathtub will continue to rise as long as the flow from the faucet exceeds the flow out from the drain and will go down only when the flow from the faucet decreases enough for the drain to remove more water than is flowing in.

Stranded assets: assets that have suffered from unanticipated or premature write-downs, or lost value, such as oil and gas corporations being forced to cease extraction.

Sunk cost fallacy: continuing on a current trajectory, such as expanding the buildings at an institution, despite evidence suggesting a course correction is necessary simply because so much has already been invested in the status quo.

Surface attitude: the manifestations of skepticism that one hears from someone, to be distinguished from the underlying attitude roots (see attitude roots).

Theory of change: (of local action) is an idea about how engaging in local action undergirds wider national-level policy shifts.

Tipping point: the point at which small changes arising from global heating might become strong enough to cause large changes that could be abrupt and irreversible and lead to cascading effects.

Transactional struggle: working toward concrete legislative and legal victories.

Transformational struggle: shifting hearts and minds, often as a prelude to the transactional struggle.

Use value: refers to the tangible aspects of a tradeable object that can satisfy some human requirement or need or which is useful.

Vested interests: one's personal stake in an undertaking or state of affairs biases one to change one's beliefs to be more convenient aligned with that interest.

Wet-bulb temperature: the temperature read by a thermometer covered in a water-soaked cloth over which air is then passed – if relative humidity is 100 percent then the wet-bulb temperature equals the air temperature: at lower humidity the wet-bulb temperature will be lower than the air temperature because of the cooling from evaporation.

Worshiping at the Church of Technology: how the promise of technofixes appeals strongly to some members of the wealthy elite, and decision-making class, since it seems to promise a climate "solution" without entailing any change to their lifestyles or wider socio-economic-political structures.

Acknowledgments

Huge thanks to my brilliant editor, Jeanne Barker-Nunn, who taught me how to write a book instead of a science paper; to my brilliant illustrator, Milena Gavala, for her beautiful figures and charts; and to my wife, Nicole Lanouette, for her huge support for this project and for co-parenting.

Thanks to Russ Poldrack for encouraging the project from the start; to Matt Lloyd of Cambridge University Press for getting it off the ground; to Dan Harding for copy-editing; to Karl Gerth for his cheerleading and tips about the writing and publishing process; and to Brian Tokar, a long-standing climate justice advocate, who was a huge resource for discussion.

Thanks to Patrick Bond for connecting me with African climate activists, and to all who did interviews with me for the chapters: Dipti Bhatnagar, Brian Tokar, Masada Disenhouse, Mark Jacobson, and Kelly Fielding.

Thanks to all who gave substantial feedback on specific chapters or parts of chapters: Kelly Fielding, Brian Tokar, John Kerr, Sebastian Holt, Cameron Brick, Anna Castiglione, Larry Edwards, Karl Gerth, David Badre, Aaron Thierry, David Romps, Leo Kleiman-Lynch, and Wolfgang Knorr.

Thanks to all my comrades – students, staff, and faculty – in the Green New Deal at UC San Diego climate action and justice movement: together we engaged in collective action, for free, willingly, and will keep doing it. Thanks also to the hundreds of undergraduate students of PSYC185, Psychology of the Climate Crisis, going back to 2019, for helping me to develop the material: good luck and keep your courage.

Royalties to the author for the first two years will be evenly split between donations to international climate justice organizations and to research groups on collective action.

Introduction

Increasing numbers of people have become aware of the problem of global warming (referred to in this book as global heating) as extreme weather events and sea levels have continued to increase, private investment in renewable sources of power and in electric vehicles (EVs) and appliances has grown, and institutions and governments around the world have pledged to take action to reduce their greenhouse gas emissions. Yet despite the overwhelming consensus among scientists and intergovernmental bodies about the danger that global heating poses to the environment and to organized life as we know it, more than thirty years of talk about the urgent need for action has led to very few tangible results. Global emissions continue to rise, and few people familiar with the situation feel confident that the world will manage to meet the internationally agreed goal of limiting global heating to 1.5°C, or even to 2°C. Why this is the case – how it is that, despite all that we know about the causes and harms of global heating, so little effective action has yet been taken, and how that can be changed – is the central question of this book.

This book is a result of my own journey to answer that question. I first came to this topic not as a climate scientist or political scientist but as a cognitive neuroscientist whose primary research was concerned with uncovering the mechanisms of human cognition. Although I first became concerned about the problem of global heating in the early 2000s, the demands of my career and of parenting got in the way of my actually doing very much about it for a long time. My first direct involvement with the issue came in 2015 when I chose to join the fossil fuel divestment movement at the University of California, the institution for which I worked. My increasing puzzlement and frustration with the seemingly low level of concern and action that I encountered as part of that work led me in 2018 to ask my department for permission to teach a course that has since come to be called the Psychology of the Climate Crisis. Only as I began to prepare for that course and pored over the actual data published by the UN-led Intergovernmental Panel on Climate Change (IPCC) did I begin to fully understand the depth of the threat we faced and the breadth of issues relating to my question. Among these new discoveries was that many of the predicted emissions pathways set out by the IPCC are misleading since they assume that huge amounts of carbon will pulled out of the atmosphere decades in the future, thereby encouraging decision makers to keep on emitting now and hope for somehow capturing that carbon later. Another was that both the historical responsibility for global heating and its effects are unevenly distributed around the world. This reality had particular resonance for me as I realized that the plants, animals, and human population of the fairly rural part of southern Africa in which I had grown up were experiencing roughly double the level of global heating despite having done almost nothing to generate those emissions and having hardly any resources for adapting to these changing conditions. As I shifted more of my research and teaching time to the climate crisis and continued to engage in collective action with others to push my university to switch away from burning large amounts of fossil fuels and using large banks that are financing fossil fuel extraction, I also began to learn and think more about the history and social science of collective action itself. This book therefore represents the sum of my own reading, teaching, research, and experiences as an activist and organizer – and part of my attempt to have a good answer when, sometime in the future, my children ask me what I was doing during this critical time in the history of the human race.

Unlike many of the existing books on the problem of global heating, which typically focus on only one major aspect of the issue, this book is intended to provide readers with a broad view of the topic to help them understand how we have arrived at this point and formulate their own personal and collective course of action in response to it. To generate the kind of major social mobilization that will be required to produce effective action, this book argues that more people must understand the causes and history of the problem of global heating, overcome skepticism about the threat it poses and the effectiveness of possible action, and convert that belief into actions that will hasten technical, political, and social responses that are also just.

Part I of the book addresses the first of these dimensions: understanding the problem of global heating and the obstacles to responding to it. Chapter 1 first offers a historical examination of the causes and discovery of global heating and the development of the scientific consensus that it is human-caused and can be curtailed only by cutting greenhouse gas emissions, and it describes how those developments nonetheless failed to lead to extensive action. Chapter 2 next explains the mechanism by which human activities have produced increasing levels of gases that have accelerated what has come to be called the greenhouse effect and have led to unprecedented heating in Earth's climate. Chapter 3 then lays out in more detail the grave current impacts of global heating, including such extreme weather events as heat waves, droughts, floods, and hurricanes; the alarming projected scenarios if substantial cuts in emissions do not begin soon; and how the methods and assumptions used by many economists have contributed to the failure to cut emissions. Chapter 4 completes this portion of the book by examining the role played by contemporary capitalism in both massively escalating emissions and creating structural, ideological, and psychological barriers to efforts to cut them.

Part II of the book addresses how to change climate change skepticism into belief so as to engender a higher level of support for curbing emissions than currently exists in most countries. Chapter 5 introduces the major causes of skepticism, including misinformation and worldviews and values, and offers some possible strategies for countering these influences. Chapter 6 next discusses ways in which scientists, activists, journalists, policy makers, and others concerned about global heating can most persuasively communicate climate science findings to the general public. Chapter 7 then explores ways in which people's perceptions of the actual risks posed by climate change can be elevated adequately to motivate them to engage in individual and collective action to counteract it.

Part III of the book then examines how to move believers to take actions that will be effective in combating global heating. Chapter 8 first argues that any such steps or program can be effective only if they adopt a justice lens and reject proposed technical and market fixes that threaten to perpetuate the same inequities, corporate agendas, and extractivist mentality that created the climate and ecological crisis in the first place. Chapter 9 then addresses the kinds of steps that must be taken to make such a renewable energy transition feasible from a technical, political, and social point of view and produce meaningful action and confrontation across many levels of government and realms of society. Chapter 10 lastly turns to the sociology of social movements, the psychology of collective action, and the histories and tactics of prominent grassroots groups, to examine how to grow and empower such a movement and increase broader advocacy.

PART I

Background

History, Science, and Capitalism

1

The History of Human-Caused Global Heating

> Whether one chooses to ignore, suppress, deny or agonize over the knowledge of what is happening, it is there, in the air, heavier by the year. And yet the descendants of the Lancashire manufacturers, whose dominion now spans the globe, are taking decisions on a daily basis to invest in new oil wells, new coal-fired power plants, new airports, new highways, new liquefied natural gas facilities, new machines to replace human workers, so that emissions are not only continuing to grow but doing so at a higher speed.
>
> Andreas Malm, *Fossil Capital*

In 1988, the prominent climate scientist James Hansen, director of NASA's Goddard Institute for Space Studies, testified before the US Congress that human emissions of greenhouse gases were heating our planet to dangerous levels. His warning, however, was ignored. Since his testimony, more than 50 percent of all greenhouse gases in human history have been emitted, and almost every biosphere and earth system indicator is blinking red.[1] Time is now running out to keep global heating from reaching levels that would be catastrophic for millions of species and for organized human existence as we know it.

Our current fossil fuel economy and industrial-scale agricultural practices are releasing vast amounts of carbon dioxide (CO_2) and methane (CH_4) into the atmosphere. These gases are called greenhouse gases because they lead to global heat accumulation via the accelerated **greenhouse gas effect**, as shown in Figure 1.1. The resulting global heat accumulation is equivalent to the addition of about four Hiroshima bombs of energy every second of every minute of every hour of every day, day in and day out, for decades.[2] This has already led to about 1.1°C (2°F) of heating since preindustrial times, which has resulted in a roughly 400 percent increase in such extreme weather events as hurricanes, floods, droughts, and heat waves since just the 1970s.[3] According

Figure 1.1

The greenhouse gas effect

Greenhouse gases slow the escape of infrared light into space. This traps heat in the lower atmosphere, making it, and the surface, warmer than it would otherwise be. The best analogy is a blanket. You stay warm when wrapped by a blanket because it slows the escape of your body heat.

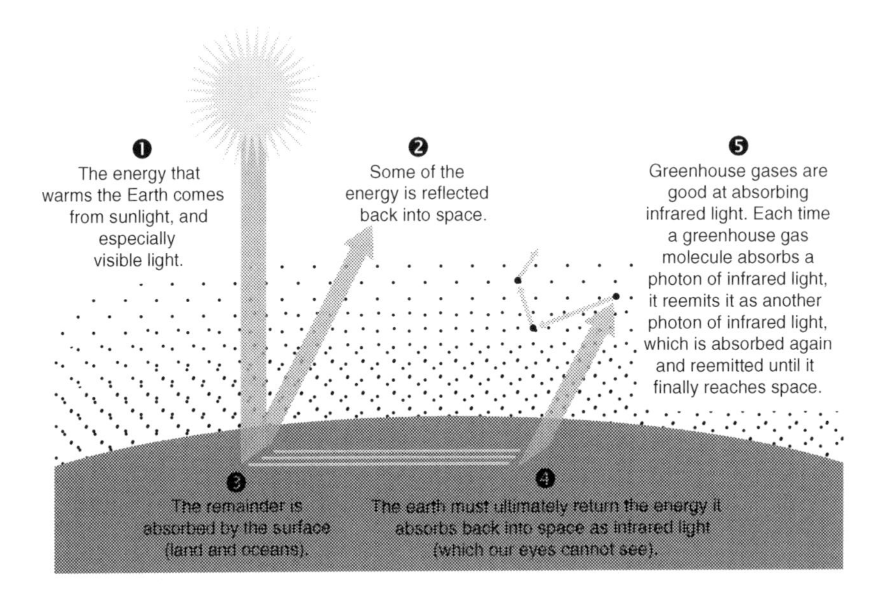

to the IPCC, which represents the international consensus among climate scientists and is backed by the world's governments, if our greenhouse gas emissions continue on our current high-emissions pathway, we are likely to experience a rise of 2.4°C (4.3°F) by some time between 2040 and 2060, and even reducing that to a medium-emissions pathway would lead to a 2.0°C (3.8°F) increase by then.[4] In the meantime, every fraction of a degree of increase in global heating will unleash dramatically worse consequences around the globe.[5]

What must be done to avert the worst of those consequences? In 2018 the IPCC declared that to have a good chance of limiting global heating to a target of 1.5°C (2.7°F) above preindustrial levels, by 2030 we would have to cut 2010 levels of greenhouse gas emissions by 45 percent.[6] Put in terms of our remaining **carbon budget**, to have even a 50 percent chance of keeping heating to 1.5°C, we can afford to emit only about 300 gigatons (Gt, or billion tons) more of CO_2.[7] Yet, counting up all the existing and currently pledged fossil fuel infrastructure and business-as-usual activities around the world indicates that we are already committed to producing 840 gigatons of CO_2. (See Box 1.1 for an explanation of the units of measurement used in this book.) It is clear that, even accounting for the temporary dip of emissions induced by the Covid-19 pandemic, we are rushing well past that target toward an increase of 2°C, which as noted above could occur in a sustained fashion in the 2040 to 2060 time frame according to the IPCC.[8]

To better understand our current predicament and why we failed to heed the warnings of the scientific consensus regarding global heating, this chapter

Box 1.1 Units and measurements

Different temperature scales: 1° Celsius of global heating = 1.8° Fahrenheit, so a 2°C increase is $2 \times 1.8 = 3.6$°F.

Converting temperature: To convert a temperature in Celsius to Fahrenheit, take the temperature in Celsius, multiple by 1.8 and then add 32. For example, 29°C = $(32 + 1.8 \times 29) = 84.2$°F.

How to weigh the gas that comes out of the chimney: Every carbon atom burned will produce one CO_2 molecule. An oxygen atom weighs 1⅓ times that of a carbon atom, so a CO_2 molecule weighs $2 \times 1⅓ + 1 = 3⅓$ times as much as a carbon atom. Power stations record how much coal they burn each year and determine the carbon content of the coal (which typically ranges from 60 percent to over 80 percent, depending upon where it is mined). So if a power station burns one million tons of coal that is 70 percent carbon, it uses 700,000 tons of carbon that produces around 2.5 million tons of CO_2.

Weights: One metric ton is 1,000 kilograms. A kilogram is 2.2 pounds; 1,000 metric tons is a kiloton; and one billion metric tons is a gigaton (Gt).

Energy: The standard unit is one joule – the amount of energy that acts on an object when one Newton moves one meter or the amount of energy dissipated as heat when one amp passes through a resistance of one ohm for one second.

Power: One watt is one joule per second. It reflects the rate at which work is done or energy is transferred. One horsepower is 745.7 watts. A kilowatt (Kw) is 1,000 watts. A medium-size car can produce about 40 Kw running at typical constant speed. A megawatt (MW) is one million watts or 1,000 Kw. This is enough to power about 400 to 900 US homes.[9] A megawatt hour (MWh) is one MW of electricity used continuously for one hour.

offers an historical examination of how global heating arose, how much we have already had and can anticipate, and the international attempts to deal with it.

The Industrial Revolution's Huge Increase in Fossil Fuel Use

As Timothy Mitchell points out in *Carbon Democracy*, before the Industrial Revolution, people used mostly renewable sources of energy, such as water or animal power and wood that "captured" energy from the Sun.[10] The renewable forms of energy that powered such activities as heating, farming, and milling in preindustrial life were weakly concentrated (not energy dense like fossil fuels) and required people to live mostly in relatively dispersed settlements near rivers, pastures, and woodland. The timescale of such energy production was also slow, depending on the life span of animals and the time it took to replenish forests via photosynthesis. But around 1800, European countries, and Britain in particular, began to replace these organic supplies with highly concentrated stores of buried solar energy, such as coal and oil – fossil fuels that were produced by the compression of the decomposed biomass of dead marine organisms from about 150 to 300 million years ago. To understand just how concentrated these fossil fuels are, consider that one liter (about a quarter of a US gallon) of gasoline used today required about 25 metric tons (about 55,000 pounds) of material from ancient marine life.

Once humans discovered a way to "free" energy from the limits of the muscles of living animals and the replenishment of woodlands, their use of such energy began to grow at an exponential rate. The amount of energy produced by this change was stupendous. The amount of coal energy put to work in Britain grew from 170,000 horsepower in 1800 to 2.2 million in 1870, 10.5 million in 1907, and 100 million by 1977. Before Britain's coal reserves were mostly exhausted about twenty years ago, they released as much raw energy as the total amount of oil provided so far by Saudi oil production.

As Mitchell explains, the enormous increase in energy made possible by this use of fossil fuels in European countries accelerated the development not only of industrial capitalism but of colonialism. First, European countries had a growing need for territory that could provide the raw materials, such as cotton and sugar, to which this enormous fossil energy could now be applied through manufacturing and production processes. Second, as more European workers became engaged in industrial production, they could no longer grow food, and thus industrializing nations needed additional territory and populations to supply their workforces with consumables. But given that faraway agrarian peoples understandably preferred to use their land and labor to provide for

Figure 1.2

Annual global emissions of carbon dioxide are rising

Fossil fuel use and carbon production have increased enormously since the nineteenth century. With them, atmospheric concentrations of carbon dioxide and temperature have also gone up.

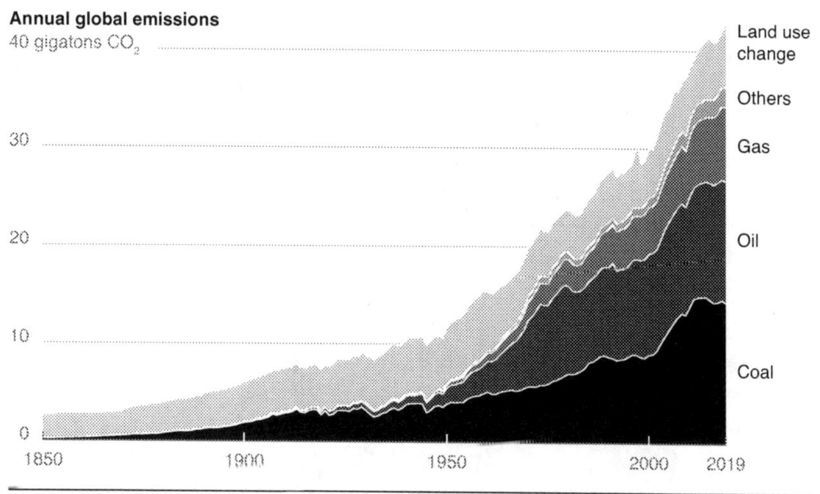

Adapted from the Global Carbon Project | CC BY 4.0

their own needs, European colonial powers devised ways to control the land and farming practices of those distant populations, such as dispossessing Indigenous populations and importing slave labor to acquire sugar and cotton from the Americas.

Thus, while coal provided concentrated energy for rapid industrialization, colonization provided the raw materials and markets for the work to be done and the goods necessary to feed the industrial workers. Since the nineteenth century, the use of fossil fuels has continued to increase enormously – not only of coal but, following World War II, of oil and then methane (so-called natural gas) (Figure 1.2). As we will see in Chapter 2, this increase in the use of fossil fuels closely parallels the increases in atmospheric CO_2 concentrations.

The Discovery of the Greenhouse Effect

Meanwhile, scientists in the late nineteenth century first began to make the links between the use of fossil fuels, increases in CO_2 concentrations in the atmosphere, and global heating. The initial step in this new understanding is typically credited to the Irish scientist John Tyndall, who demonstrated in 1859 that gases such as CO_2 and water vapor can absorb heat. (While Tyndall is generally credited with this discovery, it was in fact made three years earlier by an American amateur scientist named Eunice Foote, whose findings were largely overlooked at the time, probably due at least in part to her gender and lack of professional credentials.)[11] In 1896, the Swedish scientist Svante Arrhenius took Tyndall's findings a step further, arguing that fossil fuel burning, by adding CO_2 to the atmosphere, would raise the planet's average temperature, a phenomenon Arrhenius referred to with a Swedish term that translates as "hotbed" or "hothouse," a precursor to our term *greenhouse effect.*[12] At the time, however, few took Arrhenius' argument seriously, as they thought that changes in human activity would affect vast climate cycles only over the timescale of tens of thousands of years.

Little further progress in climate science took place until the Cold War, which led to a sharp increase in science funding for research on the weather and oceans, motivated largely by military concerns. During this period, Charles Keeling of the Scripps Institution of Oceanography in La Jolla took painstaking measurements of CO_2 from the top of the Mauna Loa mountain in Hawaii that showed that CO_2 levels were rising year by year. In the next few decades, scientists devised simple mathematical models to describe the climate and found creative ways to retrieve information about past temperatures by studying ancient pollens and fossil shells. At the time, climate scientists' calculations suggested that average temperatures were likely to rise a few degrees in the next century, a likelihood that at the time seemed too far off to inspire policy recommendations. Yet, while a few degrees may not sound like much, such an increase corresponds to enormous impacts on climate. As we will see in Chapter 2, even the less than 1°C of heating that has occurred since the 1980s corresponds to a 400 percent increase in the number of extreme weather events since then.

The inattention to climate began to change in the 1960s, however, as well-publicized ecological disasters and the publication of popular books such as Rachel Carson's *Silent Spring* created greater attention to the environment in Western societies. The huge post-World War II economic growth and increase in industrialization brought with it increased environmental disruption and risk and a growing recognition that environmental problems can also produce human health problems. Together, greater concern over such health effects among members of the well-off middle class and the context of general support for social change and social movements led to the rise of a "green politics" in European countries, as left-wing political parties began to champion environmental concerns.[13] In the USA, this new consciousness led to the bipartisan passage of extensive legislation, such as the Clean Air Act of 1970 and the Water Pollution Control Amendment of 1972, even under the conservative administration of President Richard Nixon. This growing public awareness of the harm that human activity was doing to the planet created a wider backdrop for climate science's growing focus on fossil fuel emissions. As research activity further accelerated with the use, by scientists, of international fleets of ocean-going ships and orbiting satellites, these scientists and policy makers in the USA and elsewhere began to warn that climate change was not merely a distant concern but was taking place already.

As investigative reporter Nathaniel Rich points out in "Losing Earth: The Decade We Almost Stopped Climate Change," by 1979 scientists had accumulated nearly all the knowledge that was needed to understand global heating.[14] That year a report published by a research team led by Jule Charney, a major figure in meteorology, titled *Carbon Dioxide and Climate: A Scientific Assessment*, distilled much that was known about ocean, sun, sea, air, and fossil fuels and attempted to estimate something they called **climate sensitivity**, or how sensitive the climate is to increases in CO_2. Although the researchers noted that the thermal inertia of the ocean would lead to a lag on the order of decades between the release of CO_2 and the resulting temperature rise, they calculated that atmospheric CO_2 concentrations would double their preindustrial levels sometime in the first half of the twenty-first century and predicted that, as a result, the average global surface temperature would increase by 3°C (with 95 percent confidence that it would occur between 1.5 and 4.5°C). This estimated 3°C rise in temperature was particularly striking, as the last time the world had been that warm was about twenty million years ago, when trees grew in Antarctica and sea levels were eighty feet higher, as will be shown in Chapter 2. As the report made clear, human beings had altered the Earth's atmosphere through their massive burning of fossil fuels, a problem that could be reduced to a simple axiom: the more CO_2 in the atmosphere, the warmer the planet.

The Thwarted Opportunity to Act in the 1980s and 1990s

As Rich and others have pointed out, at the start of the 1980s, with the main scientific question settled beyond debate, attention shifted from the diagnosis of the problem to its predicted consequences and the need to act. In 1980, US President Jimmy Carter signed the Energy Security Act, which directed the National Academy of Sciences to undertake a comprehensive study to analyze the social and economic effects of a warming planet. Even the major oil company Exxon, anticipating that legislation to restrict hydrocarbons might be passed following Charney's report, created its own dedicated CO_2 research program to understand how much its activities had contributed to the problem. Indeed, the company had been studying the CO_2 problem for decades, as had the American Petroleum Institute, the oil industry's largest trade association. In anticipation of the results of the National Academy of Sciences study, Exxon began to spend substantially on global warming research, including funding outside scientists. In 1982, Edward David, Jr., the president of Exxon's research division, boasted that the company would create new global energy systems to save the planet. Indeed, the CO_2 issue was now receiving major attention not only from scientists and policy makers but from the energy sector of the economy, which also began to make heavy investments in nuclear and solar power.

When the eventual National Academy of Sciences report, titled *Changing Climate*, appeared in 1983 during President Reagan's administration, it echoed the report of Charney's group in calling for immediate action to solve this pressing existential problem. Yet, that was not the way the report was represented by the chairman of the committee, William Nierenberg, a presidential advisor and director of the Scripps Institution of Oceanography in La Jolla, California. At press interviews following the report's release, Nierenberg denied there was an urgent need for action, advised the public not to be frightened by the most "extreme negative speculations," and argued that it was better to wait and see what would happen because American ingenuity would save the day. Those who knew Nierenberg were not surprised by his remarks. A devout believer in American exceptionalism and one of a group of scientists who had helped win World War II and created the booming aerospace and computer industries, he was a free market ideologue who was hugely optimistic about the saving grace of market forces and deeply pessimistic about the value of government regulation. As Rich explains, despite the evidence and conclusions of the actual report, Nierenberg's remarks, which were probably 1/500th of the length of the report, received 500 times the press coverage. This was reflected on the front page of the *New York Times*, whose headline announcing the release of the report read "Haste on Global Warming Trend is Opposed."

Historians Naomi Oreskes and Eric Conway have identified this statement as the beginning of what they call the climate change debate. As their classic book *The Merchants of Doubt* notes, "Nierenberg didn't deny the legitimacy of climate science. He simply ignored it in favor of the claims made by economists: that treating symptoms rather than causes would be less expensive, that new technology would solve the problems that might appear so long as government did not interfere, and that if technology couldn't solve all the problems, we could just migrate."[15] Yet not even all economists shared that perspective; by the late 1960s, some had begun to realize that free market economics focused on the growth of consumption was destructive to the ecosystems upon which we all depend. But Nierenberg had not put any of those economists on his panel. Instead, Nierenberg's one-sided view gave the White House the scientific cover it needed to largely ignore the impending climate crisis: a report that presented a unified view rather than the differences of opinion between social and physical scientists and which insisted that no immediate action was needed. Nierenberg, who still has buildings named after him at the Scripps Institution of Oceanography at University of California San Diego, went on to work in right-wing think tanks, where he continued to create doubt about climate change.

The effectiveness of Nierenberg's ploy was reflected in a shift in Exxon's position, which soon cited the *Changing Climate* report as evidence that "the general consensus is that society has sufficient time to technologically adapt to a CO_2 greenhouse effect" and reverted to being mainly a supplier of hydrocarbon fuels. The American Petroleum Institute also canceled its CO_2 research programs. This shift not only marked a new commitment to fostering climate change denial among powerful elites and institutions but represented a missed critical turning point at a time when a shift to non-fossil fuel energy might have been much easier to manage.

Meanwhile, as Rich points out, a new problem related to the atmosphere emerged, that of the so-called "hole in the ozone layer." In fact, there was no layer and no hole: ozone, which shields us from ultraviolet radiation, is present throughout the atmosphere, and the supposed hole was merely a descriptive metaphor for how the amount of ozone in Antarctic had begun to decline dramatically for about two months per year. Still, the reductions in ozone were real, allowing more ultraviolet radiation to reach Earth's surface and increasing the likely incidence of skin cancer in humans. This reduction of ozone was traceable to the human-made chloroflourocarbons used in refrigerators and aerosol cans, which, when released into the atmosphere, devoured ozone and also functioned as potent greenhouse gases (much more so than CO_2). Yet the huge public concern over this issue came not from people's concern about atmospheric warming but their worry about getting skin cancer. The public outcry was such as to produce alarm in dozens of American businesses that had the word "refrigeration" in their name, which formed an alliance to hound members of Congress, the Environmental Protection Agency, and President Reagan to resist pressures to outlaw these common refrigerants. But in this case, the business interests failed; every relevant government agency and every member of the US

Senate urged the president to endorse a United Nations treaty calling for such action.

And thus in 1987, four years after the *Changing Climate* report, the USA and more than three dozen other nations signed a treaty that limited the use of chloroflourocarbons. Obviously, banning chloroflourocarbons had very minor consequences for everyday life in comparison to those of getting off fossil fuels, which is likely to be incredibly difficult for just about everyone, especially industry and governments that rely on them for cheap energy to fuel economic growth. But as Rich points out, the metaphor of the ozone hole had also moved the public because it allowed people to "see" the problem in a visceral way: "Instead of summoning a glass building that sheltered plants from chilly weather ('Everything seems to flourish in there'), the hole evoked a violent rending of the firmament, inviting deathly radiation. Americans felt that their lives were in danger." As a result, "[a]n abstract, atmospheric problem had been reduced to the size of the human imagination. It had been made just small enough, and just large enough, to break through."

Inspired by the success of the international treaty on ozone, in March 1988, forty-two senators, nearly half Republicans, now demanded that President Reagan also call for an international treaty on climate change. Reagan agreed and signed a pledge with Gorbachev of Russia to cooperate.

The following year of 1988 was one of the hottest and driest in US history. Two million acres in Alaska were incinerated in wildfires, and some streets in New York melted. For the first time in its history, Harvard University was closed because of the heat. Forty percent of the nation's counties were affected, and many people began to wonder if global warming was not so far off after all. That recognition was reinforced by a hearing of the US Senate Committee on Energy and Natural Resources at which the previously mentioned James Hansen was the star of the show. Testifying about new research that showed a current warming of 0.5°C relative to the 1950–80 average, he reported that NASA had determined that the probability that this warming could be explained by natural events, rather than human causes, was only 1 percent. Hansen's dramatic testimony brought unprecedented public attention to the issue of the warming climate, with the front page of the *New York Times* reporting his declaration that "global warming has reached a level such that we can ascribe with a high degree of confidence a cause and effect relationship between the greenhouse effect and observed warming."[16]

As Rich recounts, by the end of that summer, global warming had become a major theme of the presidential campaign. While the Democratic candidate Michael Dukakis proposed tax incentives to encourage domestic oil production, it was the Republican George H. W. Bush who declared, "I am an environmentalist" and that "those who think we are powerless to do anything about the greenhouse effect are forgetting about the White House effect." By the end of the year, thirty-two climate bills had been introduced in Congress, co-sponsored by Democrats and Republicans. Meanwhile, the German Parliament created a special commission on climate change, and Canada and Norway called for a binding international treaty on the atmosphere. Even the archconservative Margaret Thatcher, who had been trained in chemistry at Oxford, declared that "the health of the economy and the health of our environment are totally dependent on each other." For its part, the United Nations endorsed the joint establishment of the IPCC by the World Metereological Organization and the United Nations Environmental Program and charged it with making a series of policy recommendations for grappling with the problem. To this day the IPCC remains the most widely recognized and respected international organization on climate, as its reports reflect the consensus of hundreds of scientists and are signed by the representatives of nearly 200 governments around the world.

The newly elected President Bush's supposed commitment to addressing global warming and growing public concern over the issue made it look like actual action on global warming might be possible. In 1989, a bipartisan group of twenty-four senators requested that Bush cut emissions in the USA even before the IPCC's working group offered its own recommendations. Yet, as Rich explains this momentum was soon stopped by Bush's chief of staff, John Sununu, who was ideologically opposed to any limitations on emissions, which to him implied imposing limitations on the economy. At a 1989 international meeting to promote policy action on climate change in the Dutch town of Noordwijk, Sununu's appointed delegate torpedoed a framework for a global treaty. With the acquiescence of Britain, Japan, and the Soviet Union, the conference abandoned any commitment to freeze emissions. Thus, what had appeared to be a decade of progress in understanding and facing the climate crisis had come to a dead end. Once again, and anticipating much of the situation today, climate action had crashed into the wall of a free market ideology that would not countenance government intervention and concerns about the primacy of economic growth.

The United Nations Framework for International Climate Policy from the 1990s to the Present

Even as efforts to take global heating seriously stalled in the USA, since the 1990s the United Nations has promoted numerous international attempts to reduce emissions. As recounted by Brian Tokar in his book Toward Climate

Justice, in 1990, the IPCC released its first assessment of the science on climate change, which declared that the amount of heating was consistent with the rise predicted by the climate models discussed above but acknowledged that its magnitude was consistent with normal short-term variability (Table 1.1 provides a brief history of IPCC assessments).[17] Even though this conclusion was much weaker than that made by NASA as reflected in Hansen's testimony, it was nonetheless enough to lead many world leaders to believe that action had to be taken and to send delegates to the UN-sponsored Earth Summit in Rio in 1992. This summit culminated in a treaty known as the United Nations Framework Convention on Climate Change (UNFCCC), whose stated objective is to "stabilize greenhouse gas concentrations in the atmosphere at a level that would prevent dangerous anthropogenic interference with the climate system." Within a few years, 192 countries had become signatory parties.

To ensure that the principle of justice would be honored in ensuing climate change agreements, this framework enshrined the important concept of **common but differentiated responsibilities**. This notion acknowledges that individual countries have different capabilities for combating climate change, recognizing that, for instance, rich countries would be able to cut emissions sooner with less dire consequences for their populace than less developed countries. It also encompasses moral considerations regarding equity and historical responsibility, such as acknowledging that raising billions of poor people in developing countries out of poverty will require the use of energy and that making energy more expensive could thus work at cross-purposes with improvements in their well-being. It also recognizes that, regardless of current levels of energy use, the vast majority of the CO_2 in the atmosphere has already been put there by rich and industrialized countries. As shown in Figure 1.3, the historical per capita emissions in the USA are more than 300 tons of carbon, or eight times that of China and hundreds of times more than African countries. Given this history, rich countries clearly must bear much greater responsibility for dealing with the emissions problem.

But as important as the UNFCCC framework may have been for encouraging international cooperation in mitigating the greenhouse effect and global heating, it included no enforcement mechanisms. Instead, it set nonbinding limits on greenhouse gas emissions for individual rich countries that were intended to reduce emissions to 1990 levels by 2000 and stipulated how subsequent international treaties (called "protocols" or "agreements") might be negotiated to specify further action. By the mid-1990s, it became clear that no country would meet the nonbinding emissions reduction targets set by the UNFCCC and that real action would require a treaty with mandatory reductions.

Table 1.1. *Key UN summits and reports*

a) Major accord or report	b) Year	c) Major findings
IPCC Assessment 1	1990	"The size of the [observed] warming is broadly consistent with predictions of climate models, but it is also of the same magnitude as natural climate variability"
Rio Earth Summit	1992	The concept of "common and differentiated responsibilities"
IPCC Assessment 2	1996	"The balance of evidence suggests a *discernable* human influence on the climate"
IPCC Assessment 3	2001	"There is new and stronger evidence that most of the warming observed over the last 50 years is *likely* attributable to human activities"
IPCC Assessment 4	2007	"Most of the observed increase in globally averaged temperatures since the mid-twentieth century is *very likely* due to the observed increase in anthropogenic greenhouse gas emissions"
IPCC Assessment 5	2014	"It is *extremely likely* that human influence has been the dominant cause of the observed warming since the mid-20th century"
Paris Climate Accord	2015	Emissions should be reduced as soon as possible to keep the increase in global average temperature to well below 2°C above preindustrial levels and to pursue efforts to limit the increase to 1.5°C, which will substantially reduce the risks and impacts of climate change
Special Report on Global Warming of 1.5°C	2018	Limiting warming below or close to 1.5°C will require reducing emissions from 2010 levels by around 45% by 2030
IPCC Assessment 6, Working Group 1	2021	"It is *unequivocal* that human influence has warmed the atmosphere, ocean and land"
IPCC Assessment 6, Working Groups 2 and 3	2022	"If global warming transiently exceeds 1.5°C in the coming decades . . . some impacts will cause release of additional greenhouse gases (medium confidence) and some will be irreversible, even if global warming is reduced (high confidence)" "The continued installation of unabated fossil fuel infrastructure will 'lock in' GHG [greenhouse gas] emissions" (high confidence)

Note: Some quotes drawn from https://insideclimatenews.org/content/growing-certainty-ipcc-climate-models-and-assessments.

In 1997, a new international treaty, the Kyoto Protocol, was signed by 192 countries with the intention of creating a schedule of binding targets for reducing emissions and a process for reaching those targets (Figure 1.4). The primary responsibility for such cuts fell on the rich countries, with the rest accepting common but differentiated responsibilities. But as so often happens, the devil was in the details, and the USA was soon objecting to mandatory cuts. At that point, then President Bill Clinton sent US Vice President Al Gore to Kyoto. Gore was credited with turning the situation around by giving rich countries and corporations an out by suggesting that the USA would sign on to the Kyoto Protocol under two conditions: that mandated emissions reductions would be limited to half of those that were proposed, and that cuts could be implemented through carbon trading. Carbon trading relied on the two approaches of **cap-and-trade** and the **Clean Development Mechanism (CDM) (carbon offsets)**, both of which turned out to be deeply problematic (see Box 1.2). Cap-and-trade ostensibly tried to cap the emissions of corporations by allowing them to trade in pollution credits, while the CDM allowed industrialized countries to invest in emissions reduction projects in the developing world and count them toward their own targets. (Such market approaches to the climate crisis are considered in more detail in Chapter 8.)

As further recounted by Tokar, the ambitions ambitions of the Kyoto Protocol were further undermined when, in 2001, the administration of George W. Bush withdrew the USA from the agreement on the grounds that it would harm the US economy. Meanwhile, some other industrialized countries continued with the legally binding protocol, whose commitment period was specified as 2008–12. Although the thirty-six countries that continued to participate in the protocols did reduce emissions, the required reductions were only 5 percent less than 1990 levels, a paltry amount, and were abetted by slowdowns created by the financial crisis of 2007–8.[18] In addition, the countries that made the greatest reductions were those of the former Eastern bloc, whose emissions had plummeted even before the deal was signed because of the dissolution of the Soviet Union. Furthermore, ten countries achieved their targets only by using carbon credits (not genuine emissions cuts), and some of the reductions were likely due to a shift of manufacturing to China. Lastly, the accounting did not include fast-rising emissions from aviation and shipping (such as for moving all those products whose emissions were generated in China). Meanwhile, global emissions increased by 32 percent from 1990 to 2010. Thus, the Kyoto Protocol can only be considered a very qualified success, and even though the USA never ratified it, the rest of the world has continued to live with its effects – a cumbersome and corporate-friendly carbon-trading system that manifestly failed to reduce emissions overall.

Figure 1.3

Countries of the Global North have contributed vastly more emissions over time

The UN concept of **common and differentiated responsibilities** refers to the dramatically different historical contributions of countries. For example, in (A) the UK contributed more than 60 percent of global emissions at the time of the Industrial Revolution, but that has declined dramatically, so that in 2019 the UK's emissions are much less than China (B). However, if one takes the cumulative emissions from the UK over history, and divides them by the current population, the per capita contribution of the UK iş the second highest in the world (C).

(A) Shares of worldwide historical CO_2 of emissions

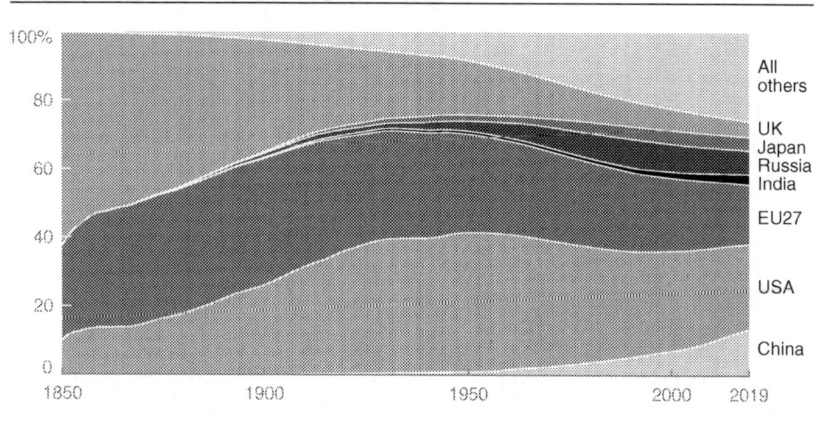

(B) 2019 per capita emissions

(C) 1751–2016 per capita cumulative emissions

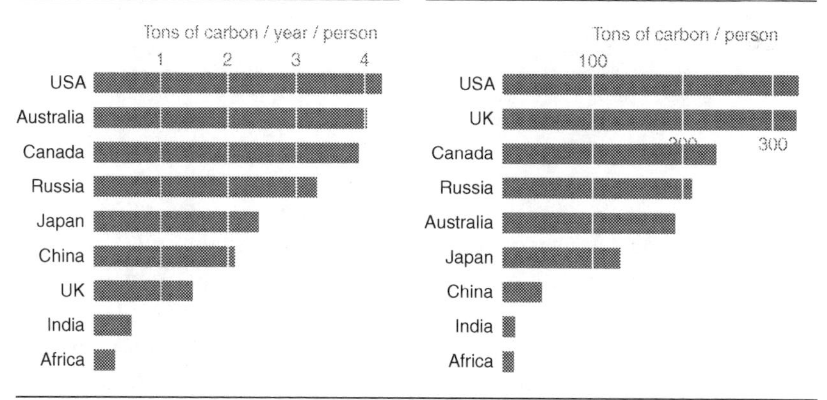

(A) Adapted from the Global Carbon Project | CC BY 4.0
(B) and (C) Adapted from 1751–2016: T.A. Boden, G. Marland, and R.J. Andres. 2017. Global, Regional, and National Fossil-Fuel CO2 Emissions. Carbon Dioxide Information Analysis Center, Oak Ridge National Laboratory, U.S. Department of Energy, Oak Ridge, Tenn., U.S.A. doi 10.3334/CDIAC/00001_V2017. April 2017 and 2016-2019: BP Statistical Review of World Energy June 2020.

Figure 1.4

Timeline of US climate reports and international negotiations and how these culminated in a deeply inadequate and non-binding framework on emissions reductions

Although the science is overwhelming and many agreements have been made over the years to limit greenhouse gases, in the end, non-binding agreements have prevailed. And these are inadequate to reach the goal of staying below 2°C.

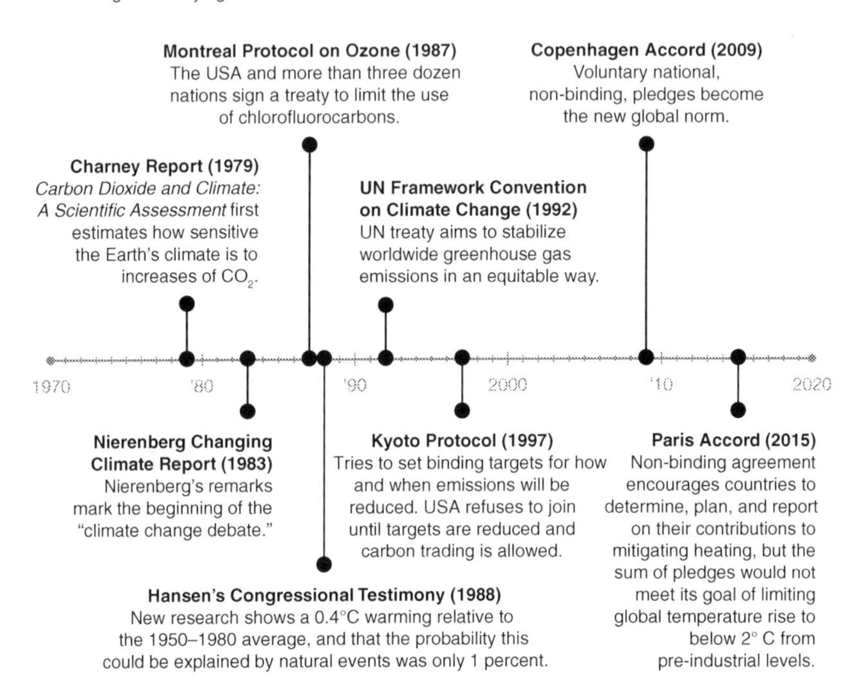

In 2009, as Barack Obama took office in the USA, climate activists and policy makers looked forward to the upcoming Copenhagen summit, which was intended to negotiate a post-Kyoto protocol. As Tokar also explains in his book *Toward Climate Justice*, outside of this UN-led process and its corporate-dominated interests, a broader climate justice movement had also been building among activists in Europe and North America and Indigenous and small farming communities worldwide.[19] This movement represented the voices of the communities most affected by the climate changes already underway and they challenged what they saw as the corporate-friendly false solutions of carbon trading and offsets, "clean coal," new nuclear plants, and industrial-scale bioenergy (see Box 1.3 for one of those voices from the Global South).

Box 1.2 Market mechanisms for emissions reduction under the Kyoto Protocol

When US Vice President Al Gore addressed the UN conference in Kyoto in 1997, he stipulated that the USA would sign the agreement if the emissions reductions were implemented under a market-based trading of rights to pollute that became known as cap-and-trade.

Under this scheme, governments set a ceiling on the maximum allowable CO_2 – the cap – for a given industry. Then, for every ton of CO_2 that a polluter reduces under this cap, it is awarded one permit to pollute that can be bought, sold, and banked. Over time, governments were supposed to ratchet down these caps, on the assumption that this would gradually make fossil fuels uncompetitive with renewable sources of energy. While many economists claim this scheme induces companies to implement the most cost-effective mechanisms to reduce emissions as soon as possible, experience has showed that cap-and-trade was often subject to fraud and manipulation. Many industries complained that the cap acted like a tax and that they were made uncompetitive by it. For example, in Europe in 2005, where the world's first mandatory trading market was established, giant utilities and smokestack industries beseeched governments for exemptions, many of which were granted.[20] In Germany, electricity companies ended up being allocated 3 percent more permits than they needed – a windfall worth about $374 billion. As governments caved in, emissions soared and the profits went to the polluters and traders. Other forms of cap-and-trade, such as that currently operating in California, might be more effective, but some critics consider even that form and indeed the overall approach to have been a failure and a distraction from what should have happened instead: genuine emissions cuts.[21]

Another market approach developed under the Kyoto Protocol was the CDM, which allows rich countries to achieve some of their emissions reductions by buying certified emissions reductions units (i.e., carbon offsets) from emissions reduction projects in developing countries. The projects were subject to approval by a monitoring board to determine that the emissions reductions were both "real" and "additional." Additionality is key to this approach, as it means that the project in the developing country would not have happened unless the rich country had paid. Yet a detailed 2016 analysis of the CDM showed that only 2 percent of the projects up until that point had a high chance of being additional, and by the Madrid 2019 climate summit the CDM market had crashed.[22] More

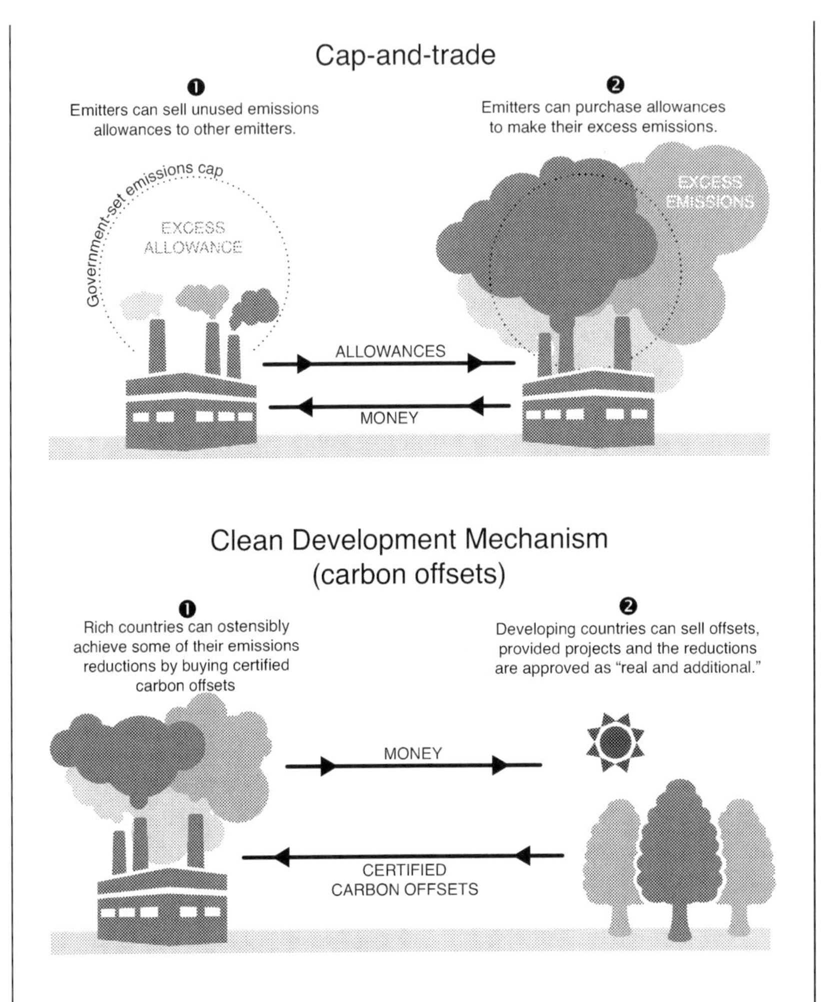

generally, however, carbon offsetting has continued to grow as Google, Apple, and other institutions such as the University of California rely on it. (A more detailed critique of offsets is presented in Chapter 8, the fundamental problem being that those who purchase carbon offsets aren't changing their planet-heating behavior.[23])

That November, North American activists held a continent-wide day of action, during which protestors in the state of South Carolina blocked the shipment of a generator for a new coal plant and in Canada blockaded the office of the finance minister. When the Copenhagen summit on climate change opened in

December, a hundred thousand protestors took to the streets with the cry of "System Change, Not Climate Change" and called for fossil fuels to stay in the ground, for Indigenous people's rights to be respected, and for reparations for ecological damage to be paid by rich countries.

During the summit, a memo put together by Denmark, the USA, and the UK was leaked to Lumumba Di'Aping, the lead negotiator for participants from the **Global South**, a term that refers broadly to the mostly low-income and often politically marginalized regions of Latin America, Asia, Africa, and Oceania that are also sometimes referred to as the Third World or Periphery to denote regions outside of Europe, North America, Japan, Australia, and New Zealand. In the UN negotiations, Di'Aping specifically represented the Group of 77 countries plus China. According to the memo, summit participants were planning to make a deal that would require developing countries to sign an agreement that gave more power to the rich, sideline the UN's role, and set a new global heating target of 2°C. Di'Aping, who had been named after the Congolese independence leader Patrice Lumumba, loudly and bravely declared that the Global South was being asked to sign a "suicide pact" (Figure 1.5).[24] Referring to the IPCC's own evidence, he explained that a 2°C rise globally actually meant a 3.5°C (6.3°F) rise for much of Africa, which he called "certain death for Africa" and a type of "climate fascism" imposed on Africa by polluters in exchange for promised fast-track funding – a carrot dangled to break the solidarity of the Group of 77 plus China. Declaring that "I would rather die with my dignity than sign a deal that will channel my people into a furnace," Di'Aping asked, "[w]hat is Obama going to tell his daughters? That their [Kenyan] relatives' lives are not worth anything? It is unfortunate that after 500+ years of interaction with the West, we are still considered 'disposables'." Indeed, the most recent IPCC report at the time had predicted that heating in parts of Africa was expected to be much more than the global average (a result that also appears in the IPCC 2021 report and in the future scenarios provided in its 2021 interactive atlas).[25]

The final Copenhagen Accord gave little comfort to the people Di'Aping represented; it agreed that global temperatures should not rise more than 2°C above preindustrial levels; that deep cuts in emissions were necessary to meet that goal; that the rich industrialized countries would set their own targets for emissions in 2020; that the world's developing countries would take steps to mitigate their emissions without having specific targets; and that flexibility should again be incorporated into climate-related policies. As Tokar explains, In the end, a handful of countries, including Bolivia, Cuba, Peru, and Venezuela, objected to the formal adoption of the accord, so that the assembled countries agreed to merely "take note" of it rather than to adopt it (and since it

Figure 1.5

"I would rather die with my dignity than sign a deal that will channel my people into a furnace"

At the Copenhagen 2009 climate summit, a memo that was being drafted by a small group from the US and richer countries was leaked to **Lumumba Di-Aping**, the lead negotiator for the Global South. He frankly pointed out that a target of 2 degrees Celsius would condemn Africa to much more heating than the gobal average. Estimates from the AR5 report of the IPCC show that under the high scenario of emissions (RCP 8.5) parts of Africa may have more than 3 degrees Celsius (5.4F) heating by mid-century compared to the 1986–2005 mean. (For definitions of AR5 and RCP8.5 please see Chapter 2.)

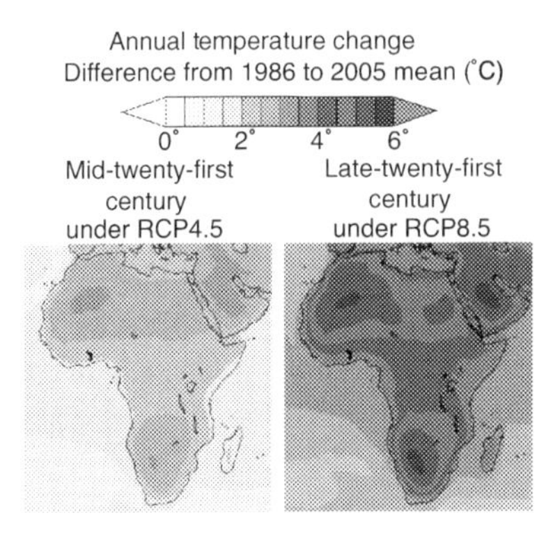

Photo courtesy of Yale University. Data reproduced from the IPCC AR5 report: Figure 22.1 (Top panel, right) from Niang, I. et al., 2014: Africa. In: *Climate Change 2014: Impacts, Adaptation, and Vulnerability. Part B: Regional Aspects. Contribution of Working Group II to the Fifth Assessment Report of the Intergovernmental Panel on Climate Change* [Barros, V.R., et al. (eds.)]. Cambridge University Press, Cambridge, United Kingdom and New York, NY, USA, pp. 1199–1265.

was a political framework and not a legal one the distinction was perhaps unimportant).

The overall outcome represented a triumph for the new agenda of the USA, which was to replace a comprehensive international treaty with a patchwork of informal, individual country commitments. Notably, the European Union, which had once called for a strong worldwide agreement to reduce greenhouse gases, now fell in line with the US strategy. The legacy of the Copenhagen Accord was to establish the notion of voluntary and nonbinding national pledges as the new global norm, which still stands today and represents a failure for emissions reductions.

Even with this evisceration of binding climate action at the international level, a new problem now emerged in many rich countries: a conservative backlash against such action and cooperation. In the USA, before about 2009 a number of prominent Republicans had openly accepted the reality of climate change and supported mitigation policies, but that all changed after the Tea Party came to prominence and the Citizens United ruling of the Supreme Court allowed the use of "dark money," such as from fossil fuel interests, to influence elections.[26] After 2009, for example, John McCain, Newt Gingrich, and Mitt Romney all adopted a skeptical position regarding climate change, and other Republicans felt they had to do the same or risk losing their positions. Similarly, in 2011 the conservative government of Stephen Harper in Canada withdrew from the Kyoto Protocol, at least in part because enacting emissions reductions would reduce the value of Canada's immense tar sands, and in 2013 the conservative government in Australia began rolling back its emissions reduction commitments.

The next major international accord on climate policy was the Paris Accord in late 2015, which had the long-term goal of keeping the increase in global temperature to "well below 2°C" above preindustrial levels. Eventually, 189 UNFCCC members, including the USA, became party to this agreement, under which each country agreed to determine, plan, and report on the contributions it makes toward mitigating global heating, although once again it included no binding enforcement mechanism. Although the Paris Accord was much heralded in the press, James Hansen called it a "fraud," describing it as "no action, just promises."[27] Within a few years, it was clear that none of the major industrialized nations were implementing the policies they had agreed to, and research pointed out that even if they had, the sum of the pledges would not be enough to keep the global temperature rise to the 1.5°C target.[28] Shortly after taking office in 2016, President Donald Trump withdrew the USA from the accord.

The next notable event in the development of the IPCC framework was its 2018 publication of a "Special Report on Global Warming of 1.5°C," a consensus document ratified by IPCC member governments that included more than 6,000 scientific references, analyses by ninety-one authors and editors from forty countries, and the input of thousands of scientists. The key finding of this report was that limiting global heating to 1.5°C (2.7°F) would require "deep emissions reductions" and "rapid, far-reaching and unprecedented changes in all aspects of society." [29] It concluded that to meet that target, by 2030 we would have to cut 2010 levels of greenhouse gas emissions by 45 percent. Much of the report was taken up by a comparison of the consequences of increases of 1.5° and of 2° on the biosphere, showing that there are quite dramatic differences between the two targets in terms of extreme weather, rising sea levels, diminishing Arctic sea ice, loss of ecosystems, and more (for example, at 2°C, 99 percent of coral reefs would be destroyed vs. 70 to 90 percent at 1.5°C). Later in 2018, the USA's own Fourth National Climate Assessment was released quietly by the Trump administration, predicting that by the end of the century climate change damage to the USA would cost hundreds of billions of dollars per year.[30]

Notwithstanding these declarations, however, governments continued to take steps to massively exploit fossil fuels. For example, in mid-2021, the Biden administration was on course to approve as much oil and gas drilling on public lands as the Trump and George W. Bush administrations, and Germany was completing the Nord Stream 2 gas pipeline from Russia that would double the methane supply from Russia.[31] This was despite the fact that at about the same time, the IPCC released its sixth assessment (Working Group 1: Science), now finally declaring that "it is unequivocal that human influence has warmed the atmosphere, ocean and land."[32]

Box 1.3 Interview with Dipti Bhatnagar

Dipti Bhatnagar is the co-coordinator of the climate justice and energy program of Friends of the Earth International, based at Justiça Ambiental/Friends of the Earth Mozambique. Originally from Kolkata, India, Dipti has fought destructive dams in India and has worked on immigrant rights and safe drinking water for farmworker communities of color in California and on climate and energy issues in Mozambique. She lives in Maputo, Mozambique.

AA: How do you feel about the climate crisis?

DB: I feel it very deeply. I feel the suffering that people are going through and that the planet is going through. This is Mother Earth that's sustaining us. It's really horrific the way that our dominant system, our economic system, is treating her. We need to be active, we need to do our part to protect the planet. And we need more people to feel it deeply and not be switched off and not sit with a sense of normalcy that a billion shellfish got roasted in their shells [in the Pacific Northwest heat wave of June 2021], not sit with a sense of normalcy about these wildfires that are ripping through the Amazon, Australia, and California.

AA: You've been an activist for twenty years. What is it about your psychological makeup and your formation that put you ahead of so many people?

DB: I grew up a normal, middle-class child playing on the streets in Kolkata. My father was in the first generation of independent India, so he saw his role as contributing to building the country. I learned to have a sense of context, to think about history, about what got us to this point, the promise of our freedom struggles. And that has been really important in my life. And then I heard Arundhati Roy [the legendary Indian author and environmental activist] speak at my college in Delhi in 1999. She spoke of what she saw in the Narmada Valley [a dam project that would displace half a million people]. That was the changing point for me. And so I went there to the Narmada Valley with my sister and became a part of the people's movement. And the respect that I have for peoples' movements and local communities and Indigenous peoples and the struggles on the ground, it comes from there. I think this is really important for students to learn as well. While we're in college, we learn how to build knowledge and to have a deep analysis, but at the same time we need to appreciate those whose knowledge isn't typically accepted or understood. I think it's critical that young people are placed in situations where they are face to face with realities that are different from their own.

AA: I want to ask you about Mozambique. It was devastated by Cyclone Idai in 2019. Do the local people understand that it was likely climate change-related and that it will escalate?

DB: Yes, Cyclone Idai was supercharged by climate change, made much more likely and intense. And there was another cyclone that year, Kenneth, and Cyclone Eloise this year. In the urban areas of Mozambique, people have heard of climate change. The government talks about it. Mozambique is one of the most vulnerable countries in the world to the ravages of climate change. Now, rural people may not use that terminology. But rural people do keenly understand and describe the changes occurring in their environment, that something is wrong, that what's happening to their fields, what's happening to their rainfall patterns, is different. Life is becoming harder. It's becoming drier. And when the rain comes, it's much heavier. There are also sea-level rise issues along the coast of this country. At the same time, Mozambicans are dealing with an onslaught of multiple interrelated crises and injustices. We have to put the climate crisis in context with all those other crises that they are facing. For example, 70 percent do not have access to electricity. People are struggling to survive. At the same time, our government is pushing coal and gas extraction, they are pushing mega-projects that are grabbing land.

AA: I want to ask about the historical responsibility for emissions. Some of the elites here don't accept it they say that India and China are emitting more than us now.

DB: First of all, historical responsibility is very real, and just because those individual elites and their governments don't recognize it doesn't mean it's not true. It's based on science. Much of the carbon dioxide that these countries emitted in building their societies is still in the atmosphere and affecting us all. They created this crisis. The concept of historical responsibility is also enshrined in the UN (Rio) Convention of 1992 [common but differentiated responsibilities]. Of course, actors in the Global North have tried very hard to get away from it. At Copenhagen, President Obama introduced the bottom-up approach, which normally sounds great because it feels like it's decentralized and building power from the bottom up, but in this case it's completely wrong because each country now offers their nationally determined contributions, how much carbon emissions reductions they feel like doing, which is not based on climate science, not based on justice, and none of it is binding. What we needed was a top-down architecture that was going to mandate emissions reductions, based on climate science and based on justice – so that would determine how much each country needs to reduce.

AA: What problems do you see in the way the Global North is responding?

DB: The rich countries continue to fund fossil fuel infrastructure in their countries and abroad while the ink is still drying on their emergency climate declarations. These are the countries that have been polluting since the Industrial Revolution and are most responsible for the climate change we are experiencing today. What's stopping them from acting? It's all about so-called "economic feasibility." They want to be seen to curb emissions while maintaining infinite growth on a finite planet. It's not going to work. This explains why they are pouring money into and pushing the rhetoric of dodgy schemes such as offsetting and carbon markets; towards inefficient and dangerous energy technologies such as mega-hydro, nuclear and bioenergy; and towards developing high-risk, unproven technofixes such as geoengineering and carbon capture and storage. In the climate justice movement we call these "false solutions" – because it poses as an alleged solution but is designed to secure profit for the corporate elite and keep unjust business as usual going, so it is not a solution at all.

AA: What would real solutions look like to you?

DB: Of course we need to quickly and justly transform our energy systems away from fossil fuels, towards renewable energy. But we also need to serve the hundreds of millions who don't even have electricity, most of whom live on this continent. We need to underpin this energy transition with just principles. How is the transition going to be done? On whose lands are the solar panels going to be set up? Where are the minerals and other materials going to come from? Friends of the Earth is working on this. It's not just any renewable energy that will be just. It's not about large solar farms in Morocco that export energy to Europe, that's not the transition we need. Ownership matters as well. Is the energy owned by a private corporation or is it in community hands? We're calling for socially owned renewable energy systems. It could be at the building level, at the village, at the city level, wherever, but the people who use the electric power must govern it and make decisions about it. Land use is also a huge factor. The corporate agribusiness model is a huge driver of climate change, also deforestation. We're calling for better, more sustainable ways to grow and distribute our food. We're calling for support for peasant agroecology. And we're demanding land rights and forest rights for the communities who have always taken care of those resources. And that's one of the big problems with the false solutions. False solutions of offsetting, carbon trading, net zero is coming to grab those lands, those forests, those resources because it wants that land, that lake, that forest to sequester carbon so that the Global North countries and the corporations can keep on polluting. And that's why the land rights are so important for local communities.

Conclusion

As this chapter has shown, anthropogenic global heating began in earnest with the burning of coal in the 1800s and has increased exponentially ever since. The basic greenhouse effect was explained as early as 1896, and the scientific evidence supporting it and the reality of global heating was well established by the time of the Charney report in 1979. Although the 1980s and 1990s offered reasons to be hopeful that the USA and international community would take action to confront the need to actively reduce emissions levels, conservative movements and free market economics seem to have so undermined such efforts as to leave them toothless and insufficient. As noted, even if all countries were to meet their nonbinding commitments under the Paris Accord of 2015, it will not be enough to limit heating to 2°C, which according to one estimate will require emissions cuts of about 5 percent per year from 2022 onward.[33] Even though doing so is possible, as evidenced by a reduction of about 6.4 percent globally and 13 percent in the USA during the Covid-19 pandemic in 2020, the economic rebound in 2021 put global CO_2 emissions on course to actually increase by 5 percent, reversing most of that decline, and governments such as those of the USA, Canada, and Germany, far from squelching new fossil fuel extraction and development, were increasing their fossil fuel investments and approving more licenses to extract such energy sources.[34] Overall, therefore, apart from temporary dips in emissions after the 2008/9 financial crisis and the Covid-19 pandemic in 2020, the total quantity of greenhouse gases emitted per year has continued its inexorable rise, notwithstanding all the billions invested in research, the nonbinding treaties, and the carbon trading.

2

Climate Science

> Human-made climate change has emerged far enough from the weather "noise" that even the public notices it . . . The science has become clearer and exposes an urgency for action that is not convenient for political operatives, but is understandable to the well-informed.
>
> James Hansen, Professor of Earth Science at Columbia University, Former Director, NASA Goddard Institute for Space Studies, 2019

As we have already seen, a planet's surface temperature is in a delicate balance between the relative amount of incoming solar radiation, reflected solar energy, and the greenhouse gases in the atmosphere. As Jeffrey Bennett points out in *A Global Warming Primer*, how that balance works can be seen in a brief comparison of the atmospheres of the Earth and Venus (Table 2.1).[1] Although the two planets are about the same size and are composed of similar amounts of rock and metal, they have hugely different surface temperatures: unlike that of Earth, the average surface temperature of Venus is hot enough to melt lead. What explains this vast difference in their temperatures? Although Venus is a little closer to the Sun than Earth, that alone cannot account for this large difference, and Venus also has brighter clouds that actually reflect more sunlight back into space than does Earth. The reason Venus is so much hotter than Earth is that its atmosphere contains about 200,000 times more CO_2, which serves as a greenhouse gas to trap heat. In contrast, the Earth is in what has been called a **Goldilocks zone** – neither too hot nor too cold to support human life as we know it. As the history outlined in Chapter 1 has shown, however, that delicate balance is being disrupted by the human shift to using fossil fuels over the past two centuries, and this chapter will help explain why and what might be done to mitigate the damage.

Table 2.1. *The tale of two planets*

	Earth	Venus
Distance from Sun	94 million miles	67 million miles
Surface temperature	15°C (59°F)	470°C (880°F)
CO_2 in atmosphere	Enough to make planet livable	About 200,000 times as much as Earth

Note: Data from Bennett, *Global Warming Primer*, page 4.

The Mechanisms of Global Heating

Although some nonscientists might find the science behind climate change complicated, its core elements are actually quite simple. Bennett memorably describes the core facts behind the current climate change and their unavoidable conclusion as Global Warming 1-2-3, which this book adapts to **Global Heating 1-2-3**:

(1) Fact: CO_2 makes planets hotter than they would otherwise be.
(2) Fact: Human activity, especially the burning of fossil fuels, releases CO_2, and adds more of this heat-trapping gas to the atmosphere.
(3) Logical conclusion: We must therefore expect that rising CO_2 levels will heat our planet.

Although this simple formalism refers specifically to CO_2, the same applies to other greenhouse gases, such as methane (aka "natural gas"), which arises from agricultural practices and extractive processes such as fracking, and also hydroflourocarbons, which are released from industrial processes.

CO_2 Is Heating Our Atmosphere

The foundational fact in Global Heating 1-2-3 is that the presence of CO_2 in the atmosphere makes planets hotter than they would otherwise be. As Bennett explains, we know this is true for several reasons. First, the specific mechanism – the greenhouse effect shown in Figure 1.1 – has been established and verified in the lab for more than a hundred years. According to the How Global Warming Works website, the greenhouse effect can be explained in as few as thirty-five words: "Earth transforms sunlight's visible light energy into infrared light energy, which leaves Earth slowly because it is absorbed by greenhouse gases. When people produce greenhouse gases, energy leaves

Earth even more slowly – raising Earth's temperature."[2] While a tiny minority of scientists disputes the degree of threat that global heating represents, no legitimate scientist has come forward to dispute the basic physics of the greenhouse effect. A second reason for Fact 1 is that if there was no greenhouse effect, a planet's average temperature would depend only on its distance from the Sun and the proportion of sunlight it receives and emits. If we calculated Earth's temperature from this alone, without the influence of greenhouse gases, we would get −16°C (3°F). But this is not correct, and we get the correct answer for our actual planetary temperatures only when we use mathematical formulas that include the greenhouse effect. A third reason for Fact 1 is that if measurements of CO_2 concentrations over the course of Earth's history demonstrate that temperatures were higher when CO_2 concentrations were higher, as discussed in more detail below.[3]

Although approximately 98 percent of Earth's atmosphere is made up of nitrogen (N_2) and oxygen (O_2), molecules of those gases do not absorb the infrared light that is sent back into space by Earth and so do not contribute to the greenhouse effect. (For molecules to absorb infrared light, they need to be able to vibrate and rotate in a particular way, but molecules with only two atoms, especially when these two atoms are the same, such as N_2 and O_2, do not move in that way.) Instead, it is CO_2, methane (CH_4), water vapor, and other greenhouse gases in our atmosphere that prevent the infrared light from escaping directly into space, and without them our planet would be frozen over. Thus, either too little or too much of those gases would prevent Earth from being a Goldilocks zone.

Although water vapor is also a greenhouse gas, and our atmosphere contains about ten times more water vapor than CO_2, CO_2 plays a much more critical role in setting Earth's temperature. This is because – unlike water vapor, which cycles quickly into the atmosphere through evaporation and out of it through rain and snow – once CO_2 concentrations increase, they remain high for thousands of years. Nonetheless, water vapor can act as an **amplifier of climate change** initiated by other factors, such as increasing CO_2. If greenhouse gases raise the global temperature, the atmosphere can hold more water vapor, trapping more heat and making temperatures rise even more; if CO_2 levels were to drop and global temperatures to decrease, however, there would be less water vapor in the atmosphere, trapping less heat and making temperatures drop further still.

Human Activity Is Causing Increases in Greenhouse Gases

The second central fact in Global Heating 1-2-3 is that human activity has been adding CO_2 and other greenhouse gases to our atmosphere. As Bennett explains, scientists have shown that levels of CO_2 have been rising in several ways. One of these is by sampling air at many locations, most notably from the top of Mauna Loa mountain in Hawaii, where such measurements have been taken since the 1950s (Figure 2.1).[4] Scientists have also developed ways to estimate what atmospheric CO_2 concentrations were like much further back in history. One such method is drilling ice cores in Antarctica to depths of nearly two miles, resulting in columns of ice made up of snow that accumulated over about 800,000 years. Analysis of air bubbles in that ice has shown that CO_2 concentrations have risen and fallen many times in Earth's history, demonstrating a range of natural variation that took place long before humans began to burn fossil fuels (Figure 2.2). Yet that analysis also shows that the amount of natural variation over most of those 800,000 years fell between 180 and 290 parts per million, a measurement dwarfed by the massive and very rapid rise that has occurred since the Industrial Revolution. Indeed, the concentration of approximately 414 parts per million in 2021 was about 47 percent higher than the preindustrial level of about 280 parts per million, which at that time was one of the highest peaks in the whole record. If we extrapolate that level of increase into the future, we can anticipate that the concentration will double from the 280 parts per million level to 560 parts per million within just a few decades.

Figure 2.1

Atmospheric CO_2 measured from Mauna Loa Observatory

The graph shows direct measurements of the amount of CO_2 in the atmosphere. Such measurements are similar at other sites in the world. The short-term squiggles arise from the respiration of plants and trees which absorb CO_2 in the spring and summer and release it as they decay in the fall and winter. The units on the y-axis are parts per million (ppm). This refers to the number of CO_2 molecules among one million total molecules of air. It is now close to 420 ppm, which is about 0.042 percent. While CO_2 represents a tiny proportion of the molecules in the Earth's atmosphere, it has an outsized influence on global heating via the greenhouse effect.

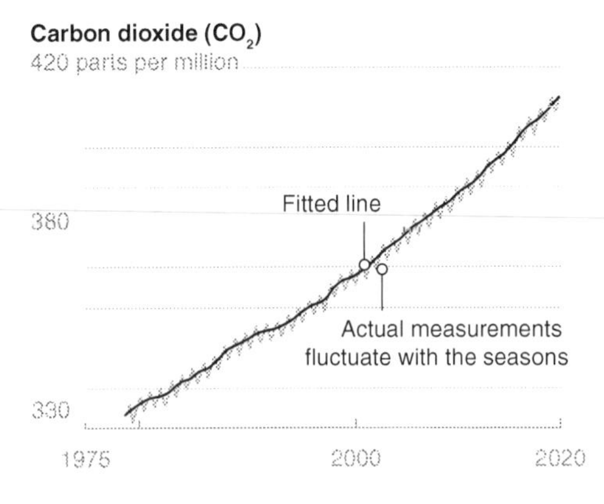

Adapted from the Scripps Institution of Oceanography Keeling Curve. C. D. Keeling, S. C. Piper, R. B. Bacastow, M. Wahlen, T. P. Whorf, M. Heimann, and H. A. Meijer, Exchanges of atmospheric CO_2 and $^{13}CO_2$ with the terrestrial biosphere and oceans from 1978 to 2000. I. Global aspects, SIO Reference Series, No. 01-06, Scripps Institution of Oceanography, San Diego, 88 pages, 2001. http://escholarship.org/uc/item/09v319r9

Figure 2.2

Carbon dioxide levels are higher now than at any point in the last 800,000 years

Analysis of Antarctic ice cores indicates that for the last 800,000 years, the Earth's carbon dioxide concentration has not risen above 290 parts per million (ppm). Our current carbon dioxide concentrations (~418 ppm in 2021) are therefore about 40 percent higher than the Earth's highest level over at least the last 800,000 years. Unless countries reduce their emissions, these concentrations are predicted to increase dramatically by the end of this century.

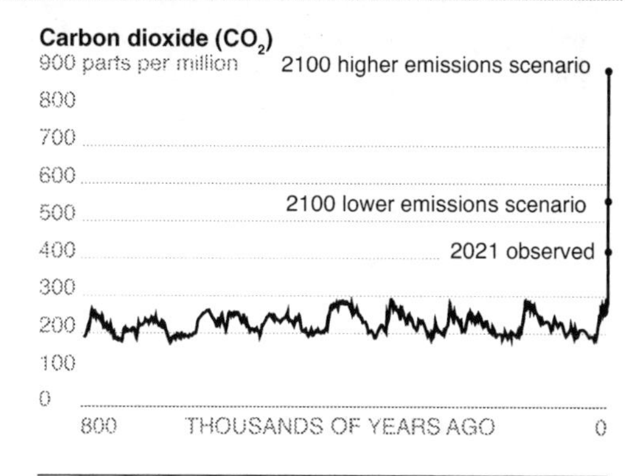

Adapted from Thomas R. Karl, Jerry M. Melillo, and Thomas C. Peterson, (eds.). Global Climate Change Impacts in the United States, Cambridge University Press, 2009. https://www.globalchange.gov/browse/reports/global-climate-change-impacts-united-states

Despite demonstrated natural variations in CO_2 levels over the history of the planet, there are many reasons to think that the huge recent increase in greenhouse gases is due to human activity. As noted in Chapter 1, chief among these is that the rise of atmospheric CO_2 coincides very closely with the increase in CO_2 from human activity that started with the Industrial Revolution around 1800 (Figure 2.3). Furthermore, because burning fossil fuels consumes oxygen, increases in CO_2 due to burning should be accompanied by a corresponding decrease in oxygen – a decrease that has indeed been measured, although one small enough not to seriously affect us. A third and also very convincing piece of evidence comes from isotope analyses of carbon from ice cores.[5] The relative abundance of the three isotopes of carbon – carbon-12, carbon-13, and carbon-14 – is different in carbon that comes from different sources, such as volcanoes or the burning of fossil fuels. Carbon-13, for instance, appears at lower levels in living organisms and thus at lower levels in fossil fuels, which are the remains of once-living organisms. When scientists measure the carbon in bubbles from ice cores, they can plot the relative abundance of carbon-13 in those ice bubbles across time. If increasing CO_2 concentrations are related to carbon from burning fossil fuels, one would expect the proportion of carbon-13 isotopes in the carbon in ice bubbles to decrease over time, which is exactly what scientists have found.

On the basis of such evidence, whether global heating is human-caused is no longer a matter of debate among climate scientists. In 2021, a IPCC report produced by 200 scientists, based on 12,000 published references and

Figure 2.3

The fact that atmospheric CO_2 levels tightly track CO_2 emissions from burning fossil fuels clearly shows that human activity relates to the build up of greenhouse gases

(A) Estimates derived from atmospheric measurements at Mauna Loa plus other sources such as ice cores. (B) Estimate of the emissions derived from burning coal, methane gas, etc. going back to the Industrial Revolution.

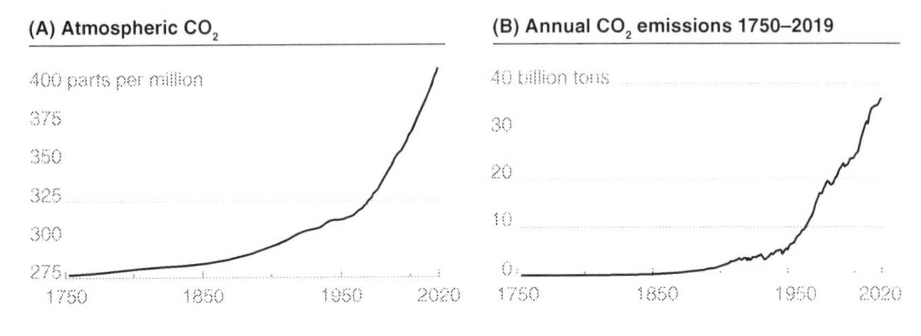

Adapted from NOAA Climate.gov | Data: NOAA, ETHZ, Our World in Data

approved line-by-line by representatives of 195 governments, declared that the evidence for this link is now "unequivocal."[6]

Although some skeptics outside the scientific community continue to argue that this heating is due to natural causes rather than human activity, their arguments are based on misunderstandings of the basic science involved. One frequently heard argument is that global heating arises from variation in the intensity of the Sun. To understand why this claim is simply incorrect requires understanding something about the influence that the Sun has on our climate, or what scientists refer to as the Sun's **climate forcing** effect. A climate forcing is a perturbation of Earth's energy balance, which is measured in watts per square meter (W/m^2) and can be either natural or anthropogenic and either positive or negative. A positive forcing, such as the impact of greenhouse gases, will cause global heating, while a negative forcing, such as a volcanic eruption that injects huge amounts of gas and dust into the Earth's stratosphere, will block sunlight and lead to global cooling. The variation in the amount of energy emitted from the Sun, known as the sunspot cycle, changes its brightness by only about 0.1 percent, creating a range of positive forcing of only about 0.24 W/m^2 , while the increase of CO_2 from the preindustrial level to 2018 corresponds to a forcing of about 2 W/m^2.[7] What this means is that any change that can be attributed to increased forcing from the Sun is only about a tenth of the change that can be attributed to the rise in CO_2 levels since the Industrial Revolution.

Other evidence that the extent of global heating cannot be explained by the Sun's influence comes from a historical comparison of the Sun's irradiance (the amount of energy it emits) with changes in temperature (Figure 2.4). If heating was caused by changes in the Sun, then we would expect the irradiance to increase along with temperature. While that did happen until about 1950, the two measures have since gone in opposite directions, with global heating accelerating and irradiance decreasing. Overall, therefore, the science clearly shows that global heating is not due to variation from the Sun. Nor are there any other natural candidates that can help explain the current degree of global heating; the contribution from volcanoes, for instance, appears to be less than 1 percent of that from human activity.[8]

Another way that scientists have addressed the question of whether natural factors alone can explain the heating is by comparing models that include only natural processes with models that include both natural processes and human activities to see which best explains the measured temperature increases. This research employs complex **earth system models** run on supercomputers that represent the weather as a grid of three-dimensional cubes, each representing one small part of the planet (Figure 2.5). Earth system models simulate the interactions of the atmosphere, oceans, land surfaces, ice, and the biosphere to estimate climate conditions and the movement of carbon through the earth

Figure 2.4

Changes in energy from the sun are extremely unlikely to explain global heating

A historical comparison of (A) the Sun's irradiance (the amount of energy it emits) with (B) global temperature change shows that, while the curves generally match up until 1950, they have since gone in opposite directions. This is precisely the period of time when global heating has accelerated. This strongly argues against the "sun spot" theory of global heating.

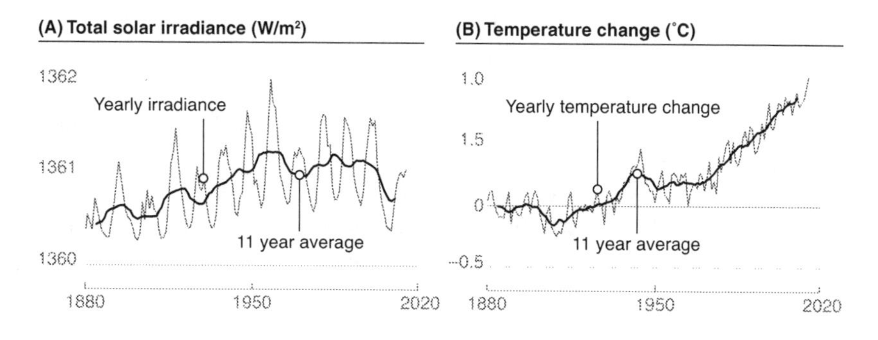

Adapted from Skeptical Science "Solar vs. Temperature" (CC BY 3.0). Available online. https://skepticalscience.com/graphics.php?g=5

Figure 2.5

Predictions about heating and other aspects of climate are made with earth system models

Earth systems modeling runs in a computer program. The planet is divided into a 3-D grid of "cubes" that straddle the Earth's surface and vertical space. Equations are then calculated which include variables such as temperature, wind speed, ocean currents, moisture, salinity, and pressure across space and time. Together, the equations simulate our atmosphere, surface vegetation, ocean biogeochemistry, and land and sea ice to provide insights into how carbon moves through the Earth's system, and how temperature and precipitation may change.

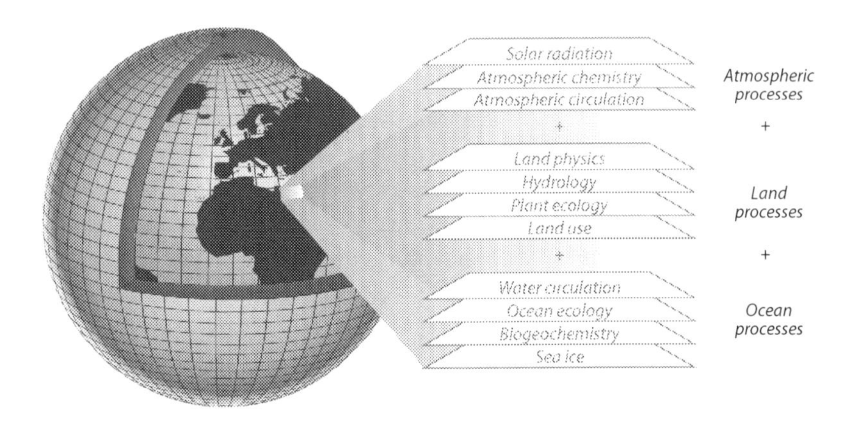

system. In addition to such variables as temperature, air pressure, wind speed, and humidity, these models include a representation of the **carbon cycle**, in which CO_2 is absorbed into the oceans, taken up into plants through photosynthesis, and then released again through the respiration of animals and the decaying of dead plant and animal matter, in order to predict how conditions are likely to evolve across time. Such models have been able to reproduce the actual climate data over the last century very well and, especially relevant here, it has been shown that when these models are run leaving out human activities, natural factors alone cannot predict or account for the actual measured temperatures (Figure 2.6).

Another reason some people argue that global heating is not actually occurring is because the weather has been colder or wetter where they live, an argument that confuses *local* effects with *global* effects, and very likely also confuses *weather* with *climate*. Global heating and climate change refer to changes in long-term trends, usually over decades, in atmospheric conditions, whereas the weather typically refers to short-term and local variability. Thus, overall increases in average global temperatures do not rule out significant

Figure 2.6

Climate models also provide evidence that global heating is caused by humans

Scientists create climate models (see Figure 2.5) that instantiate what is known about the physics of the atmosphere, land, and ocean and that can estimate temperature at each time step. If one runs such a model across time that includes natural factors (such as ocean cycles and volcanic activity) and human influence (i.e., increases in fossil fuel burning) then the model's estimate of temperature is very good; but if one only includes natural factors then the estimate is wrong. This provides further evidence that human influence underpins global heating.

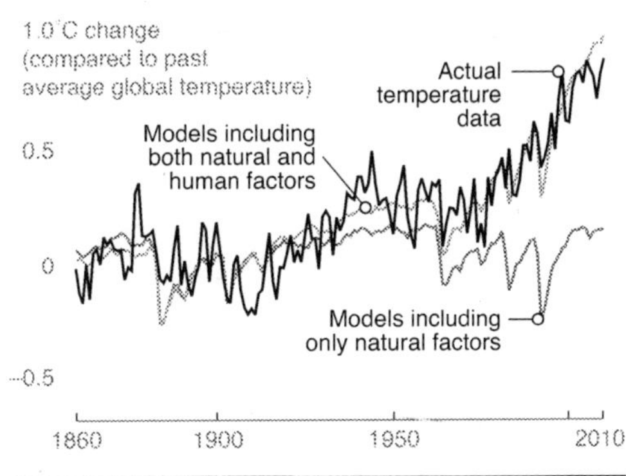

Adapted from Jeffrey Bennett, from Bennett, J. (2016) A Global Warming Primer. Big Kid Science.

regional and yearly variations in weather. Further, unusually cold and snowy winters periods – corresponding to the polar vortex that has occurred in the last few years, in which cold air from the Artic plunges down into Canada and the USA – can themselves be extreme weather events related to climate-induced disruption.[9] Misleading graphs referring to regional instead of global changes, or which cherry-pick short-term trends, are sometimes used by those trying to foster climate denial, but they should not obscure the essential point that long-term and global trends clearly show that our planet is heating up, and at an accelerating rate.

In summary, there is substantial evidence for the conclusion that human activity is responsible for rising CO_2 levels and global heating. And this, along with the fact that CO_2 makes the Earth hotter than it would otherwise be, leads to the inevitable conclusion that we must expect that rising CO_2 levels will translate into more heating. And that is exactly what has been observed by multiple independent research groups, such as the National Oceanic and Atmospheric Administration, NASA, the UK Meteorology Office, and the Japanese Meteorology Agency. As shown in Table 2.2, their measurements show that 2020 was the second hottest year since records began to be kept in the 1880s, and that as of 2021, each of the last three decades has been hotter than the one before, nineteen of the twenty hottest years have occurred in the last nineteen years, and the past six years have been the hottest on record.[10]

Table 2.2. *Recent years ranking among hottest recorded years and amount of warming over 1850–1900 average*

Year	Rank	Warming in °C	Warming in °F
2020	2	1.27 ± 0.04	2.29 ± 0.08
2019	3	1.25 ± 0.04	2.24 ± 0.08
2018	6	1.11 ± 0.04	2.00 ± 0.08
2017	4	1.18 ± 0.04	2.12 ± 0.08
2016	1	1.29 ± 0.04	2.33 ± 0.08
2015	5	1.15 ± 0.04	2.08 ± 0.08
2014	8	1.01 ± 0.05	1.82 ± 0.08
2013	13	0.95 ± 0.05	1.70 ± 0.08
2012	16	0.92 ± 0.04	1.65 ± 0.08
2011	18	0.91 ± 0.04	1.64 ± 0.08
2010	7	1.03 ± 0.04	1.85 ± 0.08

Note: Data from http://berkeleyearth.org/global-temperature-report-for-2020/. The numbers show the mean temperature in a year, e.g. 1.27°C with the 95 percent confidence range around that, indicated by $\pm$ 0.04, which means plus or minus 0.04, i.e. the temperature was 95 percent likely to range from 1.23°C to 1.31°C.

Human Activities Produce Multiple Greenhouse Gases

To understand what needs to be done about our current predicament, we also need to recognize that CO_2 is produced in several other ways than burning fossil fuels and that it is not the only greenhouse gas about which we need to be concerned (Figure 2.7). CO_2 is also released from land use: for instance, the massive deforestation to clear land for agriculture and livestock has released huge amounts of CO_2 into the atmosphere and also reduced the number of trees that can absorb excess CO_2 from the air (Figure 2.8).[11] Furthermore, while healthy soil can hold about 70 percent of land-based organic carbon, less CO_2 is stored when intensive farming practices lead to its being repeatedly plowed and compacted by heavy machinery or livestock. Another substantial source of CO_2 (about 5 percent of current levels) is the massive global production of cement, which is made from heating carbonate materials such as limestone, which then release some of their CO_2.

Yet it is not only CO_2 levels that have been increasing as a result of human actions. Methane levels have more than doubled over the last 150 years, largely from leaks within the oil and gas industry, the waste sector, and agriculture (such as the methane released from the belches of livestock).[12] Methane is also released from microbes in water-logged rice paddies. Another greenhouse gas, nitrous oxide, arises from fertilizers and animal waste and has increased by around a third over the past 150 years.[13] As noted in Chapter 1,

Figure 2.7

Human activity leads to rising levels of multiple greenhouse gases, not only CO_2

Human activity increases concentration of (A) CO_2, (B) nitrous oxide, (C) methane, and (D) the halocarbons. While the concentration of specific halocarbons (chloroflourocarbons, CFCs) was increasing steeply in the 1970s and 1980s, it began declining in the 1990s after the Montreal Protocol, when countries began working together to repair the "ozone hole."

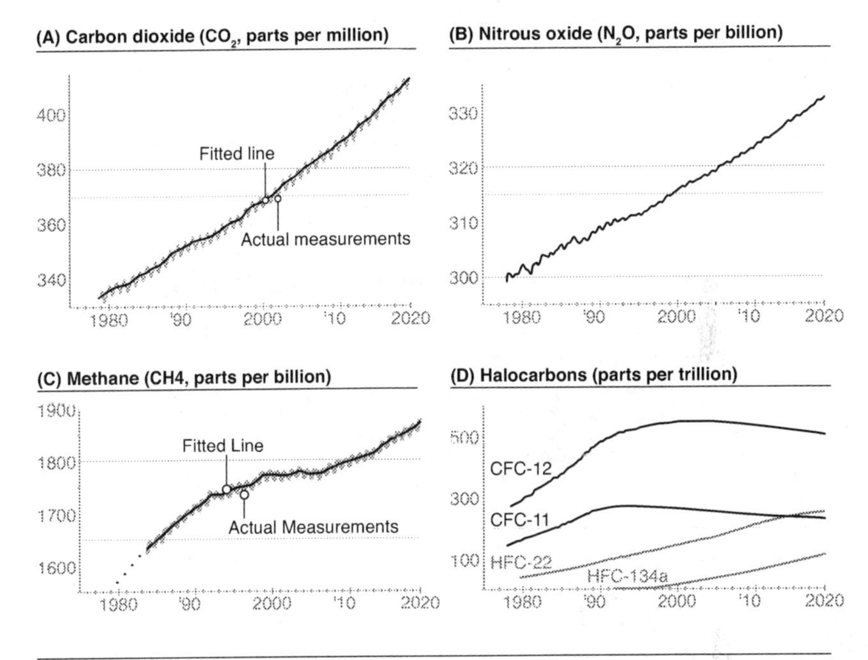

Adapted from the NOAA Annual Greenhouse Gas Index (AGGI). https://gml.noaa.gov/aggi/aggi.html

hydroflourocarbons are another class of especially potent greenhouse gases that come from industrial/chemical processes.

While, as Figure 2.7 shows, the proportion of methane, nitrous oxide, and hydroflourocarbons in our atmosphere is much less than that of CO_2, they all absorb much more heat than CO_2 – in some cases, thousands of times more. This ability to absorb heat is known as the **global warming potential** of a gas (Table 2.3). The difference in the global warming potential of various gases is important in that it means we cannot focus solely on reducing CO_2 emissions, as essential as that will be. For example, the fossil fuel industry and its political allies often refer to methane as "natural gas" and as a "clean" energy source as

Table 2.3. *Global warming potential (GWP) of greenhouse gases compared to* CO_2

Greenhouse gas	GWP after 20 years	GWP after 100 years
Carbon dioxide	1	1
Methane	72	25
Nitrous oxide	289	298
HFC-23	12,000	14,800
HFC-134a	3,830	1,430
SF6	16,300	22,800

Note: Data from IPCC Assessment Report 4. HFCs are hydroflourocarbons produced by chemical plants; SF6 is sulfur hexafluoride, a human-made industrial gas used as an electric insulator.

Figure 2.8

Global greenhouse gas emissions, broken down by groups of gases and sectors

(A) In the USA, the bulk of greenhouse gas emissions are from CO_2. Most of these emissions come from fossil fuel and industrial uses, but some are from forestry and land use. Methane is also a substantial contributor, for example from animal agriculture and from leaking pipes and well heads from fracking extraction. (B) Energy is the largest contributor to greenhouse gas emissions. This includes electricity/heat, transportation, manufacturing/construction, fugitive emissions, building, and other fuel combustion.

(A) 2018 greenhouse gas emissions by groups of gases

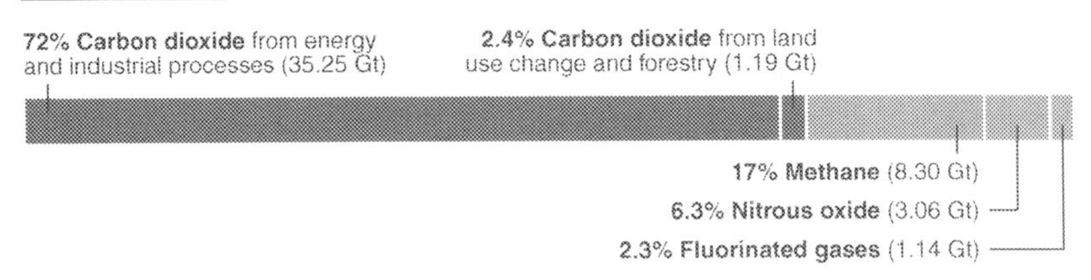

(B) 2018 Greenhouse gas emissions by sector

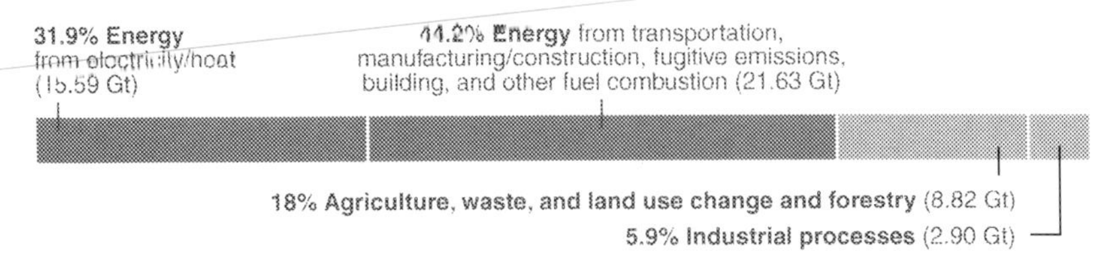

Adapted from Climate Watch. 2020. Washington, DC: World Resources Institute. Available online at: https://www.climatewatchdata.org Land-Use Change and Forestry and Agriculture data from FAO 2020. FAOSTAT Emissions Database. CO_2 emissions from fuel combustion from CO_2 Emissions from Fuel Combustion. OECD/IEA 2019 compared to coal (such as on buses bearing the slogan "Clean Natural Gas" on their sides). But because of the leakage that occurs at methane wellheads and pipelines and its much greater global warming potential, "natural gas" might actually be as bad for global heating as coal.[14]

Global Heating Will Continue unless CO_2 Levels Drop

Figure 2.9 dramatically illustrates our predicament: CO_2 levels have shot up, and we know that temperature, with some delay, is going to follow. But while we can demonstrate that temperatures have already increased 1.1°C since preindustrial times, how fast it is going to increase and to what level in the future is less certain. To try to predict the degree and speed of that rise, it is useful to return to the concept of climate forcing, which refers to the physical process of how certain factors affect Earth's energy balance and climate. As already mentioned, current calculations of this anthropogenic climate forcing from the CO_2 "blanket" that traps the heat radiation that travels from Earth toward space indicate that the CO_2 increase from the preindustrial level of

Figure 2.9

While atmospheric CO_2 and temperature have moved up and down together over the last 800,000 years, CO_2 has shot up since the Industrial Revolution, and temperature is going with it, with a delay

The change in global mean temperature on the timescale of thousands of years can be estimated using data from ice cores from many locations around the world, while prior CO_2 levels can be measured from ice cores. The results show the tight control of CO_2 on global temperature. This figure suggests that the eventual warming for about 407 ppm CO_2 will be about 3.5°C, including both fast and slow climate feedback processes.

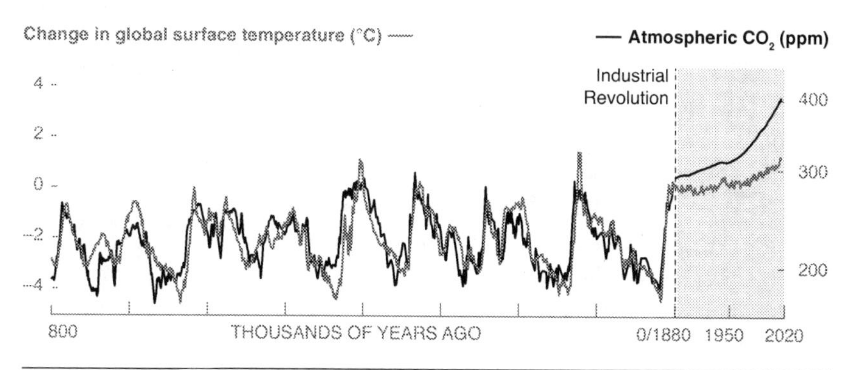

Reproduced from Makiko Sato and James Hansen, "Climate Change in a Nutshell", 18 December 2018. Available online at: http://www.columbia.edu/~jeh1/mailings/2018/20181206_Nutshell.pdf.

280 parts per million to the 2018 amount of 407 parts per million amounts to a forcing of about 2 W/m^2. But to predict future climate changes, scientists also have to take climate feedbacks into consideration, such as how global heating increases water vapor, which increases heating further (a positive feedback – the heating reinforces itself), and how more heating might produce more clouds that might thicken and thus reflect more sunlight to space and thus reduce the heating (a negative feedback – the heating leads to an effect that reduces heating).

Scientists describe the climate's response to forcing as climate sensitivity, which as we saw in Chapter 1 is the predicted change in temperature in response to the change in forcing factors. More specifically, they define climate sensitivity as how much global heating is likely to occur as a result of a doubling of CO_2 concentrations from the preindustrial level of 280 parts per million to 560 parts per million. As noted in Chapter 1, the Charney group's 1979 report concluded that our current climate sensitivity was 3°C above preindustrial levels, with a 90 percent degree of confidence of falling between 1.5°C and 4.5°C. Since then, however, research attempting to better account for the likely effect of feedback loops has yielded a narrower range of 2.6° to 4.1°C, with a best estimate of 3.1°C.[15] Although the good news is that the earlier high estimate of 4.5°C appears less likely, the bad news is that the lower range is now 2.6°C instead of 1.5°C.

The climate sensitivity parameter is an index of how much heating there will be as CO_2 increases, but how fast this will happen depends on human behavior regarding emissions. As Bennett points out, analyses of Antarctic ice cores have shown that earlier natural fluctuations in CO_2 levels, while significant, took place extremely gradually over tens of thousands of years (Figure 2.9).[16] These large and very slow changes in CO_2 levels were brought about by variations in the way the Earth travels around the Sun. Earth's less than perfectly symmetrical orbit means that its position relative to the Sun changes, driving these slow temperature cycles, also known as **Milankovitch cycles**. These changes in the Earth's position relative to the Sun affect the absorption of sunlight. For example, a small increase in the amount of sunlight being absorbed by the oceans causes them to warm slightly, which results in some of the CO_2 that is dissolved in the oceans being released into the atmosphere, which in turn leads to more ocean warming – another positive feedback loop. These slow changes in CO_2 therefore also change temperature, which over Earth's long history has driven transitions into ice ages (glacial periods) and out of them again (interglacial periods).

The critical point here is that these natural changes, which arise from a non-perfectly symmetrical Earth's orbit and are modulated by positive feedback cycles, occur over tens of thousands of years. They are completely different from the extremely rapid increase in CO_2 that we have seen over the last two centuries due to human actions, an increase that is now about a hundred times faster than those previous natural increases.[17]

While direct measurements of CO_2 from air bubbles trapped in ice cores go back only 800,000 years, scientists have found ways to indirectly estimate atmospheric CO_2 levels as far back as three million years, such as by measuring isotopes in carbonate sea shells.[18] Such evidence indicates that CO_2 levels today are higher than at any point going back three million years. Way back then, global average temperatures were 2–3°C higher than our preindustrial times and sea levels were sixteen meters higher,[19] showing again how sensitive the Earth system is to changes in greenhouse gas concentrations. We can be sure that if enough time passes, our current high CO_2 concentrations will also yield a dramatic change in temperature, as we will see later in this chapter.

Climate change skeptics point to the natural changes that we can observe in Figure 2.9 to suggest that the changes that human activity is causing today are nothing to worry about. But as that figure also shows, our current temperatures are already approaching their highest point in 800,000 years. Given that CO_2 concentrations in our atmosphere are about 47 percent higher than at any other time in that history and rising rapidly, we should be very concerned about how much higher the temperature will rise. As the figure shows, CO_2 levels and temperatures moved quite tightly together throughout this history, suggesting that the more recent uncoupling of these levels as CO_2 levels shot up is eventually going to be accompanied by temperatures also going up dramatically. And whereas the figure makes clear that the fluctuation of warm and cool periods happened over thousands of years, the changes occurring now are happening over mere decades.

Bennett also points to how some skeptics about the human cause of global heating or how grave a threat it poses argue that if we go even further back in geohistory, to about fifty million years ago – the Early Eocene era – CO_2 levels and temperatures were much higher than they are today, overlapping with the dinosaur era when much life flourished (Figure 2.10). Indeed, evidence from that period suggests that CO_2 levels may have been over a thousand parts per million; this was apparently due to a greater shifting of tectonic plates in land masses that caused much more volcanic activity, releasing huge amounts of CO_2 into the atmosphere, which also massively increased the temperature relative to now. But far from giving us reassurance, this should make us even more concerned that our current heating could be much more devastating than many assume. During the era of the dinosaurs there were no ice caps in either the Arctic or Antarctica, which suggests that we may be headed to that ice-free

Figure 2.10

Unmitigated emissions will create a climate analogous to the Eocene geohistorical period (about 50 million years ago)

Six geohistorical states of the climate system were analyzed as potential analogs for future climates. They are situated in a time series of global mean annual temperatures for the last 65 million years. Major patterns include a long-term cooling trend, periodic fluctuations driven by changes in the Earth's orbit, and recent and projected warming trends. Temperature anomalies are relative to 1961–1990.

During the Eocene, the temperature was high owing to CO_2 levels of about 1,000 ppm, and these arose from major volcanic eruptions caused by crashing tectonic plates over millions of years. According to the modeling, this is our future under the high-emissions scenario (A, RCP8.5). Under somewhat mitigated emissions (B, RCP4.5) we will experience a climate like the mid-Pliocene about 3 million years ago.

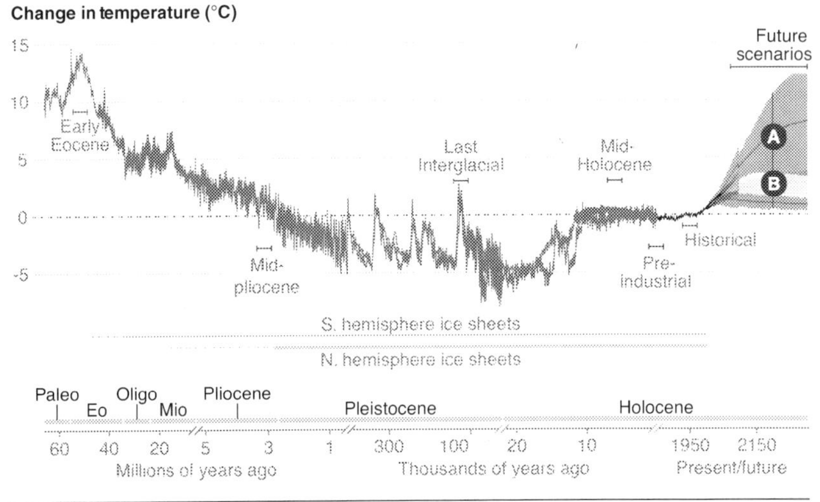

Adapted from K. D. Burke, J. W. Williams, M. A. Chandler, A. M. Haywood, D. J. Lunt, B. L. Otto-Bliesner, Pliocene and Eocene provide best analogs for near-future climates. *Proceedings of the National Academy of Sciences* Dec 2018, 115 (52) 13288-13293. DOI: 10.1073/pnas.1809600115

state too. If that happens, sea levels would rise to such an extent that every coastal city on the globe would be underwater. When the dinosaurs roamed, sea levels were as much as two hundred feet higher than they are today. And while it is true that some life forms managed to adapt to and thrive despite CO_2 concentrations at that level, most of our plants and animals – including human beings – are not so adapted, and the changes we are witnessing are happening much too fast for such natural adaptation to keep up with.[20]

Examining the geohistorical record is also interesting in a different way: it can give us some insight into what our climate future might look like. Figure 2.10 suggests that, under what the IPCC identifies as a high-emissions scenario, we will experience mid-Pliocene conditions by 2030 and after that Eocene conditions. This is a frightening prospect – just a couple of hundred years of human activity will have returned the Earth to a state that it had not experienced for fifty million years, and at a speed that is likely outside the range of our ability to evolutionarily adapt.[21]

What Can Be Done to Limit the Impact of Global Heating

The scientific consensus about the causes of global heating has led thousands of scientists to the sober conclusion that the only way out of our current climate predicament is to cut emissions.[22] Nonetheless, some skeptics have argued that taking immediate and serious action to lower concentrations of greenhouse gases in the atmosphere is unnecessary and overly costly. Take, for example, the arguments of Bjørn Lomborg, a political scientist from Denmark who in response to the IPCC's 2021 report admitted that "climate change is a real problem" but urged readers not to buy into what he described as "scare stories on climate impacts," arguing that the UN has a "long history of claiming catastrophe is around the corner" and that "although climate change in total has negative impacts, we rarely hear about the positive impacts, such as a profound global greening of the planet, equivalent to two or more continents of green, each the size of Australia."[23] This section first explains why such arguments against taking immediate and concerted action to reduce emissions fail to hold up against the science and then examines the likely outcomes of various steps the world can take to prevent global heating from getting much worse.

Dubunking the Supposed Benefits of Global Heating

Some climate change skeptics argue that global heating actually has some benefits, such as producing longer growing seasons for crops. In the continental USA, for instance, climate change has already added about twelve more days to the season for planting, cultivating, and harvesting crops.[24] But while this longer season has the potential to increase crop yields, taking advantage of it will also require more water, which in some places is getting scarcer, as in the mega-drought in the southwestern USA that is itself attributed largely to climate change.[25] Producing these additional crops will also require more nitrogen, insecticide, pesticide, herbicide, and fungicide, which together make the viability and economics of a longer growing season not quite so clear.[26] Furthermore, a McKinsey report on the world's six major breadbaskets found that if we stay on a high-emissions pathway, by 2030 the probability of a 15 percent decline in corn and soybean yields in any given year will double compared to now, and research on wheat and corn has shown that any benefits from the longer growing season are likely to be counteracted by the impact of more heat stress on the plants.[27]

In a 2020 report, the Russian government touted other supposed benefits from climate change, including decreased energy use in cold regions, expanding agricultural areas, and increased navigational opportunities in the Arctic Ocean.[28] Yet the same report also noted that Russia is heating 2.5 times faster than the planet as a whole, which poses such risks as collapsing permafrost (which destabilizes the frozen ground upon which infrastructure is built), increased likelihood of infections and natural disasters, and species being pushed out of their habitats. Other reports have flagged the likelihood that Russia will face increased migratory pressure from climate refugees fleeing the impact of climate change in Central Asia and northeastern China.[29]

The supposed benefits of climate change also have to be set against the colossal environmental and economic impacts of climate change that will be examined in more detail in Chapter 3. For example, in a 2014 report titled *Risky Business*, a bipartisan group of business leaders including Michael Bloomberg, Henry Paulson, and Tom Steyer concluded that by 2050, between $66 billion and $106 billion worth of existing US coastal property would be below sea level.[30] Consultants at Morgan Stanley estimated that climate-related disasters in just the three-year period of 2016–18 cost the world $650 billion,[31] about $415 billion of which was in North America and amounted to 0.66 percent of **gross domestic product** (GDP).[32] Another research study estimated the cost of wildfires in California in 2018 alone as $149 billion.[33] The costs of such escalating extreme weather events are predicted to increase enormously over the next decades, for both the USA and the wider world.

The claim that climate change will confer a few benefits, such as a hypothetical increase in some crop yields, also needs to be set against the source of such claims. This includes Tony Abbott, former prime minister of Australia,[34] backer of coal, and repealer of a carbon tax, who said "[c]limate change itself is probably doing good; or at least, more good than harm ... There's the evidence that higher concentrations of CO_2 – which is a plant food after all – are actually greening the planet and helping to lift agricultural yields."[35] Another source of these claims is David Koch, the oil industry magnate, who argued that, due to climate change, "the Earth will be able to support enormously more people because a far greater land area will be available to produce food."[36]

Evaluating the Expense of Responding to Global Heating

Some people who are willing to accept that global heating is human-caused and is harmful nonetheless object that it is just too expensive to deal with. For example, The skeptic Bjørn Lomborg mentioned earlier received up to $200,000 in a single year from a right-wing funder,[37] has frequently asserted that the cost of acting now is much too high to avoid impacts that he claims will not cost very much decades down the line. For instance, in 2014 he wrote: "If we insist on 2°C, we will pay an extra $60,000 billion dollars, but only prevent a stream of $100 billion damages that begins in 70–80 years."[38]

Although the costs of mitigating and preventing the worst effects of global warming are undeniably high, we need to be very skeptical of the claim that they are just not worth paying now because damages later are not likely to be as grave as the IPCC and other experts claim. First, while it is true that the carbon emissions that humans inject into the atmosphere today will not have their full effect on the climate for decades, we will not need to wait seventy years to accrue "a stream of $100 billion" damages. Those damages are already here. In recent years, as we saw earlier, the USA alone has exceeded $100 billion in such damages in a single year, the world's damages for a single three-year period were estimated at $650 billion (2016–18), and the insurance giant Swiss Re warned that climate damage could cost the world economy as much as $23 trillion in lost economic activity per year by 2050.[39] Second, as we shall see in Chapter 3, the standard economics approach of estimating damages that could take place later this century, which are often cited by Lomborg and others, have been beset by several profound methodological problems and absurd assumptions. As a quick example, the standard mainstream economic model for evaluating the costs of climate change in 2100 predicts that heating of 6°C, a rise that would produce an intolerable hothouse planet, would result in damages worth only 10 percent of GDP.[40]

In general, however, skepticism about the costs is not straightforward to refute because it requires taking into account a complex web of factors, not all of which can be reduced to simple numbers. Evaluating the full cost of global heating involves taking stock of the predicted impacts of climate change, the economic costs of damage, the costs and feasibility of shifting to renewable energy, issues of climate justice and intergenerational justice, and the incalculable damage done to the biosphere in terms of species extinctions. It also requires reckoning with the core assumptions and effects of capitalism and a larger vision of what kind of world we want. Ultimately, such a discussion is less about science and more about values, which will be discussed in Chapter 3 and subsequent chapters.

Alternative Emissions Scenarios for Addressing Global Heating

Any estimation of how much emissions will need to be cut to keep global heating from getting disastrously worse obviously depends upon how high and fast those emissions and the resulting temperatures will rise. And emissions are influenced by many human elements, including factors such as population growth, the intensity with which people use carbon-based energy sources, socioeconomic factors of many kinds, government policies and market-based factors regarding the rate of uptake of renewable energies, and the development of other technologies. To better understand how those elements interact, a community of experts (including climate scientists, energy modelers, and social scientists) has developed a set of **emissions scenarios** that describes a range of alternative futures.[41] Among the best known of these is a set of scenarios developed by the IPCC called **representative concentration pathways** (RCPs), each of which examines a different degree of climate forcing by the year 2100 as measured by W/m^2: 2.6, 4.5, 6, and 8.5. Each of these scenarios calculates the likely atmospheric concentration of CO_2 by the end of this century by inserting the emissions levels into a carbon cycle model that estimates how much CO_2 will be absorbed by the oceans and land reservoirs and how much will stay in the atmosphere. While this and later chapters refer to these RCP scenarios because many research results have been framed in that way, the IPCC has more recently shifted to discussing five **shared socioeconomic pathways** (SSPs), as described in Tables 2.4 and 2.5 and which have an approximate correspondence to the RCPs.

The highest of these emissions scenarios, RCP8.5, assumes an increasing use of coal and no serious effort to cut emissions so that they keep growing until 2100 (Figure 2.11). One of the lowest emissions pathways, RCP2.6, assumes that emissions peak around 2020, decrease throughout this century, and then go negative, meaning that carbon is actually removed from rather than added to the atmosphere, creating so-called **negative emissions**. This relies on technical approaches, such as the vast cultivation of crops that would then be burned to generate electricity, with the CO_2 being captured and sequestered underground, a process known as **bioenergy with carbon capture and sequestration** (BECCS), and another approach known as **direct air capture of CO_2**, which requires enormous numbers of built devices to filter the gas out of the air and again to sequester it underground. The intermediate RCP6 and RCP4.5 scenarios assume that some effort will be made to reduce emissions, leading emissions levels to peak around the middle of this century and atmospheric CO_2 levels to stabilize early in the next century, and they also rely on achieving negative emissions.

Table 2.4. *Shared SSPs currently used by the IPCC and their relation to RCPs*

Narrative	Relation to RCP	Partial description of narrative
SSP1	RCP1.9, RCP2.6	*Sustainability – taking the green road* The world shifts gradually but pervasively toward a more sustainable path; the emphasis on economic growth shifts toward a broader emphasis on human well-being; inequality is reduced within and across countries; consumption is oriented toward lower material growth and lower energy intensity
SSP2	RCP4.5	*Middle of the road* Social, economic, and technological trends follow their typical historical patterns; development and growth are uneven between countries; slow progress is made globally and nationally toward sustainable development; environmental systems degrade, but there are some improvements and overall resource and energy use declines; global population growth is moderate and reduces in the second half of the century
SSP3	RCP7.0	*Regional rivalry* A fragmented world of resurgent nationalism is concerned about competitiveness and security, and conflict forces countries to focus on domestic or regional issues; countries focus on achieving their own energy and food goals instead of broader-based development goals; investments in education decline; economic development is slow; consumption is material-intensive, and inequalities worsen; there is there is strong environmental degradation
SSP5	RCP8.5	*Fossil-fueled development* Societies worldwide show increasing faith in competitive markets and rapid technological progress (including using geoengineering to "solve" the climate problem); more globalization of markets; increased exploitation of fossil fuel resources and adoption of more energy-intensive lifestyles

Note: Data from Carbon Brief, www.carbonbrief.org/explainer-how-shared-socioeconomic-pathways-explore-future-climate-change/. The correspondence between RCPs and SSPs is drawn from the IPCC's 2021 report; see Table 2.5 of this book.

Although BECCS is discussed in more detail in Chapter 8, for our purposes here it is important to note that it is mostly unproven, would be hugely expensive, would involve continent-sized amounts of land and compete with the need to grow food, and would require enormous amounts of water.[42] Further, the current estimate of the cost of direct air capture is even higher than for BECCS. According to James Hansen and his team, if emissions remain constant at their current levels (not even taking into account the

Table 2.5. *Estimates for heating based on SSPs from IPCC 2021 report*

	Near term, 2021 to 2040		Long term, 2041 to 2060	
Scenario	Best estimate	Very likely	Best estimate	Very likely
SSP1−1.9	1.5°C	1.2−1.7°C	1.6°C	1.2−2.0°C
SSP1−2.6	1.5°C	1.2−1.8°C	1.7°C	1.3−2.2°C
SSP2−4.5	1.5°C	1.2−1.7°C	2.0°C	1.6−2.5°C
SSP3−7.0	1.5°C	1.2−1.7°C	2.1°C	1.7−2.6°C
SSP5−8.5	1.6°C	1.3−1.9°C	2.4°C	1.9−3.0°C

Note: Data from www.ipcc.ch/report/ar6/wg1/.

Figure 2.11

Representative carbon pathways (RCPs) are estimates of the level of emissions and heating we'll have depending on which socioeconomic behavior trajectory we take

(A) Anticipated carbon emissions on the different pathways. RCP 8.5 refers to a forcing of 8.5 W per meter squared by 2100, and is a high-emissions pathway – it assumes that we carry on emitting for many more decades without mitigation. In the new IPCC scheme it is referred to as the Shared Socioeconomic Pathway (SSP-5), "Fossil Fueled Development." RCPs 6.0, 4.5, and 2.6 are all pathways with some mitigation. For example, RCP2.5 assumes that emissions are reduced and also that "negative emissions" occur, i.e. CO_2 is "pulled out of the air" by carbon capture methodologies in the future. Note the y-axis is gigatons of carbon, rather than CO_2. (B) Projected temperature. Under RCP8.5 we are headed for a "hothouse" planet by 2100.

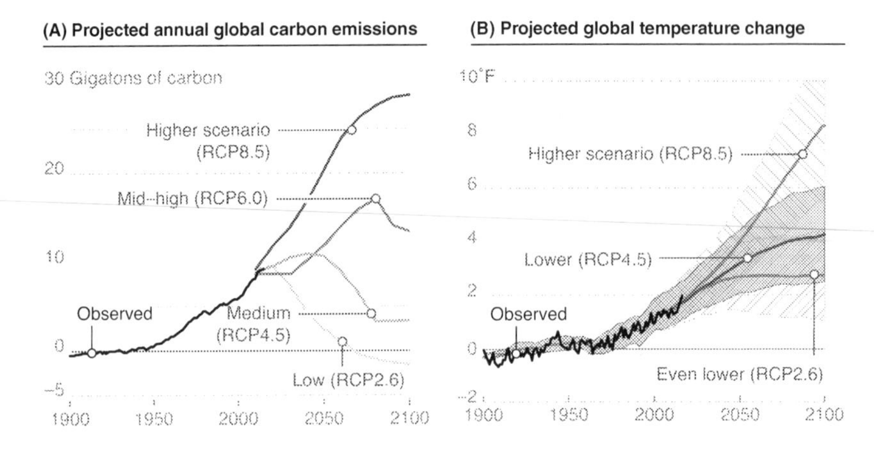

Adapted from USGCRP, 2017: Climate Science Special Report: Fourth National Climate Assessment, Volume I [Wuebbles, D.J., D.W. Fahey, K.A. Hibbard, D.J. Dokken, B.C. Stewart, and T.K. Maycock (eds.)]. U.S. Global Change Research Program, Washington, DC, USA, 470 pp.

approximately 2 percent increase per year that has generally occurred), by 2100 we would reach 547 parts per million of CO_2 in the atmosphere without any CO_2 extraction, a pathway that would produce disastrous consequences. The team calculated that bringing CO_2 levels down to 350 parts per million, a supposedly safe level, would require removing 695 gigatons of CO_2 from the atmosphere, which, taking the low-cost estimate of $40–95 per ton, would come to an estimate of around $104–223 trillion.[43] Thus, they concluded, CO_2 extraction is unlikely to provide a panacea for our climate dilemma, leaving us with no practicable current option but to immediately and rapidly cut our fossil fuel use, along with adopting improved agricultural and forestry practices to sequester carbon.

While some critics have suggested that the high-emissions pathway of RCP8.5 is misleading and unrealistic because it assumes a roughly fivefold increase in coal use by 2100,[44] others point out that the high-emissions scenario is useful quite aside from coal emissions, such as for simulating a massive release of methane from the thawing of Arctic permafrost, which would have a similar effect as a huge surge in fossil fuel use. (Indeed, in its 2021 report, the IPCC concluded with high confidence that "additional warming is projected to further amplify permafrost thaw"[45].) And whatever the source of climbing CO_2 emissions, RCP8.5 appears to be consistent with current trends in annual greenhouse gas emissions, as noted by the US government's 2018 National Climate Assessment.[46] In 2020, a comparison of the RCP8.5 pathway and estimates of current and stated policies provided by the International Energy Agency concluded, "not only are the emissions [of] RCP8.5 in close agreement with the historical total cumulative CO_2 emissions (within 1 percent), but RCP8.5 is also the best match out to mid-century under current and stated policies."[47]

Regardless of whether RCP8.5 is accurate in all its particulars, most available evidence suggests that dismissing its conclusions as alarmist would be a dangerous choice. For example, even though the economic shock of the Covid-19 pandemic led CO_2 emissions from fossil fuels to fall by an estimated 7 percent in 2020 – the largest drop since 1945 – atmospheric CO_2 still rose by about 2.48 parts per million, only a 0.32 smaller rise than expected without a lockdown (Figure 2.12).[48] In 2021, the International Energy Agency predicted that global CO_2 was set to show its second-biggest increase in history, driven largely by a strong rebound in demand for coal for electricity generation,[49] and NASA reported that global heating is accelerating and Earth's energy imbalance doubled in the period 2005–2019.[50] Such evidence strongly suggests that we are on a high-emissions pathway, and that unless we make dramatic cuts in fossil fuels soon, we will continue to stay on that path.

Figure 2.12

Fossil-fuel-related carbon dioxide increases steadily for China, India, and much of the world, with some leveling out for the USA and Europe

In 2020, the Covid-19 pandemic produced such an economic shock that CO_2 emissions (from burning fossil fuels) were expected to fall 7 percent for the year – the largest drop since 1945. However, 2021 was predicted to be a year of near record emissions growth again.

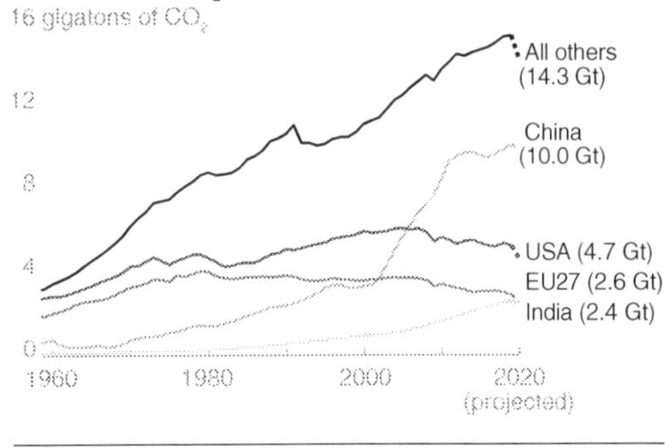

Adapted from the Global Carbon Project under the Creative Commons Attribution 4.0 International license.

Box 2.1 The Intergovernmental Panel on Climate Change

The IPCC was established in 1988 and mandated by the United Nations to undertake international assessments of scientific knowledge about climate change, including impacts and possible responses. The IPCC is governed by a group of selected government representatives from around the world, not by scientists.

The information assessed by the IPCC is integrated into an assessment report (AR) every five to seven years.[51] Each AR contains three volumes, led by a working group with several hundred members each. Working Group 1 assesses the science of anthropogenic climate change; Working Group 2 assesses the impact of climate change and how humans and nonhuman animals and plants might adapt; and Working Group 3 assesses the likely effects of mitigation efforts to slow down and prevent the worst effects. The reports on which these assessments are based are submitted by hundreds of authors and undergo a rigorous multistage review process, and the resulting summary statements are approved by the 195 governments who are part of the IPCC.

The IPCC was charged with being a technical advisory body rather than to recommend policy, but according to David Hulme's *Why We Disagree about Climate Change*, the distinction between policy and science has been hard to maintain and it is not always clear if IPCC reports reflect the views of the scientists or of the governments that approve the reports.[52] Some commentators have also claimed that the consensus-building model of the IPCC tends to produce conservative outcomes. One analysis of the language used by IPCC reports, for instance, claimed that "the main message – that our society is in climate emergency – is lost by overstatement of uncertainty and gets confused by the gigabytes of information,"[53] and another study described the reports' predictions as "erring on the side of least drama."[54] The large-scale consensus process also tends to be slow and to underweight recent evidence, which some have argued is not compatible with conveying the sense of emergency we are in; some scientists have suggested that rolling reports would be better.[55] Other critics have complained that the experts chosen to negotiate the consensus statement are not truly independent since they are nominated by governments, are biased toward the Global North, do not include representatives of fields such as sociology and anthropology, do not reflect local knowledge, and tend to represent the interests of corporations and governments.

Even given these concerns, the IPCC is an authoritative body. Its summary reports are consensus reports, every line approved by the scientists who reviewed thousands of papers and by all 195 governments who have signed up to participate in the process. When in 2021 the IPCC declared that the evidence for a human cause for global heating is "unequivocal," the wider world was on notice that the scientific debate over that aspect of climate change is over.

Our Remaining Carbon Budget

As discussed in Chapter 1, as a result of the growing recognition that the 2°C rise in global temperatures favored by some industrial nations is not safe, in 2015 a new international agreement in Paris committed signees to "pursue efforts" to keep that rise below 1.5°C. But as the IPCC recognized in 2018, even a 66 percent probability of keeping to this new target would require cutting 2010-level emissions by about 45 percent by 2030, or emitting no more than 420 gigatons of CO_2 from 2018 onward. As journalist Mark Lynas lays out in *Our Final Warning*, however, the committed emissions from the average forty-year lifetimes of existing electricity-generating coal and gas plants would themselves total 358 gigatons of CO_2 after January 2018, to which would be added another estimated 162 gigatons of CO_2 from industrial infrastructure, 64 gigatons from transportation, and 74 gigatons from the residential sector, for a total of 658 gigatons of CO_2.[56] But even that is a conservative estimate, given that more fossil fuel power generation is already planned or being constructed globally, which will emit another 188 gigatons of CO_2 over their lifetimes, bringing the total to 846 gigatons of CO_2 – more than double the 420 gigatons of CO_2 permissible to keep to the Paris target of 1.5°C.

So what can be done to meet the Paris target? Lynas points out that even if we were to cancel all new fossil fuel infrastructure and promptly do away with internal combustion engines and electrify everything, we would still be 200 gigatons over the allowable budget. His proposed solution for dealing with this – which is to shut down all existing fossil fuel infrastructure now – is a huge ambition that would run into intense opposition, including from investors, miners, and oil and gas companies who strongly influence the political process and would lose trillions of dollars in **stranded assets**. But if this is too steep an ask, there is only one alternative: we must let go of the 1.5°C target and aim at 2°C. But as noted earlier and as discussed in more detail in Chapter 3, the expected climate damages of raising the 1.5°C target even to 2°C are considerable. And even keeping heating to 2°C would require emissions cuts of 5 percent every year from 2022, when emissions are now *increasing* about 2 percent per year (Figure 2.13).

Heating Already "in the Pipeline"

As already explained, the large thermal inertia of the Earth's oceans delays the global climate response to climate forcing, meaning that some of the forcing will be realized later, which is known as **committed warming**, that is, it is "in the pipeline" to affect temperatures later. According to Hansen's calculations, about 0.5°C of additional heating is already in the pipeline and will occur over the next decades and centuries even if the atmospheric composition were to remain at today's levels. As he points out, this pipeline problem masks the actual effects of our past and present actions: "the long response time of the ocean and slow climate feedbacks allows consequences for young people and future generations to build up while most of the public does not notice much happening, as noticeable climate change is just beginning to rise above natural variability." Yet as he also points out, this delay also offers an opportunity to drastically reduce greenhouse gas emissions and avoid the most devastating effects of those human actions.[57]

To better understand what may really be necessary to avoid the worst consequences of our past and current use of fossil fuels, eighteen different research teams of scientists used modeling to try to figure out what effect a

Figure 2.13

The road to 2°C is steep; the road to 1.5°C is a cliff

Every year we delay devours the remaining carbon budget. The remaining carbon budget to keep heating below 2°C above pre-industrial levels is about 1050 Gt CO_2e. The main figure shows that, starting now, to just have a 50 percent chance of keeping heating to 2°C would require emissions cuts of about 5 percent per year globally (compare to the about 7 percent emissions reduction incurred by the Covid-19 pandemic in 2020).

Meanwhile, the carbon budget to keep global heating under 1.5°C above pre-industrial levels is only about 300 Gt CO_2e in 2022. Keeping beneath that temperature level now looks monumental, probably impossible (see inset). It would require massive cuts in emissions every year from now on, along with massive "negative emissions" to pull CO_2 out of the air (using technologies that are barely developed, hugely expensive, and deeply problematic in many ways, see Chapter 8).

Carbon mitigation curves (2°C)
For a > 50% chance of staying below 2°C, remaining budget: 1500 Gt CO_2

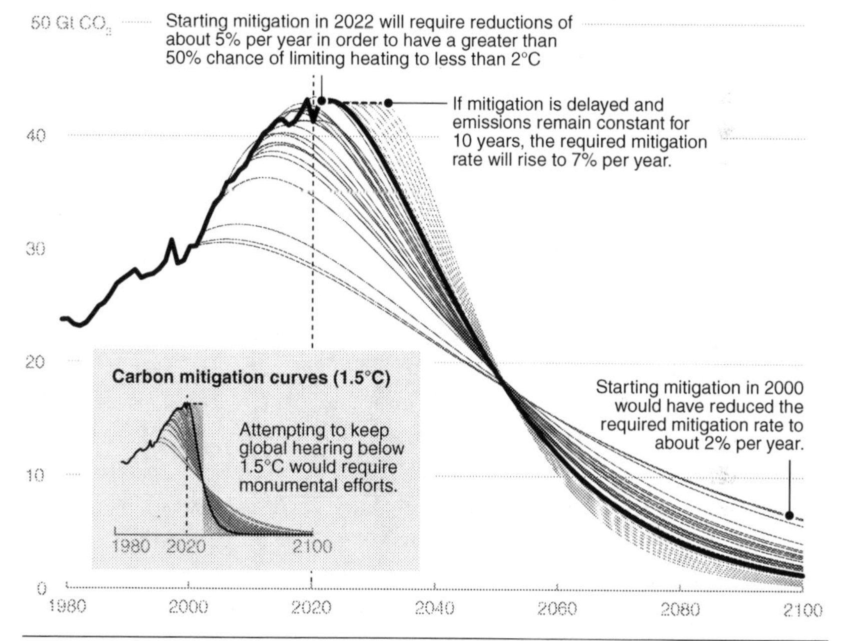

Adapted from Our World in Data, CC-BY, in turn adapted from Robbie Andrews, based on Historical emissions to 2017 from CDIAC/Global Carbon Project, projection to 2018 from Global Carbon Project (Le Quéré et al. 2018)

complete cessation of CO_2 emissions would have on the expected change in temperature.[58] Using earth system models like those shown in Figure 2.5, they raised emissions in the atmosphere to levels as high as one trillion tons of CO_2 before cutting further emissions and running the models to see what would happen. Reassuringly, they found that when all CO_2 emissions were cut, the ocean and land absorbed the carbon remaining in the atmosphere and that after fifty years, the average change in temperature across the eighteen models was close to zero, although a few of the models showed substantial heating continuing for several centuries. However, whether these models have it right is somewhat uncertain, as other studies have concluded that instantly cutting all emissions would still lead to a rise ranging from 1.1°C to 2°C in temperature.[59] One source of such uncertainty is whether the **carbon sink** – the absorption of atmospheric CO_2 by land and oceans – will continue to operate as it does now or diminish across time.[60] And of course, as these researchers recognized, there will be no sudden cut in emissions in the real world; instead, we might cut emissions slowly over a few decades, all the while incurring new heating in that time frame.

Thus, while some studies suggest that cutting emissions now might prevent us from having to incur much of the heating currently in the pipeline, different studies provide different results and, as theoretical constructions, they make assumptions that may not be or remain valid. What they do help make clear, however, is that cutting emissions is necessary no matter whether some heating is baked in or not, as we simply cannot let temperatures get to levels that will make life on Earth untenable this century.

Conclusion

This chapter has laid out conclusive and irrefutable evidence that shows that human activities such as burning fossil fuels and industrial agriculture have produced increasing levels of greenhouse gases that in turn accelerate a greenhouse effect that leads to more heating. The resulting rate of increase in atmospheric CO_2 and heating over the last 200 years and especially over the last few decades is unprecedented in the Earth's climate record, which has been traced back tens of millions of years.

How much heating we will actually see in the future thus also depends on human behavior. As the evidence in this chapter has shown, our past and current emissions and current energy policies are consistent with the high-emissions scenario of RCP8.5 (SSP5), and continuing with our current practices will make it impossible to meet the 1.5°C target proposed by the Paris Accord. Even reaching the 2°C target will require cuts of emissions of about 5 percent per annum starting in 2022. Yet without substantial cuts soon, billions of people who now live in coastal areas, depend on freshwater supplies from the Himalayas, or work outdoors in agriculture will see their lives increasingly disrupted by global heating. These impacts are the topic of Chapter 3.

3
Climate Impacts

> This tectonic challenge is man-made. It is a civilizational, moral and existential challenge – to humanity today, tomorrow, and for the future generations. If not addressed properly, the effects of this ecological challenge will be catastrophic to all future generations. Be they from the west or from the south, be they white, black, yellow or in-betweens.
>
> Lumumba Di'Aping, Lead Negotiator for the Global South at Copenhagen COP15 in 2009 (these remarks were made in 2019).

As Chapter 2 showed, the evidence that human activity is generating greenhouse gases that are heating our Earth is conclusive and irrefutable. This chapter examines some of the diverse and serious current physical and economic impacts of global heating and the projected scenarios regarding those impacts if more substantial efforts to cut emissions are not soon undertaken.

Current Impacts of Global Heating

An overview of some of the key impacts of global heating over the past several decades is provided in the charts in Figure 3.1. The first two charts demonstrate the chief effect discussed in the previous chapters: that dramatic increases in greenhouse gases such as CO_2 correspond to an increase in surface temperature. Together, the charts show myriad warning signals of serious impacts on our planet, including reductions in Arctic and Antarctica sea ice and glacier thickness, increasing ocean acidity, and rising sea levels. And as chart (H) shows, the number of extreme weather events is now about four times greater compared to the 1980s. As a result of the enormous increase in energy that arises from the accelerated greenhouse effect discussed earlier, the planet is

Figure 3.1

The planet is blinking red

Climatic response time series from 1979 to the present. The rates shown in the panels are the decadal change rates for the entire ranges of the time series. The annual data are shown using gray points. The black lines are local regression smooth trend lines.

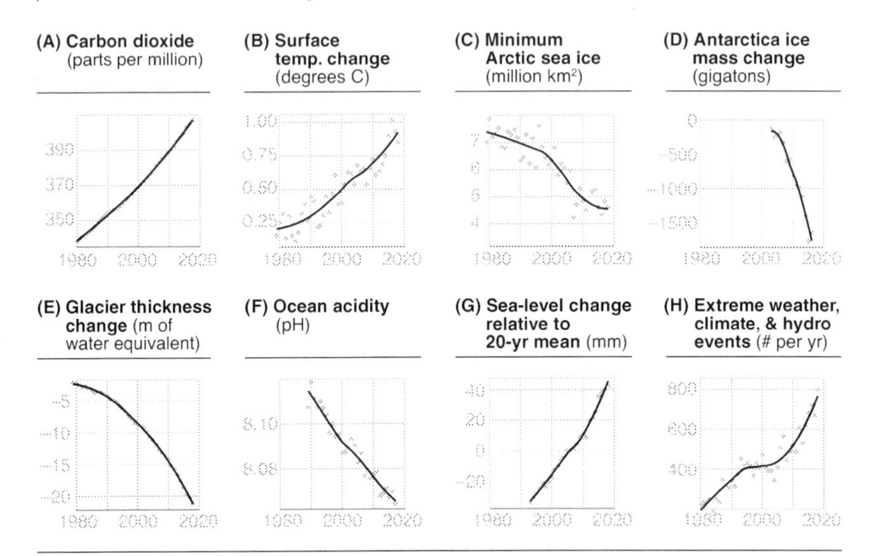

Adapted from William J Ripple, Christopher Wolf, Thomas M Newsome, Phoebe Barnard, William R Moomaw, World Scientists' Warning of a Climate Emergency, BioScience, Volume 70, Issue 1, January 2020, Pages 8–12, https://doi.org/10.1093/biosci/biz088

seeing more frequent and extreme heat waves and drought, heavier rainfall, and more intense storms. In general, across the world, wetter areas are getting wetter and drier areas are getting drier.[1] Scientists and meteorologists are also concerned that global heating may be disrupting the jet stream – a worldwide air current – which could further increase the likelihood and intensity of extreme weather events, including extreme cold periods known as polar vortexes and the explosive "bomb cyclones" that have impacted the northern USA in recent years, even as the Earth is getting hotter overall.[2]

Although scientists cannot attribute any particular extreme weather event to climate change, they can estimate how much more likely and how much more intense a particular event is made by climate change, something known as **extreme event attribution** (Box 3.1). As explained in Box 3.1, one can think of the climate as a distribution of particular weather events, such as hot days; what climate change does is to shift the distribution to the right, increasing the

Box 3.1 Extreme event attribution, or how we know an event is made more likely or more intense by climate change

The distinction between *weather* and *climate* is an important one. Weather is the short-term local conditions of the atmosphere, whereas climate is the average of weather across time and space. We might think of climate as the distribution of possible events, whereas weather is one event plucked from that distribution. What climate change does is shift the distribution. In the figure of temperatures below, global heating shifts that distribution to the right. Any particular weather event selected from the distribution after that shift is likely to be hotter than it would have been before the shift because it was drawn from a distribution with a higher mean temperature.

While it is not possible to say with high certainty that a particular extreme weather event was caused by climate change, a subfield of climate science called extreme event attribution can often determine if climate change influenced the likelihood or severity of the extreme event. According to a 2020 review article, such attributions involve the following steps.[5] Step 1 is to define the event of interest, such as the 2013 drought in California or floods in India. Step 2 is to calculate what is termed the counterfactual climate, or the likelihood that the extreme weather event would have happened in a climate without human influence. One way to do this is remove the long-term human-caused trend, such as temperature, from the climate variable and then determine how likely is it that the observed temperature would have happened from the remaining data.

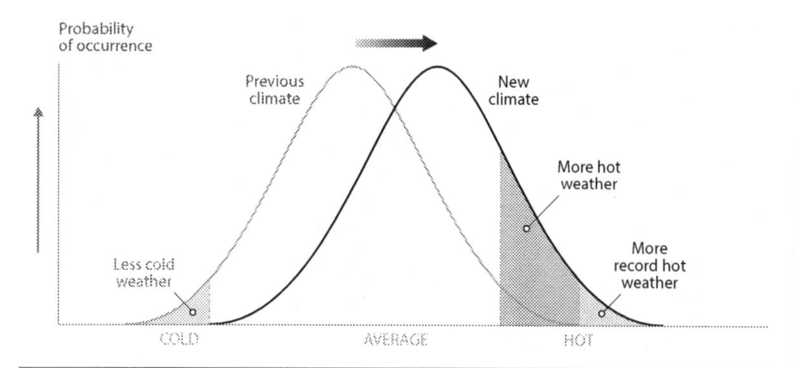

Adapted from T.R. Karl, G.A. Meehl, C.D. Miller, S.J. Hassol, A.M. Waple, and W.L. Murray (eds.). A Report by the U.S. Climate Change Science Program and the Subcommittee on Global Change Research. Washington, DC. Available at: https://downloads.globalchange.gov/sap/sap3-3/sap3-3-final-all.pdf.

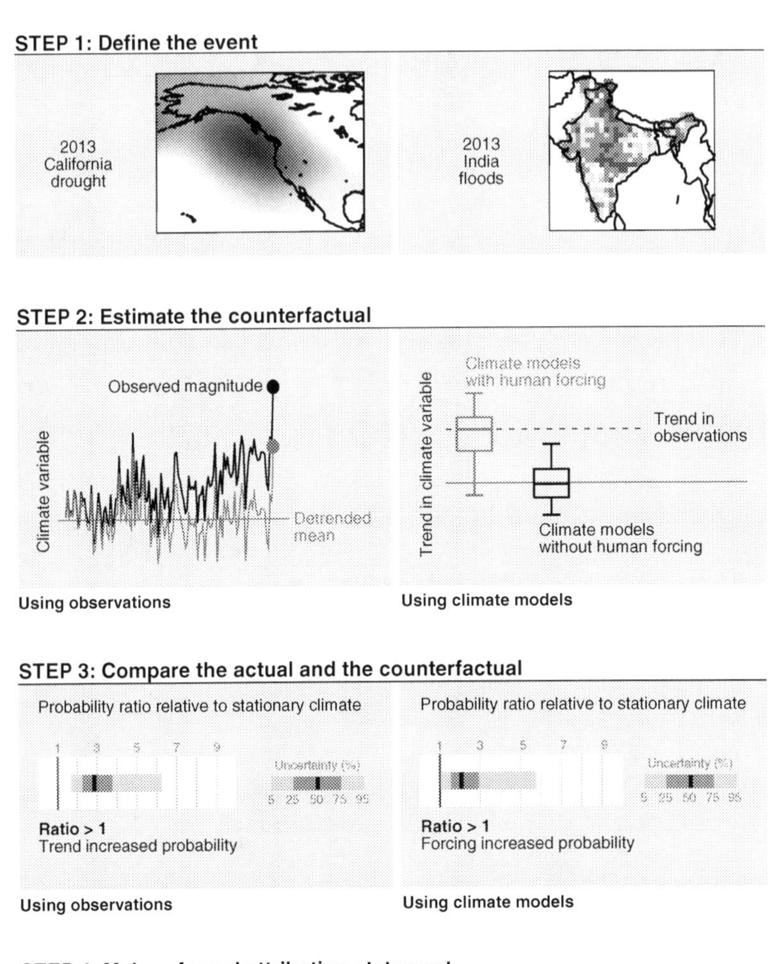

Another approach is run a climate model without human forcing and estimate the likelihood that the drought or flood would have emerged from that model. Step 3 is to compare the likelihood that the extreme event emerged from actual, human-caused conditions to the likelihood it would have emerged from the counterfactual conditions. Step 4 then involves making a formal attribution statement identifying the degree of likelihood that the event occurred because of climate change. A common way to do this is to describe an event as a one-in-a-hundred-year event, which is not to imply that such an event will happen every hundred years or that if it does happen it will not happen again for another hundred years, but simply to communicate that an event of that size is likely to be equaled or exceeded on average only once in every hundred years.

average heat of the entire distribution. Following that shift, any selected event is more likely to be hotter. To use the analogy of an athlete, if a runner is on steroids, she is more likely to run a particular race and to run it more strongly. The same goes for climate change: it is, as the Colorado Center for Atmospheric Research observed, like "weather on steroids."[3] Overall, scientists have estimated that about two-thirds of extreme weather events are made more likely or more severe by human-caused climate change.[4]

Meanwhile, as mentioned in Chapter 2, the economic impact of extreme weather events is also increasing enormously – according to one estimate, amounting to $650 billion in damages worldwide in 2016–2018 alone.[6] As Figure 3.2 shows, in just the USA, the number of weather- and climate-related events per year that produce damages of $1 billion or more each has continued to increase, amounting to twenty-two such events in 2020 alone. The effects of many of these extreme events are also widespread. For example, research on the California wildfires of 2018 found that the overall cost of damages was estimated at $149 billion – about 1.5 percent of California's annual GDP – and that about $89 billion of that figure came from indirect losses, about half of which were in industry sectors and locations outside of California, impacting far-flung supply chains and jobs.[7] Moreover, the cost to health was estimated at $32 billion, although it is not easy to measure the disruption to people's lives from increased asthma and cancers arising from inhaling tiny particles nor the social impacts of the devastation of towns such as Paradise, California, in purely economic terms.

Impacts on Weather Events, Oceans, and Melting

As Emily Grossman's freely available online book *Emergency on Planet Earth* and many other sources make abundantly clear, global heating has produced myriad extreme weather events across the globe.[8] One of these is heat waves, or very hot temperatures that last for several days, which are more likely to

Figure 3.2

Billion-dollar weather- and climate-related events in the USA, 1980–2020

In 2020 there were twenty-two separate billion-dollar weather- and climate-related disaster events. The 2020 costs exceeded $95 billion. Overall, the USA has sustained 291 weather- and climate-related disasters since 1980 where the cost of each exceeded $1 billion (accounting for inflation). The total cost of these was more than $1.9 trillion. The graph shows that the number of events and the cost of them is increasing with time.

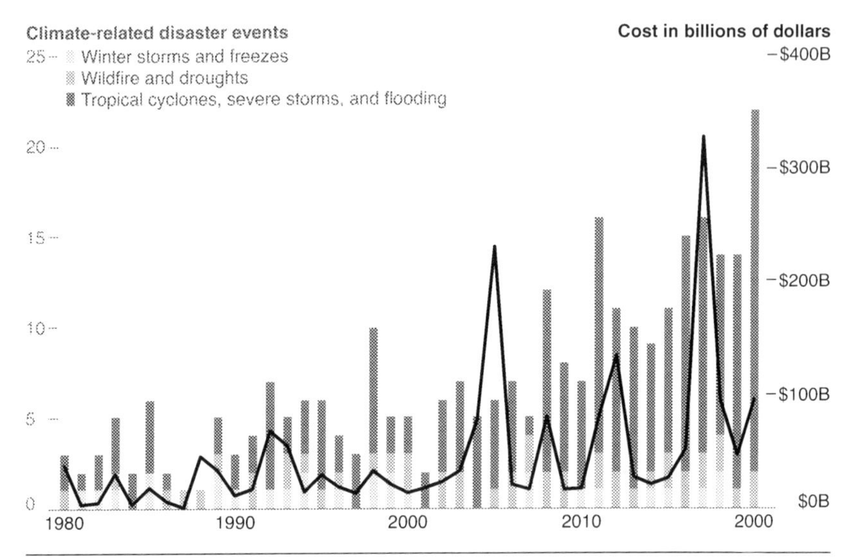

Adapted from NOAA National Centers for Environmental Information (NCEI) U.S. Billion-Dollar Weather and Climate Disasters (2021). https://www.ncdc.noaa.gov/billions/, DOI: 10.25921/stkw-7w73

occur and be stronger in parts of the world where the land is already dry (as dry soils cannot absorb as much heat as moist soil).[9] An analysis of hundreds of studies found that 95 percent of recent heat waves were made more likely or more intense by climate change.[10] In one dramatic heat wave in Europe in 2003, approximately 70,000 more people died than in a usual year, with most of these deaths attributable to the extreme temperatures.[11] Whereas that particular heat wave was made at least twice as likely by climate change, since then the likelihood of such an event has become ten times as likely.[12] Consistent with this escalation, 157 million more people were exposed to heat waves in 2017 than in 2000.[13] In 2019 alone, nearly 400 temperature records were broken across twenty-nine countries, records that were a hundred times more likely to occur because of human influence on the climate.[14] As climate scientist Sophie Lewis remarked when Australia recorded its hottest day ever that year, the problem regarding these record temperatures is "not just the frequency that we're breaking them, it's the margin" – that the level of these temperature increases is so much higher than previous levels of variability.[15] In the summer of 2021, a record-breaking heat wave scorched the Pacific Northwest of the United States, causing temperatures to soar more than 17°C (30°F) above normal for several days, and causing the death of hundreds of people and over a billion sea creatures.[16] This extreme weather event was declared by scientists to be "virtually impossible without human-caused climate change" (perhaps a 1 in 1,000 year event).[17] In April and May of 2022, a few months before summer begins, a blistering heat wave affected more than 10 percent of humanity in India and Pakistan for several weeks, devastating crops and pushing the wet-bulb temperature in Jacobabad, a city of 200,000, to 33.1°C – just below the typical estimate for human survival.[18] Scientists showed that climate change made the event more than a hundred times more likely, and suggested that high temperatures that used to occur about every 300 years may now happen every three years.[19]

Apart from heat waves, warming is also leading to more drought, or prolonged periods of drying, and to more intense drought. One reason for this is that the increased surface temperature of the Earth means that the air is generally warmer, and warmer air can hold more water vapor, leading to less condensation to form clouds that could produce rain. An analysis of data from tree rings suggests that the current drought in the southwestern USA may be the worst in 500 years.[20] In 2018, an intense drought in Cape Town, South Africa, led to severe water restrictions, and the city came close to what was dubbed Day Zero when the water supply would have to be turned off entirely.[21] In 2019, India experienced an intense drought followed by a torching heat wave that similarly led the city of Chennai to very nearly run out of water.[22] Drought also has many secondary effects, including civil unrest and conflict. For example, some climate experts have suggested that the severe drought that affected Syria between 2007 and 2010 was a key factor in the outbreak of the civil war there in 2011 because too little water forced rural populations to flee to the cities, where food riots led to political turmoil.[23]

This exacerbation of hot and dry conditions also increases the likelihood and spread of forest fires. In the two decades since 2000, human exposure to wildfires has doubled.[24] In the USA, large forest fires have become five times more frequent since the 1970s, scorching six times as much land and lasting much longer.[25] Recent examples include the Australian conflagration of 2019; the California fire seasons of 2018, 2020, and 2021; and the forest fires in Greece and Turkey in 2021. In 2019, the largest wildfire disaster in Russian history destroyed more than 52,000 square miles of Arctic forest in Siberia,[26] an event of particular concern because Arctic fires have an accelerating effect on global heating by burning the peat in the soil (a long-term carbon store) and thereby releasing even more greenhouse gases.

Even as global heating increases the drying of the Earth's land and forests, it also leads to greater flooding. Warmer air and warmer oceans produce more evaporation from the oceans, and warmer air can hold more water vapor (a 7 percent increase for each additional 1°C of heating). When that air eventually cools, clouds form and produce more frequent rainfall and often much greater downpours.[27] A 2015 report attributed about 18 percent of extreme rainfall globally to global heating, although in some parts of the world that rainfall was much heavier.[28] In the northeast region of the USA, for instance, the number of downpours has increased by 71 percent since the 1950s.[29] In 2019, following the wettest consecutive twelve months in US history, major floods in the Midwest significantly delayed the planting of soy and corn and reduced yields.[30] The following year, 2020, saw record-breaking flooding around the world. In Indonesia, 175,000 people were displaced early in the year by the most intense rains to ever fall there.[31] Japan experienced record-breaking rainfall, with over 100 millimeters (3.9 inches) falling in one hour, forcing the evacuation of more than a million people.[32] Rainfall in parts of China broke records, affected forty million people, and caused thirty-three rivers to rise to their highest level in history, incurring billions of dollars in damages.[33] In addition to causing structural damage to houses, roads, and bridges, and affecting crop yields, flooding can cause hazardous chemicals to leak from thousands of vulnerable old mines and industrial sites, such as those at toxic waste sites in Houston following Hurricane Harvey.[34] The huge economic impacts of flooding have led some insurers to warn that flood coverage could become unaffordable for almost everyone in such areas.[35]

Increased air and water temperatures also supercharge hurricanes (also called cyclones or typhoons) and the wind and water damage they create.[36] The proportion of tropical storms that have rapidly strengthened into hurricanes has tripled over the past thirty years,[37] changes that, the IPCC 2021 report declares with medium confidence, "cannot be explained by internal variability alone."[38] Moreover, the added intensity is diverting these storms from their typical route, putting more coastal cities in their path. Since 2013, five of the seven hurricane regions studied by scientists had their strongest storm on record – something extremely unlikely to occur by chance alone.[39] A dramatic increase in the ocean temperature of the Western North Pacific has intensified storms such as Typhoon Haiyan, which killed 8,000 and displaced four million people when it struck the Philippines in 2013.[40] When

Hurricane Harvey hit the USA in 2017, it broke records for rainfall (with some parts of Houston receiving over five feet) and incurred up to $125 billion in damages, one of the costliest storms ever.[41] Just weeks later, Maria, a Category 5 hurricane, devastated the islands of Dominica, St. Croix, and Puerto Rico, killing 2,900 people. Although it received less international attention, in 2019 Cyclone Idai, which the environmental justice campaigner Dipti Bhatnagar referred to in Chapter 1, struck Mozambique, Zimbabwe, and Malawi, with enormous impacts on housing, such as in the Mozambican city of Beira.[42]

Global heating also has effects on the world's oceans, as more than 90 percent of the increased heat is stored in them. Indeed, 2015–19 recorded what were then the five warmest years of ocean temperatures since 1955.[43] This extra heat causes the water to expand, resulting in sea-level rise. A recent report in *Science* magazine concluded that under a business-as-usual emissions scenario, within a century marine systems would likely suffer mass extinctions on a par with those of the distant past.[44] Hotter ocean water and air temperatures also cause the melting of sea ice, ice sheets, and mountain glaciers; while the melting of sea ice does not raise sea levels (just as a melting ice cube does not raise the water level in a glass), the melting of land-based ice sheets and glaciers does. In 2020, the land-based Greenland ice sheet was losing ice seven times faster than in the 1990s.[45] In the Arctic, the area covered by summer sea ice has shrunk 40 percent since 1979, a loss equivalent to the land area of Alaska and California combined.[46] At the other end of the planet, the ice sheet in Antarctica has lost three trillion tons of ice (equivalent to ninety million times the weight of the Empire State Building) in the past twenty-five years and is losing ice six times faster now than it was thirty years ago.[47] As Stefan Rahmstorf, a professor of physics of the oceans, has pointed out, an especially concerning issue is that these changes are happening faster than predicted.[48] In Antarctica, the ice sheet in particular is losing mass at an accelerating pace, leading some scientists to worry that it could reach a **tipping point** in which sea-level rise becomes ten times more than what we are seeing now[49] (tipping points are discussed later in this chapter).

Although until now, sea levels have been rising about 31 millimeters every ten years,[50] even this seemingly small rise is already having an impact, particularly at high tide. Rising seas have already displaced hundreds of thousands of people in low-lying parts of Bangladesh and Vietnam,[51] and in 2019 five Pacific islands were entirely lost to rising sea levels.[52] In Venice, Italy, in 2019, the worst floods in fifty years left St. Mark's Square submerged, an event that has happened only six times in 1,200 years, the last four in just the past twenty years.[53] In the USA, many coastal areas are already experiencing "nuisance flooding" at high tide, which has increased about 100 percent relative to thirty years ago and could become near-routine for some places in the next fifteen years.[54] According to the 2021 IPCC report, by 2100 the likely global mean rise from a 1995–2014 baseline is 0.44–0.76 meters for RCP4.5 and 0.63–1.01 meters for RCP8.5.[55]

Impacts on Human Health and Food and Water Supplies

In addition to its impacts on extreme weather, global heating is already affecting human health and food and water supplies. According to a recent series of reports from dozens of organizations on the health effects of climate change published in the leading medical journal the *Lancet*, "a rapidly changing climate has dire implications for every aspect of human life, exposing vulnerable populations to extremes of weather, altering patterns of infectious disease and compromising food security, safe drinking water and clean air."[56] As those reports found, 800 million people (about 11 percent of the world's population) are already vulnerable to climate change impacts such as floods, sea-level rise, droughts, and heat waves.

As the reports indicate, some of that impact on human health arises directly from the extreme weather events discussed earlier. Heat waves cause heat stress and premature deaths; storms and floods produce injuries and loss of life, shut down hospitals, and trigger disease outbreaks such as cholera; wildfires exacerbated by extreme weather dramatically worsen air pollution; and all these events lead to mental health problems. Extreme weather and soil degradation impact food production and access to water, leading to dehydration and malnutrition.[57]

Global heating also increases the spread of disease. The northeastern USA, for example, experienced a 27 percent increase in suitability for an outbreak of pathogenic vibrio bacteria between the 1980s and 2010s. Higher temperatures also increase the geographic ranges at which mosquitoes can thrive, and increased rainfall provides more sources of stagnant water in which they can breed. Mosquitoes spread pathogens such as malaria, Zika, West Nile virus, Chikungunya, and dengue, and almost 700 million people contract a mosquito-borne illness every year.[58] Nine of the ten most suitable years for dengue have occurred since 2000.[59] Higher temperatures also promote the spread of tropical diseases to parts of the world that typically have not previously had them. In 2017, for example. a few cases of Zika virus were reported in the USA from presumed local mosquito-borne transmission; in 2018, Italy reported cases of malaria in people who had not traveled outside the country and dengue fever was seen in France and Spain.[60] In the US southwest, climate change is driving a large spike in Valley fever, which is contracted by inhaling a fungus.[61] Some scientists also think that new diseases could emerge, such as the Spanish flu virus or other pathogens that may have been buried in the permafrost and could be released as it thaws.[62] There has also been an apparent increase in zoonotic viruses that jump from animals to humans, of which SARS-COV2 might be one example. While such "jumps" are driven in large part by habitat destruction, animal exploitation, and human population growth, climate change is thought to also contribute by inducing animals to migrate and creating new points of contact with both pathogens and humans.[63] Given that there are at least 10,000 virus species that have the capacity to infect humans, with the vast majority still circulating in nonhuman animals, this acceleration by climate change of cross-species infection is very concerning.[64]

Furthermore, the burning of fossil fuels not only generates CO_2 but also creates air pollution in the form of carbon monoxide, nitrous oxides, particulate matter, and unburned hydrocarbons.[65] The amount of such pollution is bad enough when burning gasoline but even worse when burning diesel: according to the California Air Resources Board, using a garden leaf blower for one hour generates as much smog-forming pollution as driving a 2017 Toyota Camry from Los Angeles to Denver.[66] Air pollution is already the world's largest environmental cause of disease, responsible for about seven million premature deaths each year and about 142 million disability-adjusted-life-years (about the same as smoking).[67] Air pollution is made much worse by climate-related forest fires, such as the 2018 and 2020 California fires, which polluted air as far east as Massachusetts.[68]

Climate change and the fossil fuel practices that spawn it also have big implications for food production and access to water. Most of what we eat depends on healthy soil, yet industrial agricultural practices are severely degrading soils.[69] Rising temperatures also impact food production. Despite some skeptics' claims that some parts of the world might see better crop yields as a result of global heating, rising temperatures have already led to global declines in the yields of many cereal crops over the last thirty years: 4 percent for corn, 6 percent for winter wheat, 3 percent for soybeans, and 4 percent for rice.[70] When heat waves affect grain yields, price hikes can affect consumers across much of the world. In 2010, for example, the Russian heat wave contributed to a global wheat shortage and a 90 percent increase in international grain prices.[71] The Australian Bureau of Agriculture has estimated that recent effects of climate change have reduced crop farmers' profits by 35 percent.[72]

Global heating also affects water supplies beyond the effects of drought by accelerating the melting of glaciers. The drainable basins of mountain glaciers cover about 26 percent of global land surface outside of Greenland and Antarctica and are home to almost a third of the world's population, who depend on downstream melted glacier water for drinking, irrigation, hydropower, and much else.[73] The glaciers in the Himalayas are now melting twice as fast as between 1975 and 2000, losing the equivalent of about 3.2 million Olympic-sized swimming pools per year.[74] As glaciers shrink more, especially in the dry season, substantial impacts will be felt by hundreds of millions of people.

Impacts on Plant and Animal Life

The climate crisis addressed by this book is also an ecological crisis. The warming of the oceans, for instance, is damaging coral reefs, with huge knock-on effects on associated marine life,[75] and changes in the North American seasons are enabling bark beetles to flourish and decimate millions of hectares of trees.[76] Warming is also having severe ecological effects on wildlife, including species that support human life in myriad ways.[77] Insects pollinate crops, microbes and fungi recycle dead matter and enrich our soils, coastal mangroves buffer us against storm surges, and as many as three billion people depend on wild-caught or farmed seafood for protein. Yet wildlife is declining globally at rates never seen in human history. According to the International Union for the Conservation of Nature, approximately a quarter of the 116,000 species assessed by scientists are threatened with extinction, including 25 percent of all mammals (Table 3.1). As other experts have pointed out, what is particularly striking about these figures is the speed with which they are occurring: species are going extinct at rates 100–1,000 times faster than was typical in the past.[78] Some scientists have gone so far as to call this decline the "Sixth Mass Extinction,"[79] referring to a period in which the Earth loses more than three-quarters of its species in a geologically short time period, something

Table 3.1. *Estimated percentage of animal life at risk of extinction*

	Proportion of species at risk of extinction (%)
Amphibians	40
Conifers	34
Reef corals	33
Sharks and rays	31
Selected crustaceans	27
Mammals	25
Birds	14

Note: Data from International Union for the Conservation of Nature, www.iucnredlist.org.

that had previously happened only five times in the last 540 million years. Although this current extinction can be attributed to multiple causes, including habitat change, exploitation (fishing, hunting, logging), invasive species, and pollution, climate change is a primary or aggravating cause of much of this trend.

Our current ecological crisis is caused not only by global heating but by a mode of economic activity that relies on relentless growth and consumption. One example is deforestation, which in tropical regions has been driven mainly by the demand for four commodities: beef, soybeans, palm oil, and wood products.[80] In the Amazon, which is home to one in ten of all known species, a combination of drying due to climate change and burning to make space for beef cattle and crops has resulted in the destruction of about one-fifth of the entire rainforest, which is disappearing at the rate of three football fields per minute.[81] The destruction of the rainforest, driven by seemingly insatiable economic growth and appetites, makes matters worse for the climate: not only do the felled trees no longer capture CO_2, but burning them releases huge quantities of carbon back into the atmosphere. In fact, such deforestation is blamed for nearly one-fifth of all greenhouse gas emissions.[82]

Projected Future Impacts of Global Heating

As discussed in Chapter 2 and shown in Figure 3.3, climate models suggest that if we remain on our current actual high-emissions pathway, global temperatures may surpass the 1.5°C target by 2030, and that even if we switch immediately to a medium-emissions pathway, we might surpass it by 2032. Although it appears extremely unlikely that we will limit heating to 1.5°C, it is certainly not out of the realm of possibility that we could take action to keep it to 2°C or some small fraction above that, and as Table 3.2 shows, even half of a degree makes a huge difference. To give just a couple of examples, with a 2°C increase, the UK Meteorology Office estimated that up to one billion people would face heat stress risk, a nearly threefold increase from a 1.5°C rise; while a 2°C rise is likely to destroy 99 percent of the world's coral reefs, whereas at 1.5°C that loss is likely to be between 70 and 90 percent.[83]

Likely Physical Impacts by 2050, 2070, and 2100

If we remain on the current high-emissions pathway RCP8.5, the impacts of global heating by 2050 will be dramatic. To give an example, the consulting

Figure 3.3

Without rapid emissions reduction, the 1.5°C target will be passed around 2030

Global average surface temperatures in any year are driven by a combination of long-term global heating plus short-term natural variability (such as volcanoes, changes in the sun's irradiance, and El Niño and La Niña ocean events). These short-term influences could make us exceed 1.5°C quite soon – the World Meteorological Organization estimates a one in four chance by 2025. But to estimate when long-term global heating exceeds 1.5°C scientists need to use climate models, the latest being an ensemble of models called CMIP6.

In (A) several different CMIP6 models were run for the RCP4.5 (also called SSP2) scenario. Each model shows the warming from the year 2020. In (B) such models were run for several different scenarios, RCPs 2.6, 4.5, and 8.5 (also referred to as SSP1, SSP2, and SSP5). Each dot represents an individual model, while horizontal bars show the range across all models. For SSP2–4.5 (medium-emissions), the range is between 2026 and 2042, with a median of 2032. For SSP5–8.5 (high-emissions scenario), the range is between 2026 and 2032 with a median of 2030.

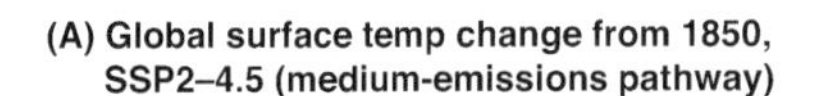

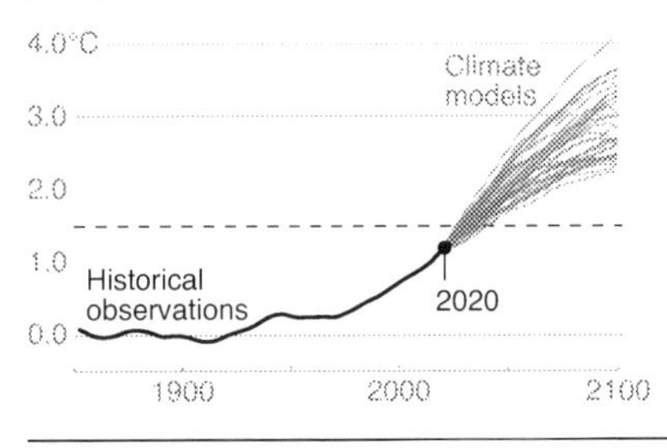

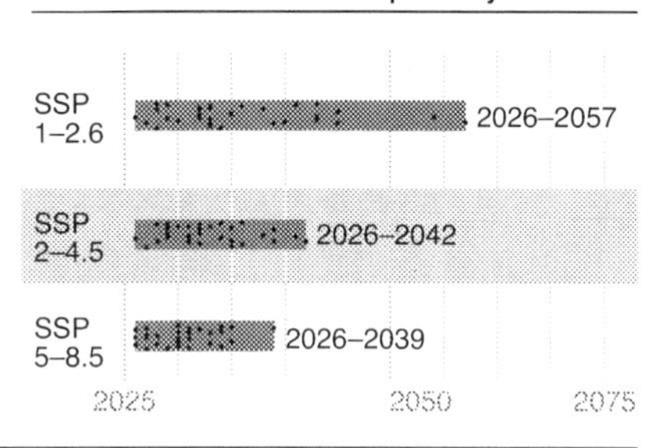

Adapted from Carbon Brief courtesy Zeke Hausfather https://www.carbonbrief.org/analysis-when-might-the-world-exceed-1-5c-and-2c-of-global-warming

firm McKinsey, drawing on IPCC data, found that by 2050 the average annual likelihood of people in a subregion of India experiencing a lethal heat wave could be as high as 40 percent.[84] (A lethal heat wave is defined as a three-day period with a maximum **wet-bulb temperature** – the air temperature measured by a thermometer covered in water-soaked cloth – of greater than 35°C; one degree higher and people start dying even in the shade.)[85] But as shown in Figure 3.4, even without life-threatening heat waves, projections by mid-century under a high-emissions scenario indicate that in parts of India, 25 percent of summer heat days will pose extreme danger, while under a mid-level emissions scenario that percentage would still be 10 percent. To give another example, in 2020 McKinsey found that 25,000 homes in Florida already experience flooding more than fifty times per year, or almost once a week, with obviously negative effects on their resale value; they estimate that by 2050 this

Table 3.2. *Comparison of impacts at 1.5°C and 2°C, based on 2018 IPCC report*

	1.5°C	2°C
Extreme heat Global population exposed to severe heat at least once every five years	14%	37%
Sea ice–free Arctic Number of ice-free summers	At least 1 every 100 years	At least 1 every 10 years
Species loss, vertebrates Vertebrates that lose at least half their range	4%	8%
Species loss, insects Insects that lose at least half their range	6%	18%
Permafrost Amount of Arctic permafrost that will thaw	4.8 million square km	6.6 million square km
Crop yields Reduction of maize harvest in tropics	3%	7%
Coral reefs Further decline in coral reefs	d) 70–90%	99%
Fisheries Decline in marine fisheries	1.5 million tons	3 million tons

Note: Data adapted from World Resources Institute, www.wri.org/blog/2018/10/half-degree-and-world-apart-difference-climate-impacts-between-15-c-and-2-c-warming.

Figure 3.4

Under RCP8.5 (high-emissions pathway), summer heat stress reaches dangerous levels for more than 25 percent of days in some parts of India after 2046

Climate models were run for RCP8.5 to predict summer temperatures over India in the 2045–2065 time frame. The Danger Level was classified as 41–54°C with "[s]unstroke, muscle cramps, and/or heat exhaustion likely[with] Heat stroke possible with prolonged exposure and/or physical activity," while Extreme Danger Level was classified as over 54°C where "[h]eat stroke or sunstroke is highly likely." Further, it was estimated that by the end of the century such heat stress would correspond to a lost of 30 to 40 percent of work days, especially in the eastern region.

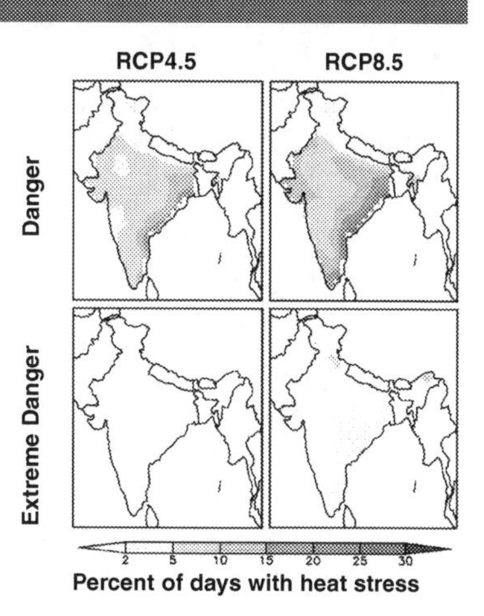

Adapted from Koteswara Rao, K., Lakshmi Kumar, T.V., Kulkarni, A. et al. Projections of heat stress and associated work performance over India in response to global warming. Sci Rep 10, 16675 (2020). https://doi.org/10.1038/s41598-020-73245-3 |

Figure 3.5

The expansion of extremely hot areas by 2070 could have significant implications for human migration

Expansion of extremely hot regions under RCP8.5. Currently, mean annual temperatures > 29°C are restricted to the small dark areas in the Sahara. By 2070, such conditions are projected to occur throughout the stippled area and could impact up to 3 billion people.

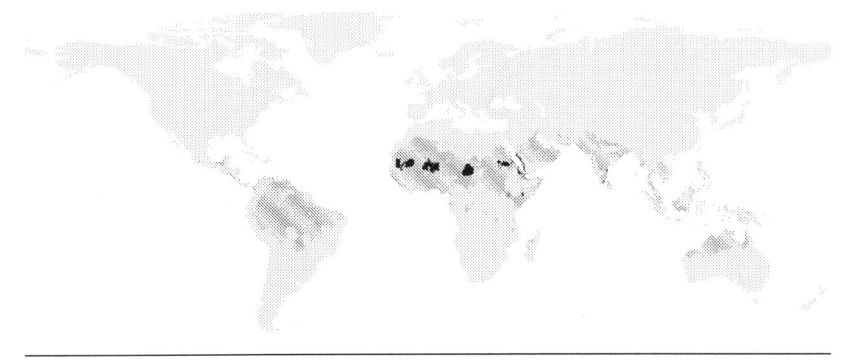

Adapted from Chi Xu, Timothy A. Kohler, Timothy M. Lenton, Jens-Christian Svenning, Marten Scheffer. "Future of the human climate niche." Proceedings of the National Academy of Sciences May 2020, 117 (21) 11350-11355; DOI: 10.1073/pnas.1910114117

number would rise to 100,000 homes, with a current combined value of $50 billion.[86] They anticipated that insurance premiums will increase by about 50 percent, with the most high-risk properties seeing a much higher jump. And even these estimates are out of date; in 2021 new rates were announced by the government-subsidized National Flood Insurance program to better reflect the actual risk of flooding due to climate change – these suggested that the cost of insuring some high-risk properties in Florida and elsewhere would now triple.[87]

Moving to 2070, the impacts of heating under a high-emissions scenario would be even more enormous. As noted by an analysis of projected mean annual temperatures around the world titled "Future of the Human Climate Niche," for millennia human populations have resided, produced crops, and raised livestock in a narrow part of the climatic envelope available on the globe in which the mean annual temperature ranges from 11°C to 15°C (52°F to 59°F).[88] Yet if we remain in a high-emissions scenario over the next fifty years, the geographic position of this temperature niche will shift, as shown in Figure 3.5, affecting one to three billion people. Specifically, the authors project that without migration, up to a third of the entire global population (approximately nine billion by that point) would experience a mean annual temperature of greater than 29°C (84°F), an average temperature currently

Figure 3.6

How much global heating will we have by 2100?

By 2100, global heating could reach 3°C on RCP4.5 (medium emissions and heavy reliance on negative emissions), and 5°C if we carry on with high emissions. This figure shows the range of warming from pre-industrial levels (1880–1900) to 2090–2100, using multiple runs of CMIP6 climate models under different emissions pathways. The SSP1–2.6 scenario – which is analogous to the "well-below 2C" RCP2.6 – shows mean warming of 2.0°C.

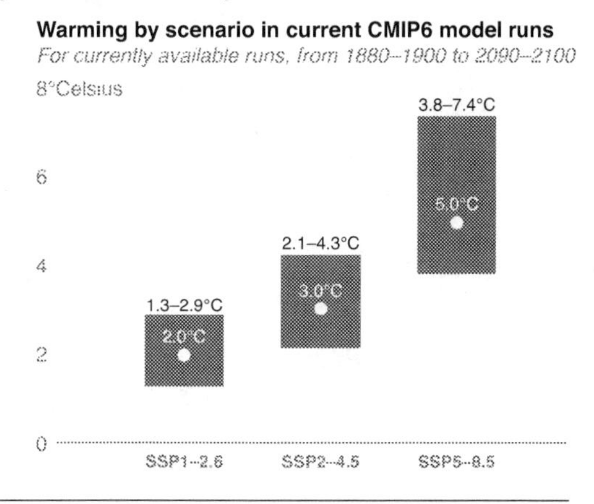

Adapted from Carbon Brief, courtesy Zeke Hausfather. https://www.carbonbrief.org/cmip6-the-next-generation-of-climate-models-explained

found on only 0.8 percent of the Earth's land surface, mostly concentrated in the Sahara. The implications of this shift for human migration and political instability are staggering.

To understand the likely impacts of global heating by the end of this century, Figure 3.6 presents a few scenarios based on the latest climate models employed by the IPCC called CMIP6.[89] As they show, under the medium-emissions scenario SSP2–4.5, we are slated to see a mean rise of 3°C by 2100, with a maximum of 4.3°C. Yet, as discussed in Chapter 2, aiming at this medium-emissions pathway will depend on negative emissions technologies such as BECCS that scarcely exist and would be hugely expensive and require massive, continent-size land use and enormous quantities of water. Taking 4.3°C, the upper bound of the medium-emissions pathway, as a likely scenario, the award-winning science journalist Mark Lynas points out that with such a rise, the US southern states would experience conditions now seen only in Death Valley, fire risk would increase by 500 percent across much of the country, and coastal areas would be beset by Category 6 superstorms.[90] At this level of heating, heat deaths worldwide would increase by 500–2,000 percent; South Asia and the North China plain would be rendered biologically uninhabitable for our species, triggering hundreds of millions of climate refugees; and southern Europe, much of Brazil, southern Africa, coastal Australia, and Central America would undergo desertification. Moreover, lethal heat waves would make it impossible to grow crops in most of the world's breadbaskets, political boundaries would be rendered obsolete as people flee their former homes, and sea levels would rise as much as three meters as Greenland thaws rapidly and most of Antarctica becomes part of the melt zone.[91]

Feedback Loops and Tipping Points

As if the current and projected impacts of climate change on the high-emissions pathway were not bad enough, there are reasons to think that heating could escalate much more through the phenomenon of positive feedback loops. One example of such a positive feedback loop is the melting of sea ice. The more ice melts, the more open ocean is available to capture the Sun's energy instead of reflecting it (reducing the **ice-albedo effect** – how ice and snow reflect solar radiation), leading to more global heating and the melting of more ice, and so on.[92] Although today's climate models do incorporate some key feedback loops, scientists acknowledge that these loops could prove to be more powerful than expected, creating a risk that global heating could end up being even more than we expect.[93]

One example of a positive feedback that some scientists think may be underestimated is permafrost thaw.[94] Permafrost is a type of soil that has been frozen for thousands of years; as it thaws, bacteria in the soil release methane and CO_2.[95] Permafrost covers about fifteen million square kilometers, or one-quarter, of the northern hemisphere, especially in Canada, Russia, and Alaska,[96] and currently stores an enormous amount of carbon, potentially holding about 1.46 to 1.6 trillion tons; almost twice as much carbon as in the atmosphere.[97] While there is little evidence so far that methane release has accelerated, there is evidence that permafrost is thawing faster.[98] Reducing further heating will be critical to preventing more such thawing; researchers have shown that the difference between a rise of 1.5°C and one of 2°C corresponds to a difference in thawing of about two million square kilometers, which could release an enormous amount of methane.[99]

Of even greater concern is that some positive feedback loops might trigger tipping points in the environment. A tipping point is the threshold beyond which a small change in temperature would cause a feedback loop to become self-reinforcing, creating an irreversible change in that part of the Earth's system and a state that would persist even if emissions subsequently go down.[100] You might think of this likelihood as being in a boat going downriver toward a waterfall: as you proceed downriver, you have opportunities to change course or to stop your movement, but if you do not, at some point you are going to go over the cliff – a tipping point in which you enter an entirely different state (this could be tantamount to losing the struggle to prevent further heating). Scientists have identified a number of such possible tipping points, such as those presented in Figure 3.7. The potential damage from a tipping point is enormous, and while the probability of any particular kind of tipping point is not known, it is not zero. And although scientists do not know just when these tipping points might occur, clearly the risk gets stronger with every increase in heating. As the authors of a recent paper in *Nature* warned, "[i]n our view, the evidence from tipping points alone suggests that we are in a state of planetary emergency: both the risk and urgency of the situation are acute."[101]

Figure 3.7

Ongoing emissions may induce tipping points – like a boat going over the waterfall

Tipping points are possible beyond which global heating could escalate beyond our control. A tipping point is a "critical threshold" that is reached when a relatively small perturbation ushers in a much more dramatic change. There is a possibility of a global cascade of tipping points, where they interact with each other; for example, through feedbacks of increasing temperature and greenhouse gas levels.

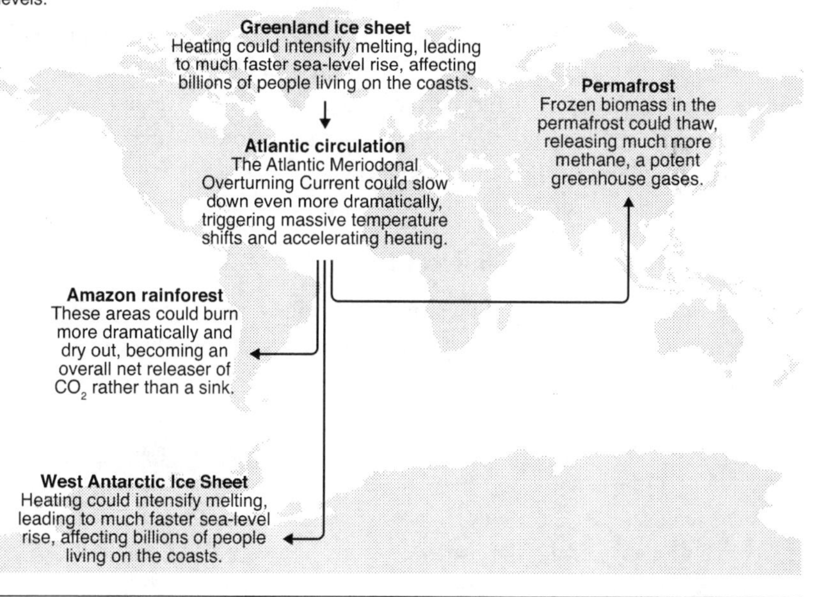

Adapted from Lenton, T. M. et al. Climate tipping points — too risky to bet against. Nature 575, 592–595 (2019), DOI: https://doi.org/10.1038/d41586-019-03595-0

In addition to the previously mentioned permafrost thaw, another of the several possible tipping points known to scientists is the Atlantic Meriodonal Overturning Current (AMOC) – a vast ocean current, equivalent to thirty times all the world's rivers combined, that supplies huge quantities of heat to northern hemisphere regions such as Scandinavia.[102] Some research suggests that this current has slowed in recent decades,[103] and the IPCC's 2021 report predicted with high confidence that "the AMOC is very likely to weaken over the 21st century for all emission scenarios." While the IPCC had medium confidence that an abrupt collapse will *not* occur before 2100, it noted that "if such a collapse were to occur, it would very likely cause abrupt shifts in regional weather patterns and water cycles, such as a southward shift in the tropical rain belt, weakening of the African and Asian monsoons, … and drying in Europe."[104] Another potential tipping point is the collapse of the Amazon rainforest (often referred to as the "lungs of the planet"), which would result in massive ecosystem loss and even more heating as billions of tons of CO_2 per year would no longer be captured. As noted earlier, large portions of the Amazon region are currently being destroyed through deforestation for agriculture, which is also being exacerbated by drying due to higher temperatures and more frequent fires. According to one estimate, if just another 20 to 25 percent of the rainforest is destroyed, large parts could flip to become a Savannah-like ecosystem, meaning it would no longer act as a major carbon sink.[105] An even bigger concern than individual tipping points is that they might start interacting, setting off a cascade like dominoes falling. As shown in Figure 3.7, for example, a further slowing of the AMOC could accelerate heating that would lead to more permafrost thaw and the Amazon collapse, which would in turn further accelerate heating.

Evaluations of Economic Risk

As a 2019 warning from 11,000 earth and biosphere scientists concluded, the climate crisis is both "accelerating faster than most scientists expected" and "is more severe than anticipated." As a result, they predicted that tipping points and "climate chain reactions could cause significant disruptions to ecosystems, society and economies, potentially making large areas of Earth uninhabitable."[106] Yet the very disturbing picture drawn from evidence from the physical sciences has unfortunately not been matched by that emanating from economists. Considering the vast economic damages and implications of climate change, to date economists have shown remarkably little interest in studying its likely impact. In 2019, a literature search of the nine most influential economics journals by economists Nicholas Stern and Andrew Oswald

Table 3.3. *Estimated impact of climate change on global GDP*

Global mean surface temperature increase (°C)	No. of estimates	Impact (% of GDP) Average of estimates	Range of estimates
<=2	4	0.3	e) −0.5 to 2.3
2.5	11	f) −1.3	g) −3.0 to 0.1
3.0	9	h) −2.2	i) −5.1 to −0.9
5.4	1	j) −6.1	k) −6.1
6.0	1	l) −6.7	m) −6.7

Note: GDP is the total value of goods and services produced in one year (here, for the whole world). Data from Tol, "Economic Impacts of Climate Change."

showed that out of 77,000 published articles, just fifty-seven were about climate change. At that time, the most frequently cited economics journal, the *Quarterly Journal of Economics*, had not published a single paper on the topic.[107] Furthermore, a 2020 analysis of the just twenty-six reports on the likely economic damages of climate change by the end of this century found that those particular estimates have been incredibly small, as shown in Table 3.3.[108] To realize how small, consider that world GDP is currently around $100 trillion, which at a growth rate of 2 percent per year would reach $500 trillion by 2100, meaning that even a loss of 7 percent (or 35 trillion), the highest value in Table 3.3, would still leave the overall predicted GDP in 2100 more than 4.5 times higher than it is today. While this "minor" impact has often been promoted by economists, and has for years been highly influential among policy makers, it is completely at odds with the considerably more dire warnings from earth and biosphere scientists. Also, as noted in Chapter 2, the worst estimate in that table, a loss of about 7 percent of GDP, was for a global heating level of 6°C – a level of heating that would be catastrophic, likely incurring the greatest mass extinction ever on Earth, greater even than the end-Permian event that destroyed 90 percent of species.[109]

The reasons for this remarkable disparity are undoubtedly complex and, as Stern and Oswald point out, due at least in part to current trends and reward systems within the academic field of economics. But as Steve Keen charges in "The Appallingly Bad Neoclassical Economics of Climate Change," they are also due to fallacies and disciplinary biases within the field of economics itself.[110] As Keen observes, most research papers on climate change in this tradition exclude almost 90 percent of the economy from their calculations – specifically manufacturing, mining, transportation, communication, finance, insurance, non-coastal real estate, retail, wholesale trade, and government services – on the assumption that these sectors will not be affected by climate change, mainly because they occur indoors or underground and so supposedly are "not exposed to the weather." Beyond the obvious absurdity of such an assumption, given the wide-reaching impacts documented earlier, Keen estimates that if this research were more realistic in modeling the wider economy, the predicted impacts by 2100 would be tens of times larger. Among the other major problems he identifies in such research is a tendency to value the expertise of traditionally trained economists over that of other social scientists and natural scientists. As Nobel Prize–winning economist William Nordhaus admitted, the small group of natural scientists he had consulted for a 1994 landmark paper on the likely damages from climate change produced estimates that were twenty to thirty times higher than those from the larger group of non-environmental economists he consulted. (For example, the natural scientists estimated that a 3°C rise by 2090 would yield a 12.3 percent loss in GDP, while the economists estimated this at only 0.4 percent of GDP.) As Keen concludes, the work of economists trained in the neoclassical tradition tends to reflect "a strong belief in the ability of 'human societies' to adapt – born of their acceptance of the neoclassical model of capitalism, in which 'the economy' always returns to equilibrium after an 'exogenous shock' – they could not imagine that climate change could do significant damage to the economy, whatever it might do to the biosphere itself."[111] Another source of distortion on some economists is the fossil fuel industry itself. In an article entitled "Weaponizing Economics: Big Oil, Economic Consultants, and Climate Policy Delay," the historian and judicial expert Benjamin Franta lays out how a group of economists hired by the petroleum industry from the 1990s to the 2010s used economic models that inflated the later costs while minimizing the policy benefits, and how their results were often reported as "independent" rather than industry-related.[112]

It is worthy of note, however, that hard-nosed assessments of firms in the business of estimating risk, who are less concerned with the doctrinaire assumptions of neoclassical economics or with the selection of appropriate experts, have been far less rosy in their assessments, as suggested by the McKinsey predictions cited earlier. In 2021 the insurance giant Swiss Re predicted that, by 2050, the world economy would lose 18 percent of GDP per year (23 trillion dollars) from climate change.[113] As a result, it soberly favored taking concerted action now to mitigate much worse damages later: "Adding just 10% to the $6.3 trillion of annual global infrastructure investments would limit the average temperature increase to below 2°C. This is just a fraction of the loss in global GDP that we face if we don't take appropriate action."[114] By 2022, the IPCC mitigation report stated that "[t]he global economic benefit of limiting warming to 2°C is reported to exceed the cost of mitigation in most of the assessed literature (medium confidence)."[115]

Conclusion

As this chapter has shown, even though is not possible to definitely attribute a particular event to climate change, climate science has clearly demonstrated that global warming increases the likelihood and intensity of particular events. Indeed, the 1.1°C of global heating we have already experienced since pre-industrial times has had dramatic effects on oceans, land masses, and many kinds of weather-related events, including heat waves, droughts, wildfires, floods, and hurricanes – all of which have serious implications for human health and food production and for the survival of other species. One need only pay attention to the news to know that the human, ecological, and economic costs are mounting and increasing rapidly. Even if it is now nearly impossible to keep heating beneath 1.5°C, it is nevertheless imperative to reduce emissions to stave off further increases, as every fraction of a degree can make a significant difference in the severity of these impacts. Against the dire picture of the likely effects of global heating provided by physical science, the methods and assumptions used by many economists and adopted by many critics of climate action have often provided an astonishingly inaccurate view of the current and future costs of our present lack of concerted effort to limit emissions and further heating. To better understand the impediments that this kind of economics and the currently pervasive form of capitalism pose for taking urgent climate action, Chapter 4 turns to the field of political economy.

4

Capitalism and the Climate Crisis

> Climate change detonates the ideological scaffolding on which contemporary conservatism rests. A belief system that . . . declares war on all corporate regulation and all things public simply cannot be reconciled with a problem that demands collective action on an unprecedented scale and a dramatic reining in of the market forces that are largely responsible for creating and deepening the crisis.
>
> Naomi Klein

As discussed earlier in this book, greenhouse emissions began to rise substantially with the huge increase in the burning of fossil fuels that accompanied the Industrial Revolution in the late eighteenth and early nineteenth centuries. But as can be seen in Figure 4.1, an even greater increase, known as the **Great Acceleration**, occurred after World War II as the result of a dramatic, continuous, and roughly simultaneous surge in growth across a large range of measures of human activity, such as GDP, transportation, fossil fuel burning, and the use of industrial fertilizers. Among the factors driving this acceleration were an increase in population and the ready availability of oil, which facilitated all sorts of transportation and personal consumption.[1] But as Figure 4.2 shows, this has since been followed by an even greater and steeper increase in emissions from the 1980s onwards, despite the growing recognition of the threat of human-caused global heating among scientists, governments, and even oil industry executives. This chapter builds upon insights from the field of **political economy** to investigate the role that contemporary capitalism has played in these massive increases in emissions and what actions might be taken to mitigate those effects.

As shown in Figure 4.3, in the mode of economic organization known as capitalism, some land, natural resources, or utilities may be owned by society as a whole, but economic activity generally remains in the hands of private ownership of the means of production. The capacity of property owners to

Figure 4.1

The Great Acceleration after World War II

This refers to the dramatic, continuous and roughly simultaneous surge in growth rate across a large range of measures of human activity (A), with impacts on multiple earth system indicators (B). OECD is the Organization of Economic Co-operation and Development, including the US, European States, Japan, Latin American states and others. BRICS includes Brazil, Russia, India, China and South Africa.

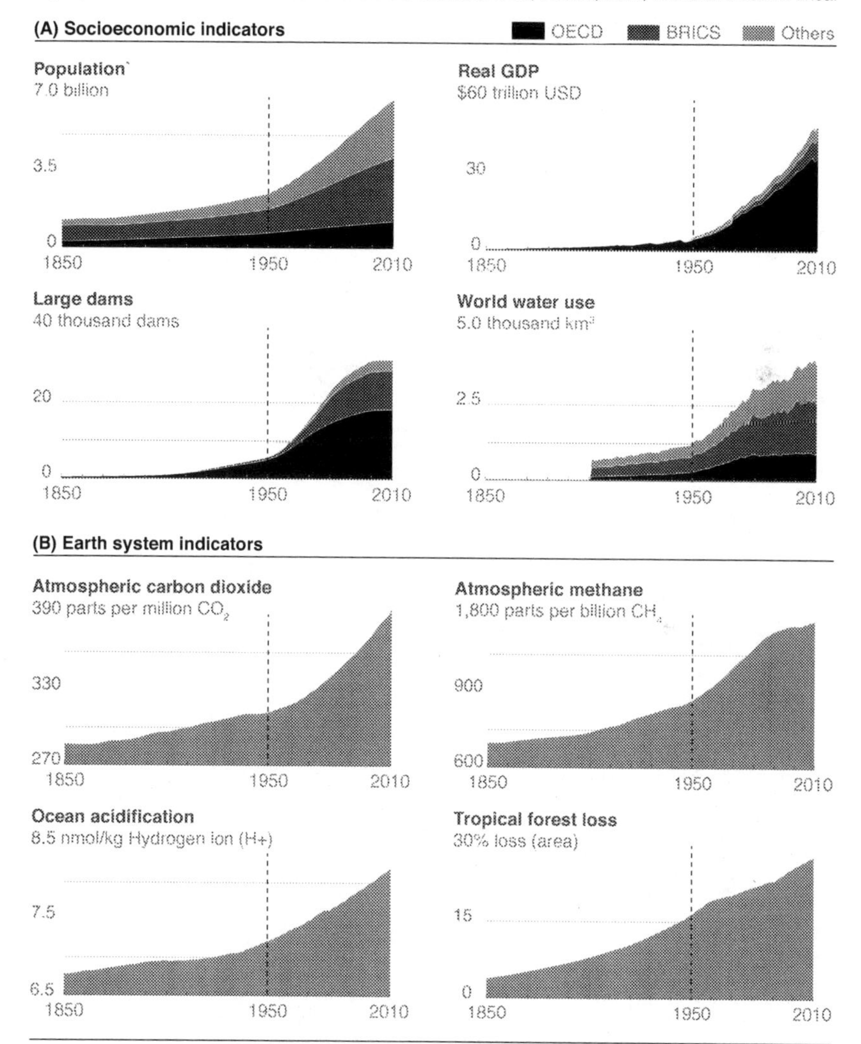

Adapted from www.igbp.net based on Steffen et al. (2014). The Trajectory of the Anthropocene: The Great Acceleration. *The Anthropocene Review*

Figure 4.2

How do emissions relate to different phases of capitalism?

(A) Emissions started rising substantially after 1900, then heavily accelerated after World War II (the Great Acceleration), even though this was a period of regulated capitalism in the USA and Europe. Emissions then heavily accelerated again in the neoliberal era (1984 onwards). (B) For England (now the UK) emissions ramped up steadily from the Industrial Revolution onward, but were eclipsed by the US around 1920. Cumulative USA emissions have skyrocketed since, joined by China, and to a lesser extent India, in the neoliberal era. Notably, Soviet Russia also made substantial emissions under "state capitalism"

(A) Yearly global CO_2 emissions

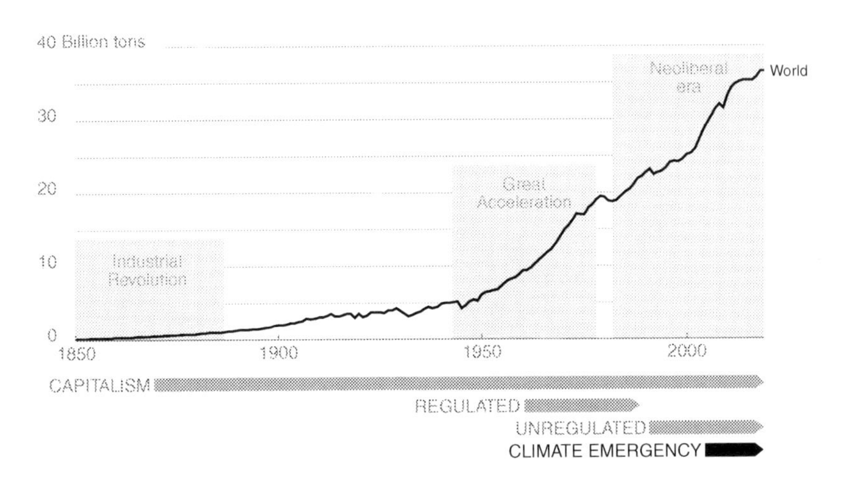

(B) Cumulative CO_2 emissions by region

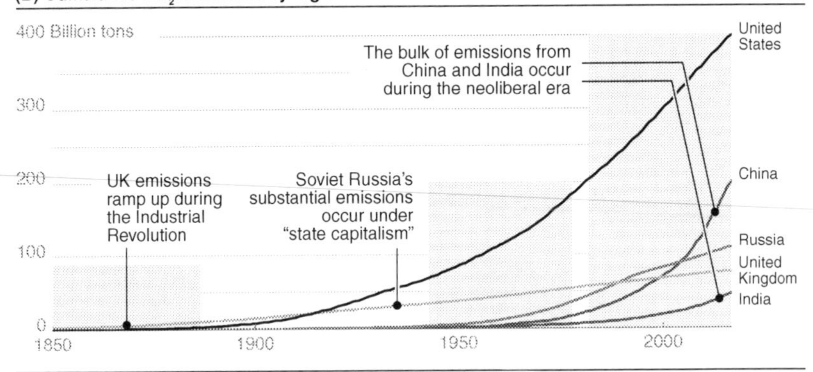

Adapted from Global Carbon Project Carbon Dioxide Information Analysis Center (OurWorldInData.org/co2-and-other-greenhouse-gas-emissions/); and Our World in Data based on the Global Carbon Project (OurWorldInData.org/co2-and-other-greenhouse-gas-emissions/)

Figure 4.3

A simplified view of key features of a capitalist system

The capitalist system is made up of several markets (land, labor, capital, goods, and services) while much economic activity is in the hands of private business. Capitalism has an ideology, including fostering the idea of a rational economic person operating according to tight rules of self-interest, and fostering the idea of competitiveness. Under capitalism, the state acts as a regulator of the various markets and it provides some public services, such as firefighting. Above all, capitalism has an expansionary compulsion – the iron law of capital dictates that more and more resources and labor must be pulled into its circuits of accumulation so more profit can be made.

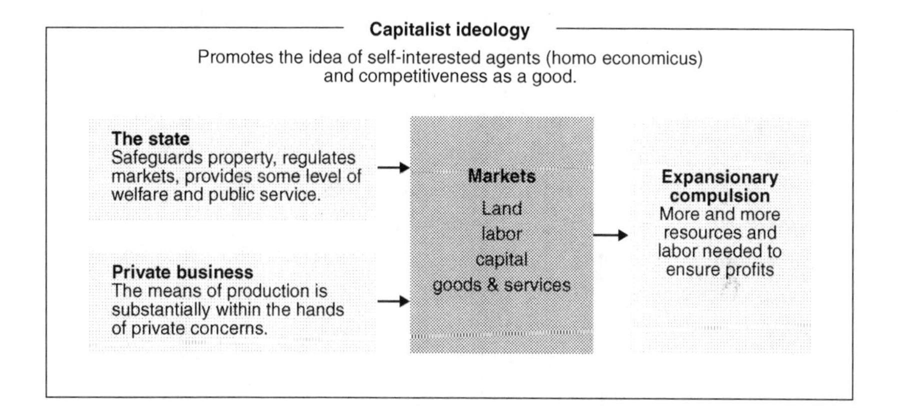

derive income from their control of the means of production (such as offices, factories, and machines) is often thought to identify them as a distinct class. Capitalism operates within various markets: the labor market includes those who do not own the means of production and whose income thus depends on selling their capacity for work, while the capital market includes a set of institutions (such as banks and the stock exchange) that funnel profits and savings to help capitalists fund their production, goods, and services.

This chapter first identifies four key features of present-day capitalism that make it difficult to cut emissions and that present structural and psychological barriers to dealing with the related ecological crisis: its expansionary compulsion, restricted role of the state, framework for international trade, and distinctive ideology. It then proposes two ways in which the most negative impacts of current capitalist practices on global heating might be addressed nationally and globally.

The Expansionary Compulsion of Capitalism

To appreciate the role capitalism has played in driving up greenhouse gas emissions and increasing ecological damage for the last several hundred years, it is helpful to understand how the capitalist system that has developed during that time is based on producing steadily increasing profits in order to continually grow and compete.

As economic anthropologist Jason Hickel points out,[2] capitalism does not refer simply to buying and selling, which has gone on for thousands of years, but to a particular way in which buying and selling happens, as illustrated in Figure 4.4. For much of human history, economies were organized around the principle of **use value**. Hickel gives the example of how an artisan might convert the commodity of wood into a bookshelf through their labor because it was useful to them or to trade or sell it for money to buy other useful things. Indeed, this is actually how most of us participate in the economy today – we buy things that are useful for us. In a capitalist model, however, the producer's goal is not to produce the chair because it is useful but in order to make a profit – to convert their investment in materials and labor into more value than they invested, providing the chair with **exchange value** beyond its use value through adding the surplus value of labor and the extraction of resources from nature. In this exchange, in other words, one is using money not simply to purchase a useful commodity but to get more money.

On a small scale, such a system can produce a fairly **steady-state economy** (near constant size and levels of consumption) in which producers and sellers make enough profit to remain in business and to sustain themselves and their families and communities. But as business enterprises began to grow in size and degree of economic risk, many of their owners began to reinvest their profits and to seek out investors willing to provide them with more capital in order to expand the production process and to yield yet more profit, resulting in a **growth imperative**, the expansionary compulsion, in which businesses must constantly expand and grow in order to compete in the marketplace and meet the expectations of their owners and investors. What makes capitalism particularly distinctive from other economic systems, therefore, is its self-reinforcing cycle, in which profit becomes capital invested to make more profit. In this way, capitalism can be said to have become unhinged from human need – some have gone so far as to compare capitalism to a virus, whose sole purpose is self-replication.

This expansionary compulsion of capitalism is a major contributor to the current climate crisis in that such growth relies not only on the skills and risk-taking of producers and sellers but on the constant takeover of more and more resources from outside its own system. In particular, it requires more and more extraction from ecological systems, including more and more energy inputs, still mostly in the form of fossil fuels that increase greenhouse gas emissions, as Figure 4.4 also shows. Yet as author Naomi Klein has noted,[3] the

Figure 4.4

Capital seeks to make more money by appropriating surplus value, with carbon dioxide as a consequence

(A) Everyday life and centuries' old trade

A commodity (C) which has use value can be exchanged for money (M), which can then be used to buy another useful commodity (C).

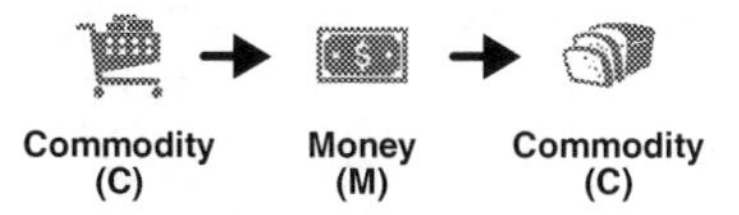

(B) Profit making under capitalism

Money (M) is used to buy a commodity (C), and that is sold to get more money (M´).

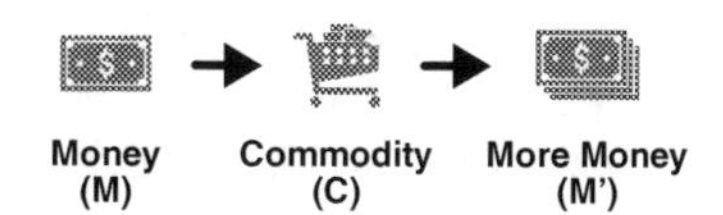

(C) What goes into (and comes out of) a commodity

In fact, the commodity the capitalist buys has several components: the means of production (MP) that has as inputs fossil fuels (F) and raw materials (RM, often from nature), and also labor power (L).

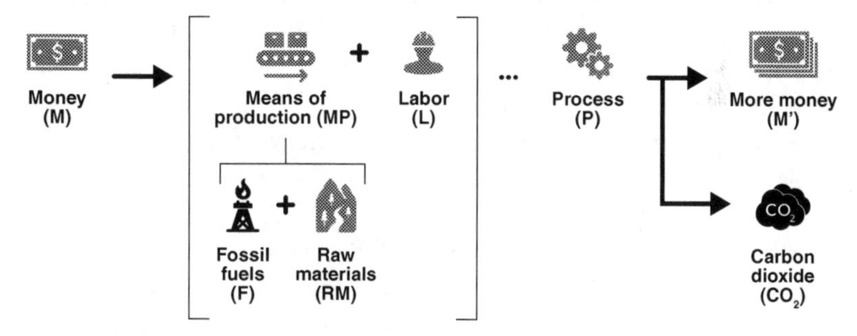

(D) The ultimate goal (and impact) of capitalism

It is not enough for the capitalist to generate profit, the goal is to reinvest that profit to make even more profit. Here, the simplified version of the overall commodity in (B) is used.

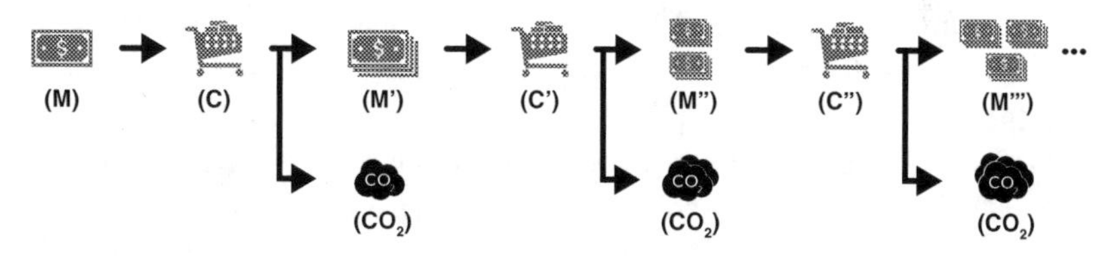

Adapted from the diagrammatic concept of Hadas Thier (A People's Guide to Capitalism, 2020, Haymarket Books, Chicago), and from the formulation of Andreas Malm (Fossil Capital, 2016, Verso, London)

accelerated increase in emissions beginning in the 1980s reflects not simply this growth compulsion but what she refers to as the very bad timing that the widespread recognition of the role of greenhouse gas emissions in global heating occurred at the very moment that capitalism took on a new form, known as neoliberalism, that aimed to shred many environmental regulations and shift greater control of the world economy to a tiny group of wealthy people.[4]

The Limited Role of the State in a Neoliberal Capitalist Economy

In most capitalist economies, the state does not play a direct role in production or distribution but instead sets necessary rules regarding property rights and markets, provides services not provided by the market (such as military, firefighting, and police), and protects the public interest through environmental regulation and social welfare benefits and services (such as health care and education). Historically, capitalist states have been generally content to limit their involvement or interference with the economy and to leave the provision of goods and services as much as possible up to the private sector, but the advent of industrialization and the movement of people from self-sufficient farms and villages into cities, forced governments to take a more active role in protecting people from the dangers of the workplace and the inevitable economic ups and downs, and expansions and contractions, of the industrial economy, which they do through regulation and public welfare services.

The role of the state within most capitalist economies expanded after World War II, when many Western states, including the USA, became what is commonly referred to as social democracies, or capitalist systems in which the state takes active steps to promote full employment, economic growth, and social welfare (including health care, education, and care of the elderly and other vulnerable citizens). The type of large-scale state intervention that developed during this period has been described as Keynesianism, referring to the economic policies of John Maynard Keynes, or as **embedded liberalism**, in which market processes and corporate activities were embedded in a web of social and political constraints in the form of strong regulations, including laws that limited pollution.[5] While this system delivered substantial economic growth in the advanced capitalist countries in the 1950s and 1960s, by the end of the 1960s the system began to break down, which was exacerbated by the fuel crisis of the early 1970s. As the reinvestment of capital no longer produced returns and the market became flooded with capital, many countries experienced a surge in both unemployment and inflation, known as **stagflation** (Figure 4.5A).

In response to this crisis, the labor movements and social movements in a number of countries began to press for more genuinely socialist alternatives to the then-current compromise between capital and labor;[6] to shift to greater

Figure 4.5

The rise of neoliberalism, and some consequences

The post-World War II period of Keynesianism, regulated capitalism, and more broadly shared prosperity came crashing down in the early 1970s with a crisis of capitalism. Some symptoms were huge inflation and unemployment (A). The ruling elites found a solution in the economic doctrine of neoliberalism – a prioritization of market principles over social and political judgments, the crushing of labor power, and widespread deregulation. From the late 1970s onwards worker productivity in manufacturing increased hugely, while real wages (in today's dollars) remained flat: this amounted to massive profits for capital at the workers' expense. From the 1990s onwards the share of wealth held by the top 1 percent rebounded to levels not seen since the early twentieth century (B). All data are from the USA.

(A) The crisis of the 1970s: unemployment and inflation rise

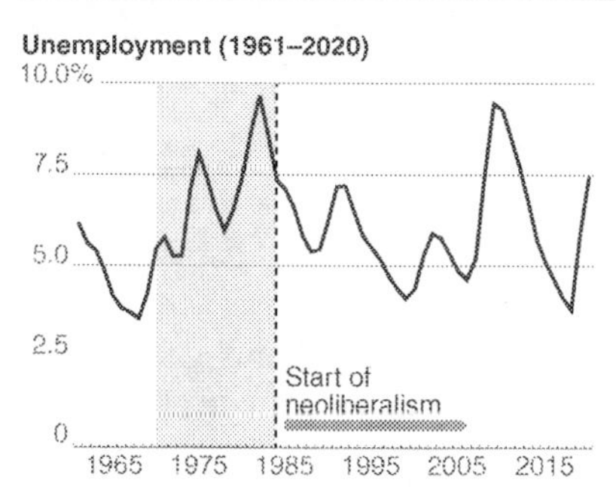

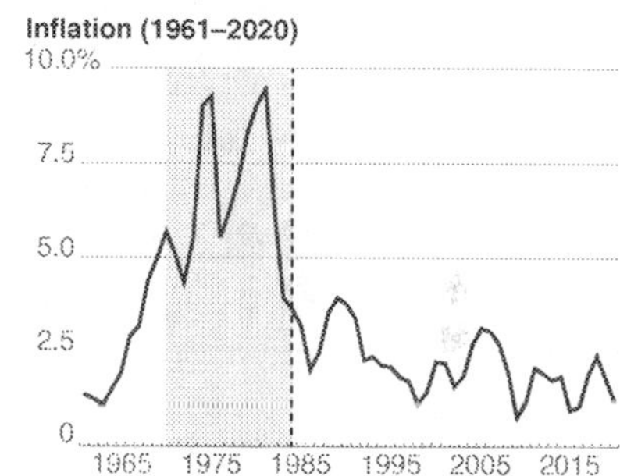

(B) Neoliberalism at work: attacks on labor and the re-concentration of wealth

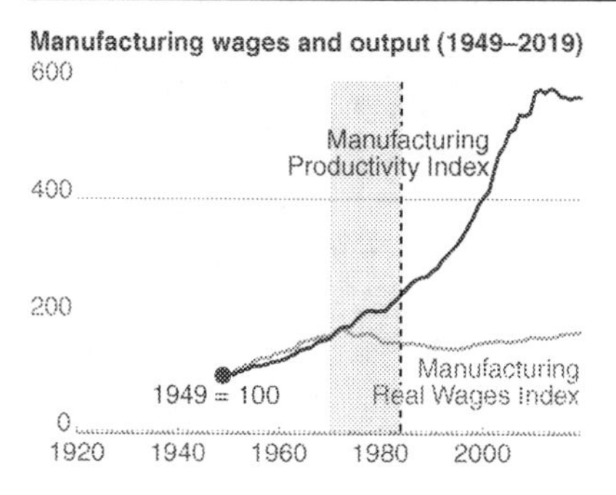

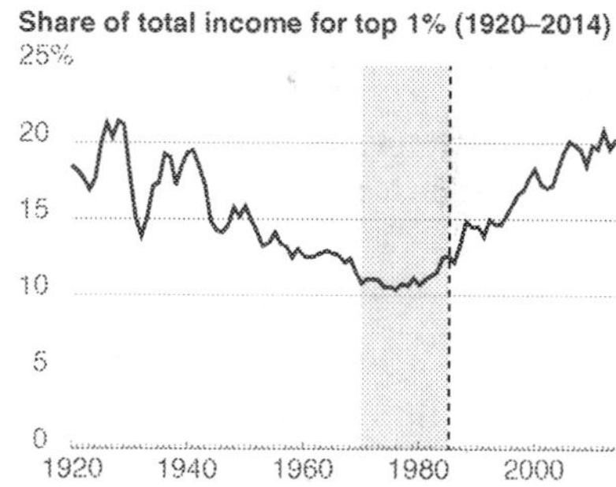

Data: (A) Unemployment data from US Bureau of Labor and Statistics: https://data.bls.gov/pdq/SurveyOutputServlet. Inflation data from the World Bank | CC-BY 4.0. (B) Income: Data from World Wealth and Income Database. Adapted from "Share of Total Income going to the Top 1%, 1920 to 2014". Online at https://ourworldindata.org/grapher/share-of-top-1-in-pre-tax-national-income?time=1920..latest&country=~USA. Wages and productivity: Data from US Bureau of Labor Statistics. Adapted from "Real wages and output per production worker in manufacturing in the US (1940–2019)", online at https://ourworldindata.org/grapher/real-wages-and-output-per-production-worker-in-manufacturing-in-the-us?country=~US"

public ownership, higher taxes, redistribution of wealth, and an emphasis on human flourishing rather than profit. In Europe, some socialist parties began to gain ground, while in the USA, social movements were calling for progressive and even radical social and cultural changes that, as historian Neil Maher has noted, were encouraged to some extent by a growing recognition of planetary boundaries: the space missions of the late 1960s carried stunning photographs back to humans of a planet, teeming with life, apparently suspended all alone in a vast universe.[7]

As economic geographer David Harvey has argued, this push for economic and social changes was viewed as a threat by many economic elites and upper classes around the world, who during the preceding period of high growth had been willing to accept a relatively smaller share of the economic pie to support raising standards of living and increased social programs.[8] In response to this perceived threat, many politicians and advocates of moneyed interests and members of powerful elites began to promote a set of ideas that came to be known as **neoliberalism**, a new version of classical liberalism's free market perspective, which prioritized market principles over social and political concerns and against the state interventionist approaches of embedded liberalism.

Neoliberalism is most widely associated with President Ronald Reagan in the USA and Prime Minister Margaret Thatcher in the UK, both of whom diluted protections for workers and the environment and oversaw the beginning of the current lopsided distribution of wealth and political influence and power in the USA and elsewhere. Soon after President Reagan managed to enact his neoliberal policies, the share of US income held by the richest 1 percent soared again to 15 percent, which was close to pre-World War II levels, as also shown in Figure 4.5B. Another eventual consequence was that capital interests' relentless attempt to reduce costs, especially labor costs, led to a dramatic and long-term reduction in real wages, even as worker productivity steadily increased through automatization. Today, neoliberalism is the dominant form of capitalism in many countries and has produced a partial retreat of embedded liberalism and social democracy elsewhere.

As Harvey explains, neoliberal theory calls for the maximally free operation of businesses and corporations within an institutional framework of free markets and free trade. To accomplish this vision, neoliberal advocates assiduously promote the privatization of assets, including property and also sectors of the economy formerly run by the state, such as railways, the post office, health care, and education. They also promote enhanced competition between individuals, cities, regions, and nations, arguing that individuals should be held responsible for their actions in the marketplace, including providing for their own welfare, education, health care, and pensions. They also call for sweeping deregulation, freeing corporations and businesses from state intervention and standards, such as banking rules and environmental controls. They justify these positions by claiming that they eliminate red tape, increase efficiency and quality, lower costs, and help individual consumers through better products and lower taxes.

The promotion of this vision by political leaders on the right (and acceptance of parts of it even by politicians on the left, as discussed in the next section) in the intervening decades has resulted in a diminishment of state control over the economy and a reduction of a shared sense of collectivism or responsibility for the welfare of the whole. This narrowing of the state's role in disciplining markets and setting national economic priorities has had myriad negative consequences, not only for the vast majority of the public who have not profited economically or socially from these shifts but for the climate and environment as well. In addition to the ecological damage resulting from the reduced oversight of natural resources around the world, such moves as the privatization of electric utilities in countries such as the USA and Chile are more likely to delay the transition to renewables.[9] These effects have been further magnified by the international trade framework created to advance the goals of neoliberal capitalism, which, as the next section shows, has dramatically escalated emissions and created legal structures that will make it all the more difficult to turn that life-threatening trend around.

The Framework of World Trade Established by Neoliberal Capitalism

The escalation of the climate crisis has also been heightened by the enshrining of the above principles of neoliberal capitalism into global trade agreements such as the North American Free Trade Act (NAFTA) and international legal frameworks such as the World Trade Organization (WTO), which together have played a central role in escalating the climate crisis and served as barriers to dealing with it. This new international trade regimen has contributed massively to the underlying causes of global heating: whereas in the 1990s emissions went up about 1 percent per year, by the early 2000s they were increasing by about 3 percent per year, and by 3.4 percent in 2019.[10] Indeed, almost half of the total emissions in history were emitted in just the thirty years between 1990 and 2020.[11]

A common assumption among mainstream economists is that such free trade ultimately enhances the economic welfare of all countries engaged in international trade through the efficiencies of something known as **comparative**

Figure 4.6

How comparative advantage becomes absolute advantage

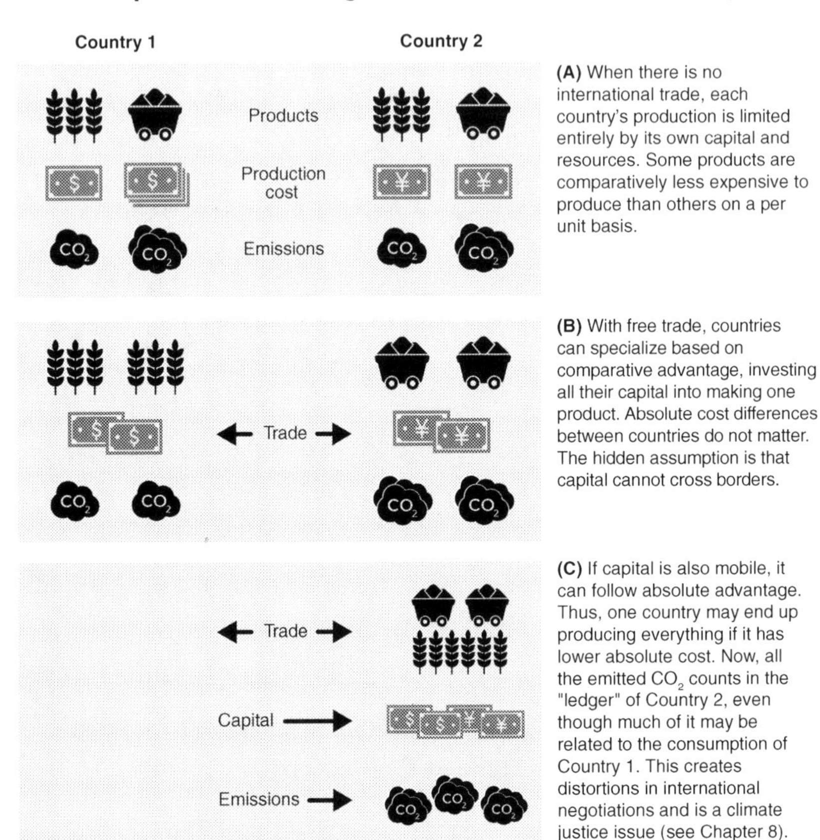

Adapted from H Daly, 1993, The Perils of Free Trade, Scientific American

advantage.[12] This notion was first formulated by the classical economist David Ricardo in the nineteenth century, who observed that countries with different customs, resources, and technologies will incur different costs when they try to make the same products. In the example in Figure 4.6, Country B can produce coal more cheaply than Country A, which gives Country B a comparative advantage, which may lead Country B to specialize in producing coal and Country A to specialize in something else, leaving all parties the better for this arrangement. What this theory overlooks, however, is that capital can cross borders, and thus one country can end up producing almost everything if it has lower absolute costs – known as **absolute advantage**.

As demonstrated by China in the period in which emissions have risen most dramatically, the mobility of capital invalidates the assumption that free trade is necessarily mutually beneficial – and certainly that it has no serious effects on emissions.

Another assumption of mainstream economics is that the efficient allocation of resources requires the internalization of all costs, which refers to holding producers responsible for paying all the costs of production, including the disposal of waste, and raising their prices to cover such costs. But this principle is undermined when producers use **externalized costs**, making someone else pay them, as when untreated wastes are released and cause illness among the public. Because profit-maximizing firms facing ruthless competition will always have an incentive to externalize costs, most developed nations have established legal and administrative structures such as the US's Environmental Protection Agency (EPA) to set and monitor the social and environmental standards of domestic industries. Yet the current neoliberal trade regime gets around such limits by encouraging industries to shift their production to countries that allow the greatest externalization of costs. A particularly dramatic example is Western manufacturers' rush in the 1990s to shift their production to China, where wages were very low, trade unions were suppressed, and the state was willing to spend seemingly limitless funds on massive infrastructure projects (including coal-burning plants) to ensure that the factories could keep cranking out products to fill container ships for export back to the rich West. And when pressure from domestic or international critics eventually led China to increase some of its standards and costs, global manufacturers simply moved their operations to even cheaper and less restrictive countries elsewhere in Asia or Africa.

Supporters of neoliberal globalization often counter concerns about such standards-lowering competition by saying that overall growth will eventually improve living standards for everyone, and indeed there is evidence that globalization has reduced poverty.[13] Yet that picture is complicated by the fact that globalization has massively escalated greenhouse gas emissions, threatening disastrous climate impacts for the whole world that certainly do not feature in the rosy neoliberal assertion that a "rising tide lifts all boats." Globalization has also dramatically increased inequality in some countries and produced other environmental consequences that will be very challenging to repair.[14] In China alone, for instance, this era of massive economic growth has generated notoriously poor air quality in its main cities, polluted as much as 80 percent of its groundwater, and contaminated about one-fifth of its arable land.[15]

Returning to the trade agreements, their primary concern was expanding trade, often in very extractive and energy-intensive sectors of the world economy, while protecting the rights of financial firms and corporations and removing regulations that govern the public good, including environmental regulations. That the decision makers who devised the current neoliberal world trade framework were determined from the outset to keep the issue of climate change quite separate from the new trade regime was already evident in the agreement produced at the Rio Earth Summit in 1992, which, among other things, explicitly declared that "measures taken to combat climate change, including unilateral ones, should not constitute . . . a disguised restriction on international trade."[16] This intention to separate trade and its impact on global heating continued in the negotiations for NAFTA and the WTO in the 1990s, right through the Trans-Pacific Partnership signed by the Obama administration and a dozen Pacific Rim countries in 2016, in which the words "climate change" did not appear even once in the 6,000 pages of the final document.[17]

These agreements produced further obstacles to addressing the impact of trade on emissions by creating legal means by which international business interests could contest local environmental efforts. For example, in 2016, a WTO legal panel ruled that a program in India that provided subsidies to promote the local production of solar panels was unfairly discriminating against foreign solar panel producers, and when India defended its support for the program by citing its UNFCCC climate obligations, the WTO determined that those obligations did not protect India's solar program from trade rules.[18] Nor was this an isolated case; one analysis of the first five years of the WTO found that it rejected the environmental argument in every environment-related dispute brought before it and that its oversight of global environmental rules consisted of narrowly defined, nonparticipatory, secret procedures that do not allow stakeholders and public advocates to offer or evaluate the relevant evidence.[19] Further, some trade agreements go so far as to allow foreign corporations to use a private tribunal system of three lawyers to challenge the laws of another country if they believe those laws are unfair, with no avenue for appeal, which has allowed multinational corporations to challenge proposed bans on offshore drilling and fracking and to undermine attempts to install local energy provided from renewables. In this vein, since 2004, mining industry interests have been able to sue national governments to maintain or expand their extractive activities more than a hundred times.[20]

This neoliberal trade regimen did not come about without protest from environmental and workers' groups. In the case of NAFTA, substantial opposition was raised by workers in the USA, Canada, and Mexico on environmental grounds as well as fears it could lead to offshoring their jobs.[21] The environmental movement itself was split on the issue. Assisted by Vice President Al Gore, some groups supported the treaty, including the Environmental Defense Fund, the Natural Resources Defense Council, and the World Wildlife Fund, while others opposed it, including Greenpeace, Friends of the Earth, the Sierra Club, and a number of smaller groups.[22] Yet as Klein explains, by making it possible for the Clinton administration to claim that 80 percent of the members of national environmental groups endorsed NAFTA, the pro-NAFTA groups were quick to take credit for helping pass the agreement. As the director of the Natural Resources Defense Council claimed, "[w]e broke the back of the environmental opposition to NAFTA. After we established our position, Clinton only had labor to fight. We did him a big favor."[23] The signing of NAFTA into law in 1993 was hugely significant, not only for trade between the three countries but because it anticipated and formed the groundwork for the WTO.[24]

As the formation of the WTO unfolded, there was opposition to it, too, in Europe and South Korea, and later very large protests reached their apogee during the WTO talks in Seattle in 1999. As recounted in the documentary film *This Is What Democracy Looks Like*, tens of thousands of students, labor union members, environmentalists, Indigenous peoples, and others took to the streets to protest the WTO policies on globalization (Figure 4.7). Such protests helped

Figure 4.7

The battle for Seattle

In 1999, tens of thousands of environmentalists, indigenous peoples, union workers, and students undertook a week-long protest in the streets of Seattle, Washington. They were protesting the World Trade Organization meeting they understood to have devastating environmental, social, and labor results. In this picture, also conveyed in a key moment in the documentary *This Is What Democracy Looks Like* (available on YouTube), police discharge pepper spray point blank at the faces of peaceful protestors.

bring worldwide attention to a darker and potentially catastrophic dimension of economic globalism, including the way it would crush environmental legislation and the later impact it would have on the huge acceleration in the growth of greenhouse emissions.[25]

Even though China, which during this period became the world's greatest emitter, is nominally a socialist state, its economic system is best described as state capitalism, meaning that its economy is driven by for-profit commercial activity, though under tight state control.[26] Furthermore, China's remarkable success in growing its economy and raising the living standards of its vast population over the past several decades is a direct result of its decision to participate in and effectively compete within the neoliberal capitalist marketplace. Indeed, as noted earlier, much of the huge growth in China's emissions is directly attributable to its serving as a vast factory for the Global North – and in that sense, China's emissions are shared by us all. Much the same was true for Soviet Russia. Despite its nominally socialist political ideology, even before the neoliberal era, its Cold War political and economic goals made Russia subject to the same growth imperative as contemporary China, leading its emissions to increase substantially throughout the twentieth century, as shown in Figure 4.2.

Yet another example of how difficult it can be to escape the competitive matrix of neoliberal capitalism is the case of Bolivia under Evo Morales, that nation's first Indigenous president. Morales was elected in 2006 on a platform to end fossil fuel **extractivism**, the exploitation and export of raw materials, mainly from the Global South, to feed industrial development in the North. Although Morales pledged to honor the Indigenous concept of "buen vivir," which defines "living well" as building a society that is in harmony with nature, he ultimately was able to dramatically reduce the proportion of the population living in poverty only by opening up Bolivia's natural parks to oil and gas extraction.[27] As fossil gas exports grew sixfold during his first term, from $1.1 billion a year to $7 billion, a representative of a Bolivian grassroots social movements group said, "[t]his is absolutely not the kind of policy we expected from president Morales after such big speeches and discourses about Mother Earth and Pachamama and how we should start living in harmony with her.... What we have seen in these nine or 10 years of government is that our natural resources and our forests – the richness of Bolivia – has been depleted much more than any other period of time."[28] As the case of Bolivia shows, the prevailing world economic order makes it difficult for any state, even a leftist one with an Indigenous-oriented mindset, to escape the international nexus of competitiveness.

The Distinctive Ideology of Neoliberal Capitalism

Although scholars (perhaps most notably Max Weber) have traced the advent of some of the psychological and social characteristics of modern capitalism back hundreds of years, most commentators agree that the roots of the distinctive ideology related to neoliberalism can be traced to a group of passionate thinkers gathered around Friederich von Hayek in the 1940s, which included economists such as Milton Friedman, who would later be hugely influential in the USA.[29] In his classic text of that time, Von Hayek argued that unrestricted competition, governmental noninterference, and individualism were essential to a successful economy, leading him to oppose collectivism, or a government's taking goods from the few to distribute to the many in the name of a social ideal.

A few years later, that underlying notion that selfishness is admirable was further popularized by the Russian-born US philosopher and novelist Ayn Rand, whose influential admirers included Alan Greenspan, chairman of the US Federal Reserve for nineteen years, US President Ronald Reagan, and UK Prime Minister Margaret Thatcher. Rand's essays and novels offered an extended argument for what she termed "rational self-interest," unfettered individualism, and laissez-faire capitalism. Her novels glorified unyielding business leaders and denigrated characters who believed they owed anything to society. In her 1957 novel, *Atlas Shrugged*, the protagonist Hank Rearden, who is standing trial for breaking a government regulation, is greeted by applause by the crowd when he proclaims, "I refuse to apologize for my success – I refuse to apologize for my money. If this is evil ... let the public destroy me. ... The public good be damned, I will have no part of it!"[30] According to the psychotherapist Sally Weintrobe, in her book *The Psychological Roots of the Climate Crisis*, decades of constant repetition of various versions of this neoliberal ideology in the mass media, political framing, promptings from social groups, and advertising supports what she terms a **culture of uncare**, by which she means encouraging individuals and groups to protect entitlement and privilege, and to refuse to accept the moral implications of their own self-focus.[31]

The right-wing ideas of von Hayek and Rand did not appear ripe for general consumption when they first emerged in the 1940s to 1960s time frame, as most capitalist countries at the time had installed a functioning welfare state to counteract the harshest aspects of free market capitalism. In the UK, for instance, the state provided free health care for everyone, family and sickness allowances, unemployment benefits, and social housing; while in the USA, under Republican President Eisenhower, investments in social spending were financed by a top marginal income tax rate of 91 percent. But when the era of embedded liberalism appeared to break down in the 1970s, Randian acolytes such as Reagan and Thatcher jumped into action, as we saw, weakening the power of labor, encouraging hugely increased consumption, shredding environmental regulations, and paving the way for globalized capitalism.

This neoliberal thinking and the accompanying culture of uncare were perhaps best epitomized by the approach taken by Thatcher, who upon her election in 1979 began a revolution in fiscal and social policies intended to overturn the social democratic state. Famously claiming that there is "no such thing as society, only individual men and women," she attempted to dissolve all forms of social solidarity in favor of individualism, private property, and family values.[32] Of this relentless ideological assault, she said chillingly, "[e]conomics are the method ... the object is to change the soul." In the UK, US, and elsewhere, popular consent for the neoliberal turn was created through many information channels and institutions and required the conversion of many intellectuals and members of the media to the neoliberal way of thinking. As a result, according to author Naomi Klein, "veneration of the profit motive has infiltrated virtually every government on the planet, every major media organization, every university, our very souls." To Klein, "neoliberalism's single most damaging legacy" has been "the belief in the central lie – that we are nothing but selfish, greedy, self-gratification machines," a vision of human relations that "has isolated us enough from one another that it became possible to convince us that we are not just incapable of self-preservation but fundamentally not worth saving."[33]

As the scholar Kate Raworth has noted, this vision of economic man (**homo economicus**) was not original to neoliberal thinkers but can be traced back to such nineteenth-century thinkers in political economy as Adam Smith and William Stanley Jevons, who described this cartoon of human nature as having four key features: that people are solitary, calculating, competitive, and insatiable.[34] Although fully aware that actual human interactions and motivations are more complicated than that, those thinkers offered the simplified vision of humans as homo economicus as part of an effort to develop economics as a science of behavior modeled on the science of physics, which described the rules that govern isolated atoms. According to both Raworth and Weintrobe, proponents of neoliberal ideas have since normalized this cartoon view into a creed, one that scholars such as Murray Bookchin had long argued is not only a simplified but an incorrect view of human nature.[35] Based on extensive historical and anthropological evidence, these scholars propose a different view of human nature as often trusting, reciprocating, and cooperative, or what Raworth refers to as **homo socialis**.

Whereas twentieth-century market capitalism is well known to have encouraged a culture of consumerism,[36] neoliberalism's culture of uncare, with its emphasis on personal choice and selfishness, has since escalated this culture to hyperconsumption.[37] In addition to the glorification and promotion of consumption in its own right, the reality of everyday life under a highly competitive form of capitalism promotes consumption in other ways as well. To stay competitive, people are pushed to increase their efficiency and time by buying all sorts of products, such as smartphones, computers, cars, and appliances, in order to manage their high workloads, secure their income, and maintain their private life.[38] Such competition also translates into other forms of increased commodification, such as buying fast food instead of cooking together, which in turn requires increased energy and material inputs. Positive feedback loops are generated as more workers must make these investments to keep up with each other, further exacerbating energy use and resource extraction and increasing environmental destruction and emissions. All of this consumption is further accelerated by capitalism's propensity to generate artificial scarcity, such as through the **planned obsolescence** of products, and also through advertising – think of the need to keep replacing a smartphone every few years. This cultural and psychological commitment to consumerism presents a significant barrier to reducing energy demand and consumption to sufficiently lower emissions, just as the wider culture of social atomization works against the kind of solidarity that is needed to join grassroots social movements, climate advocacy groups, and workers unions.

Overcoming Neoliberal Capitalism's Structural and Psychological Barriers to Dealing with the Climate Crisis

As the previous sections have shown, capitalism, and particularly its current neoliberal manifestations and philosophy, creates a number of structural and psychological barriers for lowering greenhouse gas emissions as radically and rapidly as will be needed to avoid the most disruptive effects of global heating. Yet as this chapter has shown, other economic systems have also found it difficult to survive and compete within the larger international trade arrangements imposed by the imperatives of powerful capitalist states and organizations, and even if we could be assured that another set of economic relations could improve our current situation, such changes are unlikely in at least the short term. Thus, for better or worse, we must find ways to quickly and substantially reduce emissions within the current capitalist system. This section proposes two major ways in which that system could be reformed to create a more just and healthy future, both nationally and internationally: granting governments more power and control over their economies and limiting levels of economic growth and consumption.

Assigning a Stronger Role to the State

In their 2020 *Climate Crisis and the Global Green New Deal*, Noam Chomsky and Robert Pollin argue that even though "capitalist logic, left unconstrained, is a recipe for destruction," the existential danger it presents and the narrow time frame now remaining for action suggest global heating must be addressed within a capitalist framework, which history has shown can at times "readily accommodate major initiatives of industrial policy, public subsidy, state initiatives, and market interference."[39] Indeed, some democratic capitalist states such as Spain, Norway and the UK are now generating, or planning to generate, substantial proportions of their energy needs from wind and solar.[40] Apart from these country-level examples, many local governments around the world are taking action at the regional and city levels to reduce emissions, ranging from banning new fossil extraction to making new commitments to public transportation and building electrification to bringing privatized utilities and other entities into local city control (known as **remunicipalization**).[41]

In the USA, the best historical evidence of the tools available to the federal government to influence the economy and markets is the World War II period, during which, as ecologist Stan Cox points out, Congress spent $321 billion domestically – more than the US government had spent in the entire period 1790 to 1940 – and instigated price controls, resource allocation, and rationing.[42] During this five-year period, the War Production Board quickly restructured the entire economy, halting car manufacturing in Detroit and converting those factories to military production, blocking the production of new air conditioners used solely for personal comfort, restricting civilian travel to conserve gas and coal, and regulating the production and sale of lumber, bolts, industrial chemicals, bedsprings, and coal- and oil-fired heating stoves. This board went so far as to issue standardized instructions for manufacturing many kinds of products, such as eliminating unessential materials in produced items and restricting the number and variety of manufactured items, such as clothes, shoes, furniture, and farm machinery parts. To prevent hoarding and under-the-table dealing, the government introduced fair-shares rationing in which households were issued a monthly set of ration stamps for meats, cheese, sugar, kerosene, bicycles, typewriters, and so forth. As Cox points out, even at its highest point, public approval of rationing outweighed disapproval by two to one. Although few people yet view the climate crisis with an equivalent sense of danger and national interest as citizens did the war during that era, this history shows that a mass technical reorientation of the economy and reduction of consumption can in fact be accomplished within a capitalist system and can be well tolerated when viewed as a necessary response to a perceived national crisis.

The USA has also shown great willingness for strong government action under other conditions viewed as emergencies. Although what the elements of a government mobilization to address our current climate crisis might look like is examined in more detail in Chapter 9, the US government covered tens of billions of costs and losses associated with the collapse of hundreds of companies during the savings and loans crisis of the 1980s and spent hundreds of billions bailing out banks and other companies during the 2007–8 financial crisis. More recently, the US government injected trillions into the economy during the first year of the Covid-19 pandemic, which, as Cox has pointed out, also forced a previously unthinkable level of political and economic discussion among the public, experts, and politicians about what services and products should be considered essential.[43] Meanwhile, other forms of state engagement with the threat of global heating are also possible, including institutional change. In 2021, for example, the country of Chile appeared to throw off the yoke of nearly half a century of neoliberalism dating from the Pinochet regime by convening a new constitutional convention in which a citizens assembly was charged with deciding how to embed climate and ecological principles and action into the constitution itself.[44]

Reducing Levels of Growth and Consumption

In the previously mentioned 2020 article titled "World Scientists' Warning to Humanity," more than 11,000 earth and bioscientists directly addressed the danger that the prevailing capitalist growth imperative poses to the planet. Declaring that "excessive extraction of materials and overexploitation of ecosystems, driven by economic growth, must be quickly curtailed to maintain long-term sustainability of the biosphere," they identified two steps that must be taken to accomplish that vital goal: not only do "we need a carbon-free economy that explicitly addresses human dependence on the biosphere and policies that guide economic decisions accordingly," but "our goals need to shift from GDP growth and the pursuit of affluence toward sustaining ecosystems and improving human well-being by prioritizing basic needs and reducing inequality."[45]

What they and numerous other experts are calling for is a shift away from the focus on GDP growth, sometimes known as **degrowth** or **post-growth** (the phrase used in this book), in order to reduce consumption, especially in the affluent.[46]

Some, such as the economist Robert Pollin, have taken these proposals to mean an overall reduction of GDP growth, which he argues is inadequate to reduce emissions. Pollin points out that contracting GDP by 10 percent over the next two decades would be equivalent to a reduction four times greater than during the 2007–9 Great Recession,[47] yet the overall effect of this reduction for CO_2 emissions would be to push them down only by 10 percent – from 42 billion to about 38 billion tons per year – which would be entirely inadequate relative to the IPCC target of a 45 percent cut in emissions – or a cut of 22 billion tons – by 2030. Yet post-growth proponents reply that they are not advocating reducing economic activity overall (they acknowledge that some things such as renewable energy will need to increase), but instead advocating an economic program, at least in the already-rich countries, in which continued economic growth is not necessary. For them, the aim should be to keep these economies stable while supporting strong social outcomes without economic growth – which requires a set of post-growth policies that aim to reduce consumption.[48]

The fundamental motivation for this line of argument is that if the overall aim is to reduce emissions then it will be very difficult to do that while growing the economy at the same time; specifically, it will be challenging to **decouple** emissions and economic growth (Figure 4.8A). Although data from a few countries do suggest that a shift to renewable sources of energy and improvements in technology and efficiency have allowed some such decoupling between increasing GDP growth and emissions,[49] a 2020 meta-analysis of 835 peer-reviewed articles on this topic concluded that this is not generally feasible: "large rapid absolute reductions of resource use and greenhouse gas emissions cannot be achieved through observed decoupling rates."[50] To appreciate the difficulty, consider that the few countries that did achieve absolute decoupling of GDP from emissions, such as the UK, Spain, and Romania, achieved emissions reduction rates of only about 3.4 percent per year from 2005 to 2015; yet if one aims at the IPCC goals and assumes that high-income countries grow at the typical rate, then they will need to reduce emissions at more than 12 percent per year, which is an enormous challenge.[51] Moreover, even if there are large efficiency savings from a shift to renewable energy and electrification, research data show that in a growth-oriented system, about half those savings are then shifted into further expanding consumption and production – system-wide **rebound effects** – such as occur, for instance, when

Figure 4.8

To make emissions go down, technology will not be enough – we must confront overconsumption (affluence)

(A) Coupling: growing material extraction and emissions increase together with increased economic output (GDP), while emissions are not far behind. These global data argue against decoupling GDP from growth/material extraction and emissions. They all increase together. Although technological change can act to slow global environmental impact, it is overwhelmed by increases in per capita consumption. Technology development alone is thus unlikely to reduce emissions; we will have to counteract the effects of growing consumption and affluence.

(B) Another view of coupling. Those countries with higher average annual income have higher energy use per person. There is a tight correlation between economic scale (per capita GDP, the sum total of a country's economy divided by population, which effectively indicates average annual income) and energy use (usage per person in watts). Note the logarithmic scales. Size of dot reflects population.

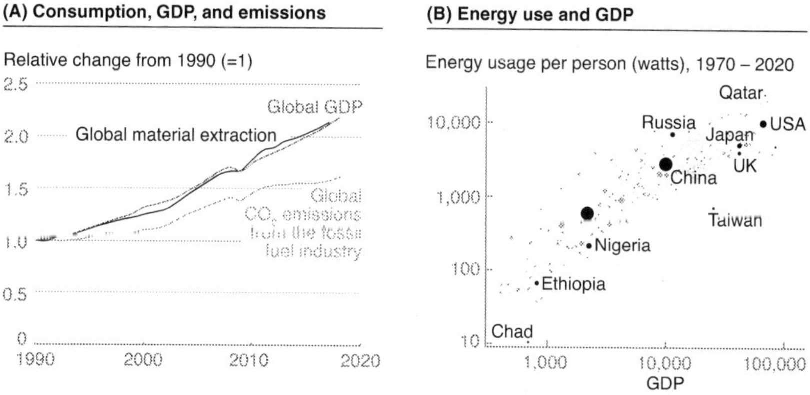

(A) Consumption, GDP, and emissions

(B) Energy use and GDP

Adapted from Wiedmann et al. (2020). Nature Communications 11:3107. Creative Commons 4.0, and from Murphy T. Energy and Human Ambitions on a Finite Planet. (https://www.nature.com/articles/s41467-020-16941-y/figures/1)

households use the savings from reduced electricity costs to buy more carbon-intensive goods or services.[52]

Thus, post-growth proponents argue that an emissions reduction energy transition must also be accompanied by a reduction in consumption, especially in affluent countries, as Figure 4.8B shows. Within those countries it is especially affluent people who drive resource use, generate emissions through high consumption, and socially propagate consumption as powerful members of their societies (the richest 1 percent across the world produced double the emissions of the poorest 50 percent).[53] Post-growth policies will also need to recognize this disproportionate effect of the affluent; it is not correct to pitch this as Global North vs. Global South – in the Global North there are many working-class people who experience limits to their access to basic necessities, often living paycheck to paycheck,[54] and in the Global South there are highly affluent individuals in countries such as Nigeria, South Africa, and India.

Such a shift away from GDP growth as the main goal would require a profound rethinking of some core assumptions because, as Jason Hickel points out in *Less Is More*, growth has a strong grip on our imaginations from being promoted by governments, economists, and opinion leaders as essential to our well-being and development.[55] Yet empirical evidence of the relationship between GDP growth and well-being is weak, and Hickel argues that how income is distributed within a nation and the level of public services it provides are far more important than growth to the well-being of a nation's population. As he observes, major improvements in the standard of living in Britain in the nineteenth century were prompted not by growth in its larger economy but improvements in public sanitation and services. Such observations are also supported by many current examples of countries such as Spain that have better health and education outcomes than the USA (such as life expectancy five years longer) despite having 55 percent less GDP per capita.[56]

One way to visualize what is truly necessary for a safe and socially just standard of living across the globe is Kate Raworth's metaphor of a doughnut-shaped "safe space for humanity." [57] As shown in Figure 4.9, this doughnut is bounded on one side by an *environmental ceiling* beyond which lies environmental degradation and the tipping points discussed in Chapter 3, and on the other side by a *social foundation* beyond which lies poverty and human suffering. According to Raworth and the signers of the "World Scientists' Warning to Humanity," defining human welfare and the responsibility of governments toward their people in terms of how they help provide that safe space rather than by abstract economic measures (such as GDP) is likely to be a necessary step in accomplishing that goal. As utopian as such an image may seem, a 2020 economic analysis demonstrated that the nine billion people expected to live on Earth by 2050 could live within these boundaries with ample water, heating and cooling, internet access, public transportation, and universal health care and education while using only about a quarter of the energy required by our current business-as-usual trajectory.[58] While such a lifestyle would surely be more modest than that used to by many people in the world's richest countries, it would hardly require going back to living in caves, as critics of the climate movement sometimes claim, and would represent a considerably higher standard of living than currently experienced by many people in poor and affluent nations alike. What post-growth advocates are thus calling for is to favor human thriving in a more equitable way and within planetary boundaries. Chapter 9 takes up the question of what such demand and consumption reduction (post-growth) policies might look like. It was striking that the IPCC 2022 report on mitigation clearly included detailed proposals on demand reduction for the first time, noting that "[d]emand-side

Figure 4.9

A safe and just space for humanity

A. The planet in balance.

The safe space for humanity is the "doughnut" space above the social foundation and beneath the ecological ceiling.

The social foundation is food and housing, gender equality, peace, justice, and a political voice.

The environmental ceiling consists in a set of planetary boundaries beyond which lies environmental degradation and tipping points.

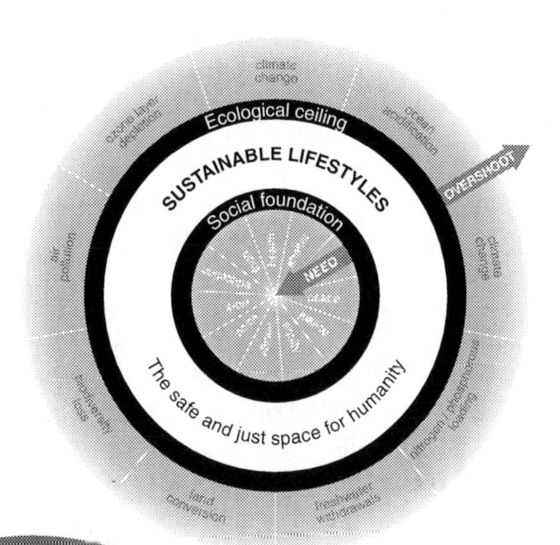

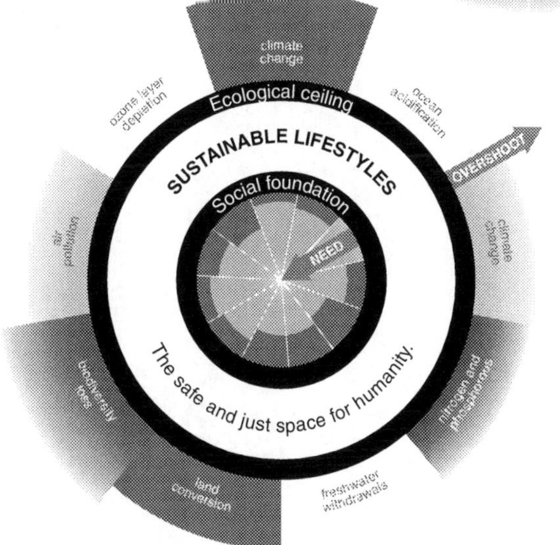

B. The planet is out of balance.

The planetary boundaries are exceeded on climate change, nitrogen and phosphorous loading, land conversion and biodiversity loss, and there are shortfalls in the social dimensions in multiple ways.

Beyond the boundary
Boundary not quantified

measures and new ways of end-use service provision can reduce global GHG emission in end use sectors by 40–70 percent by 2050 compared to baseline scenarios … and … Demand side mitigation response options are consistent with improving basic well being for all. (high confidence)."[59]

Conclusion

This chapter has examined some of the key features of contemporary capitalism to help explain why global emissions continue their inexorable rise and why governments find it so difficult to meet even their stated emissions goals. (At the current rate, for instance, Europe is expected to be twenty years late in reaching its main climate goal.[60]) As the chapter has shown, the expansionary compulsion, trade framework, and distinctive role of the state and ideology of neoliberalism have all massively escalated the climate crisis and presented powerful structural and psychological barriers to dealing with it. Yet the urgency of reducing emissions by more than half within a decade to achieve climate stabilization will require massive state intervention and reduced levels of growth on a national level and of consumption on a personal level. To achieve this will undoubtedly require increased political pressure and participation in a grassroots social movement and other forms of advocacy, which in turn will require a large-scale communication campaign to shift beliefs and overcome denial, which is the topic of Chapter 5.

PART II

From Skepticism to Belief

5

Skepticism, Misinformation, and Motivated Cognition

> It is in a certain sense unimaginable, even absurd, to think of us destroying our very climate... No wonder people resist, deny. To get beyond such denial requires you to remake your very world.
>
> Rupert Read

Achieving national action on emissions reduction will require widespread public support that is not currently forthcoming in a number of countries, such as the USA and Australia, partly because of widespread skepticism about climate change. In 2020, for instance, researchers at the Yale Climate Change Communication Group found that while 72 percent of US adults surveyed accepted that global warming is taking place, only 57 percent believed that it is human-caused, a proportion that dropped below 30 percent among conservative Republicans.[1] As illustrated by the following responses of participants in a research study by Stuart Capstick and Nicholas Pidgeon, such skepticism extends from the existence of climate change to its causes, the seriousness of its impacts, and how to respond to it:

> It's just natural and it just happens, because I suppose it's like us as a person, we change as we get older, so I guess the Earth is changing as it gets older.
>
> I'm not saying it's not happening, but ... I wouldn't have thought it was quite as disastrous as some people perhaps make us think it will be.
>
> All the world's governments can't manage to cobble anything together between them, despite all saying that they need to.[2]

Clearly, it will be difficult for US policy makers to commit to genuine emissions reductions if only just over half of all adults (and less than a third of conservatives) believe climate change is human-caused. And given the outsize role of the USA in the historic and current generation of emissions and in economic and military matters and in influencing international institutions, it is difficult to expect other countries to commit to binding agreements to limit emissions without similar US policy commitments.

Changing this skepticism to belief – the topic of this chapter and Chapters 6 and 7 – will require a better understanding of the factors that cause and propagate skepticism and what steps might be taken to halt and reverse it. As Capstick and Pidgeon have pointed out, such skepticism can be categorized into two major types: **epistemic skepticism**, which refers to skepticism about knowledge that global heating is happening, is human-caused, and is having and will have bad impacts; and **response skepticism**, which refers to whether and how one responds to that knowledge at a personal, group, national, and international level. To help us better understand why belief in human-caused climate change is so low in the USA and some other countries, this chapter offers an introduction to the causes of epistemic skepticism, discussing in particular how it is related to worldviews and values. As indicated in Figure 5.1, the remaining two chapters in Part II focus on the effect of two other major factors – science communication and risk perception – on epistemic skepticism, and the chapters in Part III discuss response skepticism in the context the technical, justice, political, and social frameworks needed to move away from fossil fuels.

What Survey Data Say about Skepticism toward Climate Change

Research in a number of fields has provided considerable information about the various factors that are associated with individuals' level of skepticism toward global heating. Among those studying this topic is the previously mentioned Yale Climate Change Communication Group, which since 2008 has regularly conducted surveys that typically include more than a thousand individuals randomly drawn from across the USA.[3] As shown in Figure 5.2, in 2020 the percentage of participants who reported believing that climate change is human-caused differed markedly by state and by counties within states. For example, although that percentage was very low in some parts of Florida, it was much higher around Miami, where the population is not only more liberal and educated but also more directly affected by rises in sea level.

The Yale studies also reveal that US attitudes have changed across time and as a function of political affiliation. As Figure 5.3 shows, belief that humans are responsible for this heating has been consistently highest among moderate and liberal Democrats, for whom it has also increased by about

Figure 5.1

Types of skepticism and relation to the chapters of this book

Epistemic skepticism concerns doubt about whether global heating is human-caused, and about whether the impacts will be serious. Response skepticism concerns doubt that action can be taken at personal, national, and international levels. Those who accept a role in acting on the climate crisis will have the greatest impact when they get organized with others (collective action).

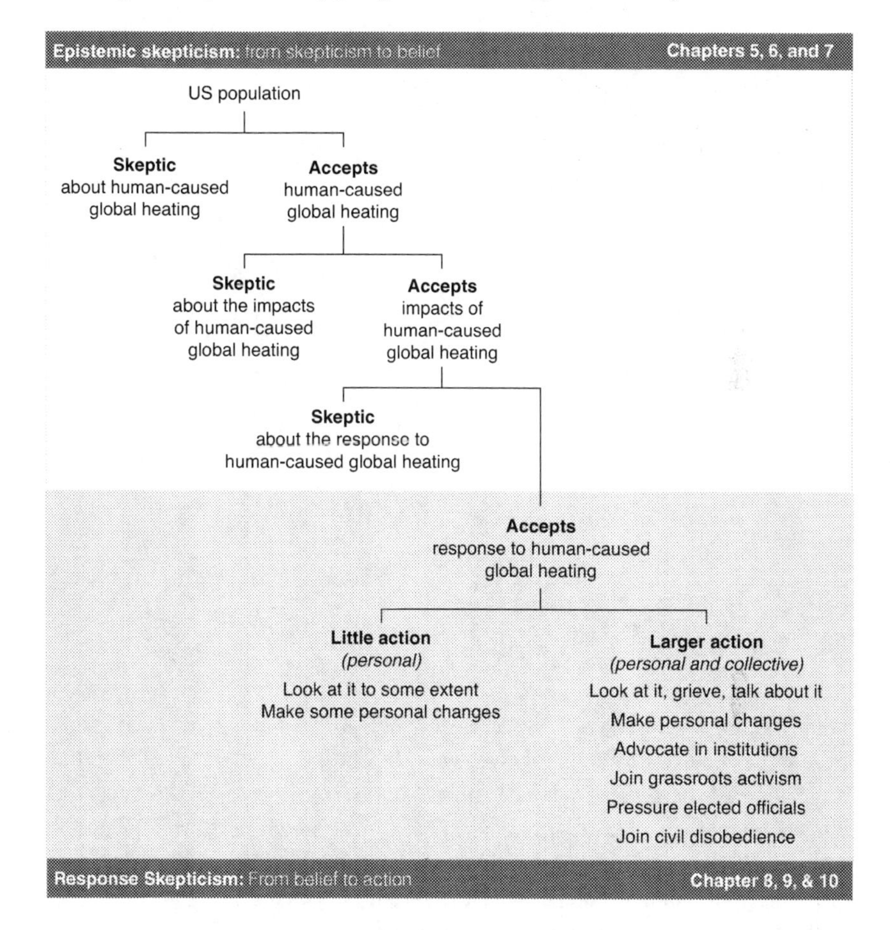

10 to 15 percent since 2008; it is lower among independents and moderate Republications; and it is lowest of all among conservative Republicans. In this last group, the belief that humans are responsible has barely changed across time, apart from a dip in belief around 2009 when Republican leaders began enunciating climate denial positions. Other researchers have found a similar situation in Australia, where 77 percent of adults polled in 2019 agreed that climate change is happening but only 46 percent thought it was entirely or

Figure 5.2

Percent of US adults who believe that global heating is human-caused (2020)

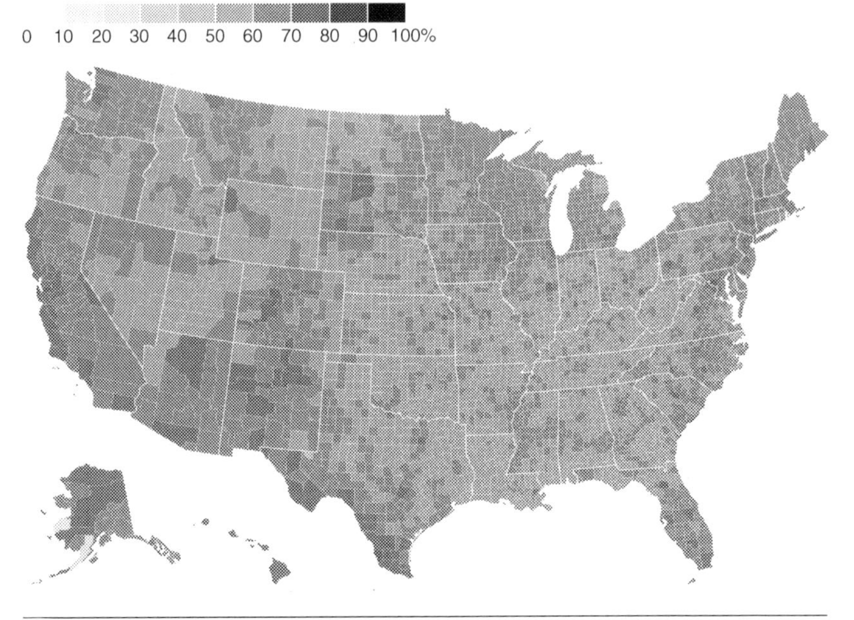

Map created using Datawrapper, online at https://app.datawrapper.de. Data from Yale Program on Climate Change Communication, Yale Climate Opinion Maps: https://climatecommunication.yale.edu/visualizations-data/ycom-us/ and Howe, Peter D., Matto Mildenberger, Jennifer R. Marlon, and Anthony Leiserowitz (2015). "Geographic variation in opinions on climate change at state and local scales in the USA." Nature Climate Change. doi:10.1038/nclimate2583

mainly caused by humans.[4] In contrast, belief in the human causes of global heating has been much stronger elsewhere. In Europe, a 2018 study found that more than 90 percent of adults in most of the countries polled agreed that climate change is at least partly caused by human activity (Table 5.1), and in China, a 2013 study showed that 85 percent of people believed fairly strongly or strongly that humans are responsible for climate change.[5]

Research has also provided some important clues as to the various demographic and psychological variables that relate to epistemic skepticism. Particularly useful in this regard is a 2016 **meta-analysis** of 151 studies from across 56 nations, the results of which can be seen in Figure 5.4. Matthew Hornsey and the other authors of this meta-analysis aggregated the data across all the studies to calculate a strength of correlation (or **r -value**) between each of seven demographic factors and beliefs regarding climate change.[6] As shown, the strongest correlation they uncovered was with *political affiliation*:

Table 5.1. *Comparison of climate change beliefs in studied European countries, 2016–17*

Country	% agreeing climate change is at least partly caused by human activity	% agreeing climate change impacts will be bad
Austria	91.8	74.0
Belgium	94.0	66.3
Czech Republic	89.5	68.0
Estonia	88.8	59.7
France	93.8	73.7
Germany	94.8	77.4
Hungary	92.7	77.0
Ireland	91.1	63.2
Italy	93.6	69.0
Netherlands	91.8	61.6
Norway	87.8	71.9
Poland	89.6	70.4
Portugal	93.6	81.1
Russia	83.8	61.8
Spain	95.7	87.9
Sweden	92.4	81.2
Switzerland	94.4	74.0
United Kingdom	91.0	66.0

Source: European Attitudes to Climate Change and Energy, 2018: www.europeansocialsurvey.org/docs/findings/ESS8_toplines_issue_9_climatechange.pdf.

Figure 5.3

Percent of US adults who believe global warming is caused by humans, across time and stratified by political views

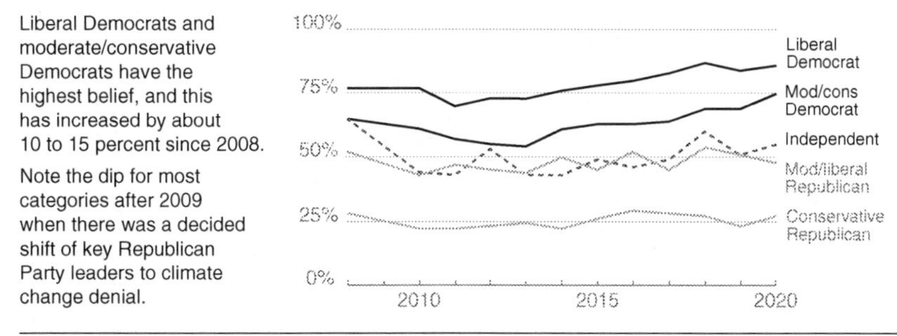

Adapted from Yale Program on Climate Change Communication (YPCCC) & George Mason University Center for Climate Change Communication (Mason 4C). (2020). *Climate Change in the American Mind: National survey data on public opinion (2008-2018)* [Data file and codebook]. doi: 10.17605/OSF.IO/JW79P and Ballew, M. T., Leiserowitz, A., Roser-Renouf, C., Rosenthal, S. A., Kotcher, J. E., Marlon, J. R., Lyon, E., Goldberg, M. H., & Maibach, E. W. (2019). Climate Change in the American Mind: Data, tools, and trends. *Environment: Science and Policy for Sustainable Development*, 61(3), 4-18, doi: 10.1080/00139157.2019.1589300

Figure 5.4

Belief in climate change is positively related to political affiliation, trust in science/scientists, and nature-loving values and negatively related to free market and other views

The figure shows results from a meta-analysis that averages over dozens of individual studies. Each bar reflects the average correlation between belief in climate change and a particular variable: it might be a negative correlation, for example increasing belief in climate change occurs for decreasing age (younger people); or a positive correlation, for example increasing belief in climate change occurs for people with nature-loving values (new ecological paradigm). Overall, the results suggest that "evidence" about climate change is noticed and remembered in ways that fit with people's political preferences and worldviews. This has important implications for science communication and trying to shift beliefs.

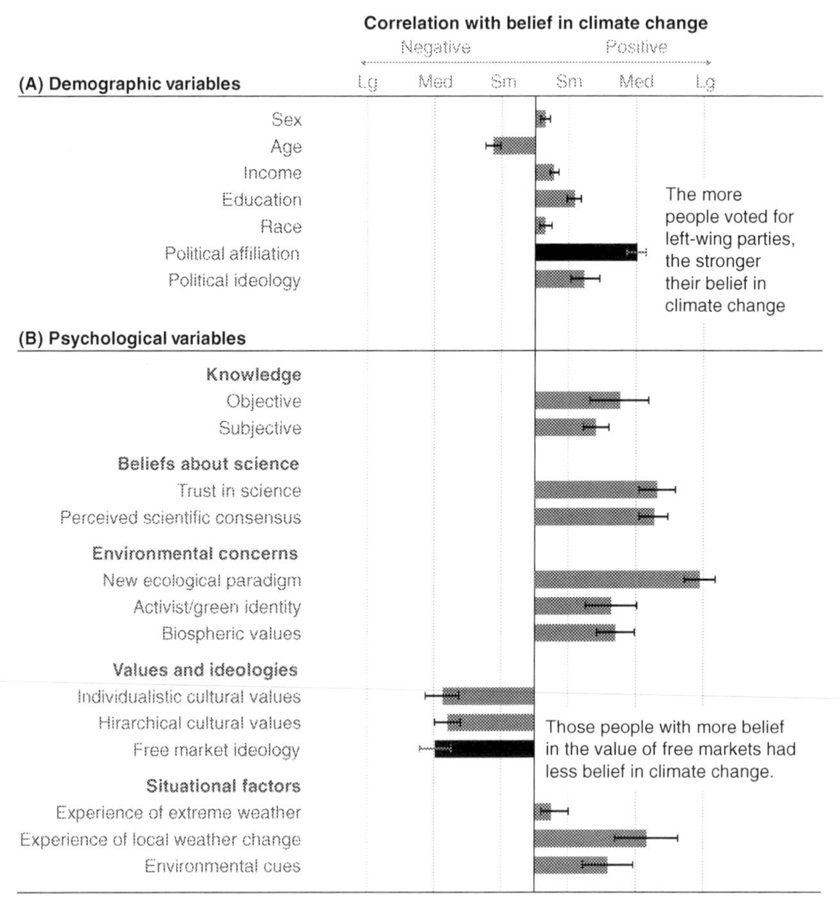

Adapted from: Hornsey, M., Harris, E., Bain, P. et al. Meta-analyses of the determinants and outcomes of belief in climate change. Nature Clim Change 6, 622–626 (2016). https://doi.org/10.1038/nclimate2943

the more likely respondents were to vote for left-leaning parties, the stronger their belief in climate change. Participants' *political ideology*, or the degree to which they reported being liberal or conservative, independent of their voting intention, was also related to a belief in climate change, although less strongly. The other demographic variables studied – sex, age, income, education, and race – had much weaker relations to respondents' beliefs regarding climate change, although, as one would expect, younger people had stronger belief in climate change.

This meta-analysis also examined the available research on the correlation between beliefs related to climate change and thirteen different psychological variables, which they grouped into five categories: *knowledge*, *beliefs about science*, *environmental concerns*, *values and ideologies*, and *situational factors*. The results regarding the knowledge category showed that belief in climate change was greater among those with greater scientific knowledge, among those believing that scientists are trustworthy, and among those believing there is a scientific consensus on climate change.[7] Regarding environmental concerns, the research found that respondents who received a high score on the New Ecological Paradigm measure of environmental attitudes also reported high levels of belief in climate change.[8] Perhaps surprisingly, the research showed a relatively weaker correlation between belief in climate change and an activist green identity, which the authors suggested might be explained by some respondents' feeling a certain amount of stigma attached to activist identities or thinking of their green concerns as a personal value rather than a social identity. Under the category of values and ideologies, the meta-analysis also found that respondents who held individualistic, hierarchical, and free market beliefs tended to have a lower level of belief in climate change. Such participants' skepticism about climate change appears consistent with a high level of trust in powerful interests and distrust of restrictions on commerce and industry. Lastly, and perhaps most counterintuitively, experience with extreme weather events was found to correlate only weakly with belief in climate change (which is taken up in Chapter 7). The main takeaways from this meta-analysis is that traditional demographic variables, such as gender and income, do not appear to be very relevant to epistemic skepticism on climate change, while factors related to worldviews, political affiliation, values, and ideologies appear to be much more important.

As Hornsey and Kelly Fielding pointed out in a different study, the tendency for conservative voters to be more skeptical of climate change than liberals is reflected in the public statements of many leading Republican politicians, such as Ted Cruz's claim that "global warming alarmists are the equivalent of Flat-Earthers," Donald Trump's declaration that climate change "is a very

Figure 5.5

There is nothing inherent in conservative ideology that makes people reject climate science

The correlation between liberal/conservative political ideology and climate change skepticism is shown for each of twenty nations. The USA has the strongest correlation (r=0.33); i.e. those participants ranking themselves as more conservative expressed more skepticism about climate change. While other large fossil fuel nations such as Brazil and Canada also showed fairly strong positive relationships, the relationship was weak in about three-quarters of countries. This suggests there is something particular in the USA for citizens to view climate change through their worldviews, while the weak relationship between ideology and skepticism in most countries, even Australia, suggests that conservative ideology does not necessarily predispose people to reject climate science. The error bars show 95 percent confidence intervals.

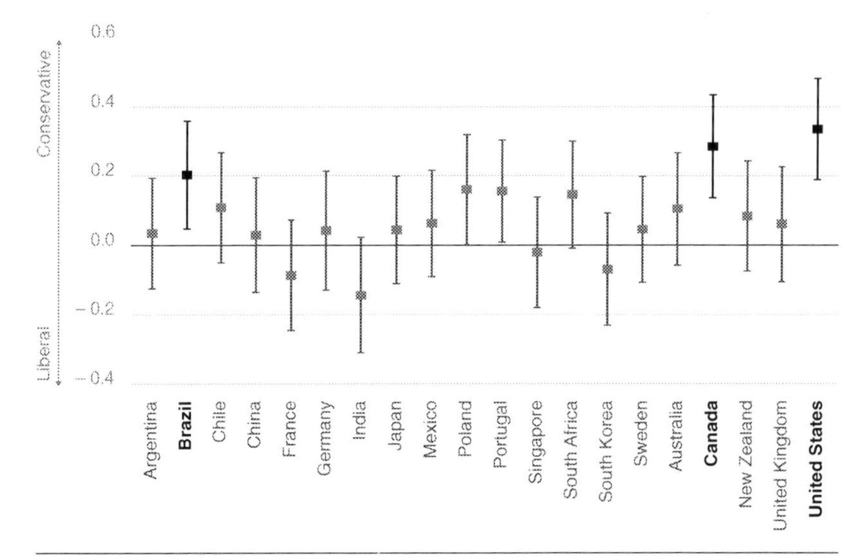

Adapted from Hornsey, M.J., Harris, E.A. & Fielding, K.S. Relationships among conspiratorial beliefs, conservatism and climate scepticism across nations. Nature Clim Change 8, 614–620 (2018). https://doi.org/10.1038/s41558-018-0157-2

expensive hoax," and Rick Santorum's charge that it is simply "an excuse for more government control of your life."[9] The mainstreaming of climate skepticism within the Republican Party is evidence that the issue has been pulled into the larger "culture wars" that currently define partisan divides in the USA around such issues as gun control, abortion, and mask-wearing. But an analysis of studies across twenty-five nations showed that even though a strong relationship between climate change skepticism and political conservatism was evident in a few countries such as Australia, Canada, and Brazil (although nowhere as strong as in the USA), that link was found to be weak or not significant in about three-quarters of the studied countries (Figure 5.5).[10] Such results suggest that there is nothing inherent in conservative ideology that leads people holding conservative beliefs to automatically or universally reject the conclusions of climate science. Indeed, in 2019 the conservative government in the UK announced a climate emergency and in 2020 it announced its intention to phase out all coal power stations by 2024[11] (although by 2022, the government had reversed its position and ordered more oil and gas drilling in the North Sea).[12]

One factor that helps to explain the strong relationship between political affiliation and climate change skepticism in the USA, Australia, Canada, and Brazil is that all of these countries are major extractors of fossil fuels. Indeed, research has found that the higher a country's per capita carbon emissions, the higher the degree of climate skepticism among conservatives.[13] This finding undoubtedly reflects the fear that a national policy to cut emissions would directly threaten the power and wealth of that industry as well as the millions of people who work in it.

Role of Misinformation

The high level of skepticism about climate change in the USA obviously has enormous political implications, as it makes it very difficult to get legislation through Congress on such issues as clean electricity standards or removing fossil fuel subsidies. Although skepticism is a characteristic of individuals that concerns their beliefs and (as will be discussed later) is therefore affected by emotion and even unconscious influences, it can also be influenced by misinformation, which is conscious- and logic-based. In the case of climate change, as this section shows, an enormous amount of misinformation has been deliberately generated and disseminated by fossil fuel interests and their political supporters. Furthermore, as this section also shows, the news media have often been inadequate, leading to a woeful public ignorance. First, however, a note is in order about the distinction between skepticism and denialism: one way to see this is that the skeptic, like a scientist, could be swayed, in principle, by more evidence, whereas the denialist persists in believing something in the teeth of the evidence.[14]

Fossil Interests

As shown in Chapter 1, research by the major petroleum companies themselves had made clear as early as the 1970s and 1980s that rising levels of CO_2 in the atmosphere were leading to potentially catastrophic effects.[15] But as a vast amount of research has since shown, the fossil fuel industry, rather than face its culpability in these likely outcomes, chose instead to embark on a public misinformation campaign so as to continue to reap huge riches from its extractive activities (Figure 5.6).[16] The fossil fuel industry contains not only coal, oil, and fossil gas companies but electric utility companies, who also feel they have a lot to lose by the need to cut emissions and the transition to renewable energy, as Leah Cardamore Stokes has shown in her book *Shortcircuiting Policy*.[17]

To shape public opinion despite knowing that the facts were against it, the fossil fuel industry began to actively attack those facts, creating an alternate universe in which rising CO_2 levels and rising temperatures were unrelated to each other and branding any scientist who said otherwise as an alarmist. In short, it built a denial machine: a sprawling network of talking heads, questionable and misleading research, and front groups, all to persuade the US public that it had nothing to worry about.

As the public interest group Climate Reality Project has pointed out, this denial machine, apart from fostering misinformation through right-wing channels such as Fox News and social media sites, has four key elements.[18] First, the denial machine runs on vast amounts of funding from very rich companies. A 2019 report found that in the three years following the Paris Accord , the five largest publicly traded oil and gas companies – ExxonMobil, Royal Dutch Shell, Chevron, British Petroleum, and Total – invested more than $1 billion of shareholder funds on lobbying and spreading misinformation about climate change.[19] And that is not even counting the hidden "dark money" that allows fossil fuel and other interests to have enormous influence over politicians since the Citizens United ruling of 2010 that awarded corporations the free speech rights of individuals, with considerable implications for electoral campaign funding.[20]

Second, the denial machine is underpinned by right-wing think tanks, such as the Cato Institute, the Heritage Foundation, and the Heartland Institute in the USA and similar groups in Europe, that were created by or funded by the fossil fuel industry, and which produce an alternate worldview regarding climate change. For example, these think tanks produce questionable or misleading research papers and other communications that have amounted to a megaphone of climate change denial. A 2016 analysis published in *American Behavioral Scientist* showed that 90 percent of skeptical or denialist climate change papers published in the USA originated from these right-wing think tanks.[21]

Third, the denial machine attempts to infuse the fossil fuel industry's self-serving position with an illusion of credibility by supporting and publicizing the views of a very small group of climate-denying scientists. In the 1990s,

Figure 5.6

Exxon: science vs. misinformation

"In the first place, there is general scientific agreement that the most likely manner in which mankind is influencing the global climate is through carbon dioxide release from the burning of fossil fuels."

James F. Black
Exxon Senior Scientist, 1978

"Currently, the scientific evidence is inconclusive as to whether human activities are having a significant effect on the global climate."

Lee Raymond
Exxon Chairman and CEO, 1997

"Present thinking holds that man has a time window of five to ten years before the need for hard decisions regarding changes in energy strategies might become critical."

James F. Black
Exxon Senior Scientist, 1978

"It is highly unlikely that the temperature in the middle of the next century will be significantly affected whether policies are enacted now or 20 years from now."

Lee Raymond
Exxon Chairman and CEO, 1997

"There is unanimous agreement in the scientific community that a temperature increase of this magnitude would bring about significant changes in the earth's climate, including rainfall distribution and alterations in the biosphere."

Roger Cohen
Exxon Sciences Lab Director, 1982

"A major frustration to many is the all-too-apparent bias of IPCC to downplay the significance of scientific uncertainty and gaps."

Brian Flannery
Exxon Position Paper, 2002

"It is now clear that for a number of years, both Bush administration political appointees and a network of organizations funded by the world's largest private energy company, ExxonMobil, have sought to distort, manipulate, and suppress climate science, so as to confuse the American public about the reality and urgency of the global warming problem, and thus forestall a strong policy response."

Dr. James J. McCarthy
American Association for the Advancement of Science, 2007

"ExxonMobil has always advocated for good public policy that is based on sound science. We will continue to do that despite criticism from those who make unsupported and inaccurate claims about our company."

Ken Cohen
Exxon VP of Public & Government Affairs, 2015

Reproduced from Paul Horn of InsideClimate News.

public relations firms paid by fossil fuel companies began promoting contrarian scientists to the media and public as experts whose opinions should be considered as equally valid to those of climate scientists, even though very few of those contrarians actually had expertise in climate science. As Naomi Oreskes and Eric Conway spell out in their book-length study, *Merchants of Doubt*, most of these supposed experts had specific anti-government regulation axes of their own to grind, just as they had in earlier campaigns such as denying that cigarette smoking causes cancer.[22] As these contrarian scientists recognized, most members of the public do not have the time or expertise to distinguish between experts; when TV announcers refer to someone as a "scientist" or "PhD," that is often enough to give their contrarian views a veneer of respectability despite the overwhelming consensus of scientific experts that human activity is causing the temperature to rise.

Fourth, the denial machine promotes the illusion that climate change skepticism has widespread public support by spreading misinformation via **front groups** with innocuous or patriotic-sounding names that are not obviously related to the fossil fuel industry. Accordingly, the fossil fuel industry has funded groups such as the American Legislative Exchange Council to spread misinformation produced by right-wing think tanks and to discreetly lobby and petition government officials on its behalf. In response, the Royal Society of the United Kingdom, the oldest national scientific society in the world, called attention to such activities as early as 2006, when it sent a letter to ExxonMobil accusing the company of funding at least thirty-nine organizations that featured "information on their websites that misrepresented the science on climate change, by outright denial of the evidence . . . or by overstating the amount and significance of uncertainty in knowledge."[23] A related approach is **astroturfing**, or masking the actual sponsors of a message to make it appear that it arises from grassroots participants. Although the California Drivers Alliance and the Washington Consumers for Sound Fuel Policy sound like real grassroots consumer movements, a 2017 report by the Union of Concerned Scientists revealed that they and more than a dozen others like them were actually operated by the Western States Petroleum Association, the top lobbyist for such US oil companies as ExxonMobil, Chevron, and Occidental.[24]

This denial machine managed to develop particularly good traction during the US presidency of former Texas oilman George W. Bush in the early 2000s. In 2007, the president of the American Association for the Advancement of Science publicly accused the Bush Administration and a network of organizations funded by ExxonMobil of actively seeking "to distort, manipulate and suppress climate science so as to confuse the American public about the urgency of the global heating problem, and thus, forestall a strong policy response."[25]

Overall, therefore, the climate change denial machine – which is most extensive in the USA but is also active in other countries such as Australia – involves a large ecosystem of actors, including fossil fuel corporations, electric utilities, petroleum and other trade associations, conservative think tanks, conservative philanthropists and foundations, front groups, public relations firms, a small number of contrarian scientists, and a vast conservative echo chamber of TV, radio, newspapers, bloggers, social media denizens, and, in the USA over the last decade, almost the entire Republican Party.[26] The aim of the denial machine to sow uncertainty about climate change in the public mind and to make global heating a theory rather than a fact has clearly been stunningly successful, especially in the USA,[27] and it continues to be successful in its more recent approaches to influence school education and research universities (Box 5.1), and to foster the ideas of "clean" natural gas and other "low-carbon solutions" such as technical carbon capture from smokestacks and hydrogen made from methane, all of which confuse the public and function to prolong the fossil industry and sustain its economic and political power. But these campaigns would not have worked out so well without the inadequate and even misleading role of much of the wider news media, as discussed next.

Box 5.1 How the fossil fuel industry colonized school and university education

In an article entitled "How the Oil and Gas Industry Has Broken Climate Education," Katie Worth provides myriad examples of how fuel-funded education programs have targeted school children in dozens of states in the USA,[28] some of which are officially sanctioned.[29] These programs work by providing free curricula, sponsored activities such as science fairs, and scholarships, and sending presenters to schools who teach children about all the ways that fossil fuels are important for humankind, about $100,000 oil rig jobs, and about how wind turbines kill birds (they do, but a miniscule number compared to the 2.4 billion birds killed by domestic cats).[30] One of the biggest players in the USA is the National Energy Education Development Plan, a $4.7 million nonprofit with sponsors of every stripe. A review of its fact sheets, science fair projects, and website links failed to reveal mention of climate change.

Even when fossil fuel industry–sponsored efforts do discuss climate change, they do so in ways that try to let the industry off the hook.

A research study in Saskatchewan Canada, which ran interviews and reviewed teaching resources, referred to a petro-pedagogy whose "teaching practices and resources work to center, legitimize, and entrench a set of beliefs relating to climate change, energy and environmentalism that align with the interests of fossil fuel industry actors"; it went on to elaborate that "this petro-pedagogy intends to restrict the imagination of possible climate solutions to individual acts of conservation that fail to challenge the structural growth of fossil fuel consumption." This fits a pattern of Big Oil having coined the term "carbon footprint" to shift the blame to individuals.[31] Of course, individual consumption reduction is also important for mitigation (as Chapter 4 discussed) but when done within a collective action framework and as part of a wider advocacy to leave fossil fuels in the ground, not as a single thing for individuals on their own.

In higher education, numerous research articles, such as Anthony Ladd's "Frackademia"[32] and research by Benjamin Franta and Geoffrey Supan has shown how fossil fuel interests have also colonized academia.[33] Supran and Franta explain that funding from Chevron, British Petroleum (BP), Shell, and other oil and gas companies is extensive in Harvard's energy and climate policy research, MIT's Energy Initiative, Stanford's Global Climate and Energy Project and the Stanford Woods Institute of the Environment, and Berkeley's Energy Biosciences Institute. They conclude that "[f]ossil fuel interests have colonized nearly every nook and cranny of energy and climate policy research in American universities, and much of energy science too. And they have done so quietly, without the general public's knowledge." Indeed, a strategy document from the late 1990s from the oil and gas industry describes support for some scientists as one specific tactic of promoting an atmosphere of delay and uncertainty.[34] Fossil fuel–financed scientists in academia have, for example, promoted talking points from the fossil fuel industry and private electric utilities which cast doubt on the feasibility of a renewable energy transition and instead tout carbon capture and hydrogen made from methane, themes that are discussed in Chapter 9.

News Media

As noted in Chapter 1, when climate change first appeared on the national radar in the 1980s and early 1990s, it was widely seen as a bipartisan issue, with even the Republican President George H. W. Bush proclaiming that he would "combat the Greenhouse Effect with the White House effect." That the opportunity to take significant and bipartisan action against climate change was lost can be attributed not only to the misinformation campaign described earlier but also to what a 2019 report in the *Columbia Journalism Review* called the news media's failure to properly inform the public about the science and the threat of climate change, which "has given rise to a calamitous public ignorance which has in turn enabled politicians and corporations to avoid action."[35] Despite early coverage of the issue, in the 1990s much of the news media appeared to lose interest in the climate issue and would later fall victim to fossil fuel industry propaganda that climate change was a matter of legitimate debate. As veteran journalist Bill McKibben has said, one of the most damaging failures of US journalism's coverage of the climate change story was to treat it like a political story rather than a science one. As a result, US print and broadcast media presented climate change stories as a disagreement between two "equally" valid points of view, thereby creating what has been termed **false equivalence**. Indeed, an analysis of 3,543 newspaper articles about climate change appearing in the *New York Times*, *Los Angeles Times*, *Washington Post*, and *Wall Street Journal* from 1988 to 2002 found that 53 percent of them gave equal attention to the views that global heating is human-caused and that it is related entirely to natural fluctuations.[36] This journalistic failure to properly cover the story had other manifestations: out of a total of six presidential debates that occurred in 2012 and 2016, the moderators did not ask a *single* question on climate change. And when the IPCC released its 2018 report expressing the consensus among representatives of governments around the globe that the world had only about twelve years remaining to cut emissions in half, only twenty-two of the fifty biggest newspapers in the USA even covered it.

Some outlets have done a better job of covering the climate crisis than others. The previously mentioned *Columbia Journalism Review* report identified the *Guardian* as an outlet that has provided particularly outstanding coverage on this issue. In addition to its UK newspaper, the *Guardian* has become a major free-access website with international climate crisis content provided by at least nine full-time climate reporters. The excellence of its coverage of climate change is undoubtedly due at least in part to the fact that it receives most of its support by subscribers and a trust, which frees it from some of the business model tensions facing other news outlets that must appeal to a broader audience and advertisers. (See Box 5.2 for a longer list of sources of good coverage of climate issues.)

As the *Guardian* example and the *Columbia Journalism Review* report suggest, another likely reason for the lack of clear, impartial, and substantive

Box 5.2 Regularly updated content about the climate crisis

Because information about fossil energy, fossil finance, extreme weather, and climate policy is constantly changing, readers who wish to be informed about these issues will want to seek out such regularly updated sources of information as the following.

News Sites

The Guardian: www.theguardian.com
Carbon Brief: www.carbonbrief.org
Inside Climate News: https://insideclimatenews.org
Skeptical about global warming skepticism: https://skepticalscience.com/

US-based Climate Journalists

David Roberts: www.volts.wtf
Bill McKibben: https://twitter.com/billmckibben
David Wallace-Wells: https://twitter.com/dwallacewells
Amy Westervelt: https://twitter.com/amywestervelt

Climate Justice Advocates

Greta Thunberg: https://twitter.com/GretaThunberg
Winona Laduke: https://twitter.com/WinonaLaduke
Indigenous Environmental Network: https://twitter.com/IENearth
Nakabuye Hilde F: https://twitter.com/NakabuyeHildaF
Dipti Bhatnagar: https://twitter.com/diptimoz
Peter Kalmus: https://twitter.com/ClimateHuman

Climate Scientists

James Hansen: http://www.columbia.edu/~jeh1/mailings/
Saleem-Ul Huq: https://twitter.com/SaleemulHuq
Katherine Hayhoe: https://twitter.com/KHayhoe
Kevin Anderson: https://twitter.com/KevinClimate
Michael Mann: https://twitter.com/MichaelEMann

Social Scientists

Jason Hickel: https://twitter.com/jasonhickel
Holly Jean Buck: https://twitter.com/hollyjeanbuck
Ben Franta: https://twitter.com/BenFranta
Julia Steinberger: https://twitter.com/JKSteinberger

Figure 5.7

While the news reports on climate change more often than ever, basic climate science is often missing

A research paper from 2019 examined how many times five basic climate science facts occur in hundreds of *New York Times* climate change articles from 1980 to 2018. With one exception, the frequencies with which these facts appear is vanishingly small. Further analysis showed that the prevalence has not generally increased over time.

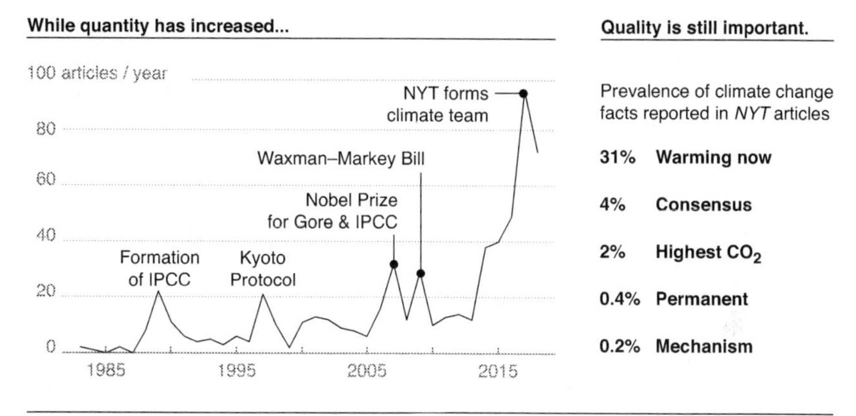

Adapted from David M Romps and Jean P Retzinger 2019 Environ. Res. Commun. 1 081002. https://iopscience.iop.org/article/10.1088/2515-7620/ab37dd/pdfhttps://iopscience.iop.org/article/10.1088/2515-7620/ab37dd/pdf

coverage of climate change in the US news media is a lack of scientific knowledge and training among those reporting on the news. This lack of expertise even among many larger news outlets is reflected in a 2019 research paper that examined more than 600 articles on climate change that appeared in the *New York Times* from 1980 to 2018. The research aimed to see how often the newspaper noted five basic climate science facts: that warming is already happening, that the greenhouse effect is the mechanism of global warming, that the scientific consensus holds that global warming has a human cause, that CO_2 levels are higher now than any other time in human existence, and that the effects of global warming are essentially permanent.[37] With the exception of the fact that warming is happening now, the frequency with which these facts appeared in the studied articles was vanishingly small, even in recent years (Figure 5.7). As the authors of the research paper noted, given that the *New York Times* is renowned for its environmental reporting, it is doubtful that other papers do better. As this analysis suggests, most newspaper articles regarding global warming fail to provide the scientific context that readers need to make sense of the problem, and most short-form social media and television reporting is even less likely to provide such vital understanding.

This is a missed opportunity. As Chapter 6 shows, reporting on the scientific consensus and the mechanism of heating increases people's belief that climate change is human-caused and that they should be concerned about it.

To be sure, in recent years, many mainstream media organizations, such as *CNN*, the *New York Times*, and *Bloomberg*, have markedly increased their coverage of climate change and its impact on extreme weather.[38] Yet even those organizations, which have the resources to support science reporters and investigative reporting, seldom critique the growth imperative of capitalism that contributes to this crisis, the outsized influence of the fossil fuel and utility companies, or the deep inadequacy of current political proposals and action; and when they do carry a climate change story it might be juxtaposed against an advertisement for more consumption or even an advertisement for the oil and gas industry.[39] Further, such media reporting has too often uncritically recapitulated US politicians' claims that effective climate action is impossible because Chinese emissions are so large, repeated myths such as that nuclear power is carbon free despite its enormous related carbon inputs, and failed to convey the extent to which proposed congressional climate action plans included a heavy component of carbon capture from fossil fuel plants that is not only unproven but maintains the current fossil fuel status quo.

Counteracting the current high level of skepticism regarding climate change and the influence of conscious, self-interested misinformation about its sources, mechanisms, and potential solutions will thus require better reporting and higher standards of journalistic integrity and public service among US news outlets. Such changes are unlikely to take place, however, without increased public demand for such information and for journalistic accountability.

Role of Motivated Cognition and Cognitive Biases

As the research survey data in Figure 5.4 showed, a major reason people reject scientific findings is that those findings are in conflict with their worldviews, ideologies, or values. Because of this, our thinking is steered toward particular conclusions by our goals and needs, something that cognitive and social psychology refers to as engaging in **motivated cognition**. Unlike intentional misinformation spread by the fossil fuel industry or misleading or inadequate news coverage that perhaps mostly effects our thinking at a conscious, logical level, motivated cognition probably has a more unconscious influence on our reasoning. Apart from motivated cognition, another type of cognitive barrier that distorts our thinking on climate-related issues is a set of inbuilt mental shortcuts known as **cognitive biases**. As this section explains, both motivated cognition and cognitive biases must be overcome if we are to change climate change skepticism.

Motivated Cognition

According to the authors Hornsey and Fielding, and as shown in Figure 5.8, understanding a person's **surface attitude**, such as skepticism about human-caused global heating, requires identifying the **attitude roots** that provide its psychological power and coherence and motivate its holder to resist new evidence contrary to that attitude and to selectively seek out other evidence that will reinforce it.[40] Table 5.2 lists several such attitude roots that are thought to be closely associated with epistemic skepticism toward climate change. In addition to several already mentioned, these include **conspiracy thinking, vested interests**, and **social identity needs**.

People who are prone to the first of these attitude roots, *conspiracy thinking*, tend to believe that powerful interests are capable of conspiring with one another to withhold scientific breakthroughs (such as slowing the development of cures for diseases so as to profit from drugs used to treat it), to fabricate events (such as the moon landing), or to withhold evidence (such as that the MMR vaccine causes autism). Conspiracy thinking surrounding climate change has been well documented, even though it requires believing that many

Figure 5.8

The surface attitudes and the attitude roots underlying anti-science beliefs

When we discuss or debate with someone, we hear the surface attitude. But it is the attitude root, which underlies the surface argument, that lends it psychological power and coherence. The root makes the surface attitude stable, and it provides motivation for the person to resist new evidence, and to seek out other evidence, in a biased and selective way, to reinforce their surface attitude

Adapted from Hornsey, M. J., & Fielding, K. S. (2017). Attitude roots and Jiu Jitsu persuasion: Understanding and overcoming the motivated rejection of science. *American Psychologist*, 72(5), 459-473. http://dx.doi.org/10.1037/a0040437

Table 5.2. *Attitude roots found to underpin rejection of climate science*

Attitude root	Definition	Link to the rejection of climate science
Hierarchical vs. egalitarian worldviews	Hierarchical values refer to the acceptance of privilege based on static social strata; egalitarian values refer to the prizing of a social order free of class and other structures	People with hierarchical worldviews tend to reject the findings of climate science because they imply policy changes that threaten elites (e.g., corporate owners, rich polluters)
Social dominance	Social dominance refers to the dominance of a ruling class and human dominion over nature	People with beliefs that social dominance is natural and valuable tend to reject scientific findings that threaten the dominance of the ruling class or human dominion over nature
Individualistic vs. communitarian	People with individualistic values prize self-reliance, independence, and freedom, while those with communitarian values prize collective interests	Believers in individualism tend to reject climate change findings because they imply imposing limitations, such as on the freedom to pollute
Free market ideology	Free market beliefs refer to an acceptance of the minimum regulation of the marketplace	Those who endorse free market ideology tend to reject the results of climate science because they imply much stronger governmental regulation
System justification vs. belief in a just world	System justifiers defend the status quo, or the prevailing socioeconomic model around us; others view that system as being unfair and want a more just world with a more equal allocation of resources	Those with strong system justification beliefs tend to reject the conclusions of climate science because they imply a fundamental change away from the status quo
Conspiracy thinking	Conspiracy believers believe that vast networks of people can hatch malevolent plots in almost total secrecy	Conspiracy theorists tend to reject climate science findings because they see them as motivated by socialist ideas or based on fake evidence or lies
Vested interests	Having a vested interest means holding an attitude because it is consistent with one's personal lifestyle	People with vested interests in high-carbon lifestyles are likely to reject climate science results because they imply changing those practices
Social identity needs	Social identities impose on people a need to align their attitudes, beliefs, and behavior with those of their group	People may reject climate science if it does not align with the beliefs of their in-group members or is embraced by a rival group

Source: Adapted from Hornsey and Fielding, "Attitude Roots and Jiu Jitsu Persuasion." *American Psychologist* 72, no. 5 (2017): 459–73.

thousands of governmental officials and individual scientists have cooperated in a vast, near-secret cover-up.[41] According to this reasoning, both government and industry have been victims of a lie spread by climate scientists, who, according to conspiracy believers, have exaggerated the threat of global heating to gain political power for themselves or their allies, to increase science funding, to promote nuclear energy, or to advance "green" or "socialist" ideological goals, such as curbing capitalism.[42]

Among the specific allegations that conspiracy thinkers have made regarding climate change are that climate scientists have manufactured or tweaked climate modeling data, misrepresented temperature data, bullied skeptical scientists, and engaged in biased editing of IPCC reports. In one widely publicized example that became known as Climategate, in 2009 hackers selected a few phrases out of context from the emails of climate scientists at the University of East Anglia in the UK that they claimed proved that the scientists were in cahoots to promote the hoax of climate change, an incident that damaged public acceptance of the overwhelming consensus that had developed among climate scientists.[43]

A second type of attitude root is having a *vested interest* in the status quo or a specific outcome. If we accept evidence that our behavior is bad for us or for the planet, we have to face the pain of either changing that behavior or living with the cognitive dissonance of acting contrary to our beliefs. One way to resolve this cognitive dissonance is to simply adjust our belief by denying the truth of the scientific message. Indeed, considerable research has demonstrated that people tend to evaluate scientific evidence differently if the outcome of that evidence is personally inconvenient.[44] This effect is illustrated by a study of smokers and nonsmokers that asked participants to interpret the information in Table 5.3, which reported rates of depression among smokers and nonsmokers.[45] When asked whether this evidence shows that smokers are (a) more likely, (b) less likely, or (c) just as likely to feel depressed as people who do not smoke, 85 percent of nonsmokers appeared to fall into the trap of simple reasoning and selected explanation (a) based simply on the raw number of

Table 5.3. *Results of an experiment on the effect of vested interests on cognition*

	Smokers	Nonsmokers
Depressed	86	51
Not depressed	43	17

Source: Mata, et al. "Strategic Numeracy: Self-Serving Reasoning about Health Statistics." *Basic and Applied Social Psychology* 37, no. 3 (2015): 165–73.

respondents in that category. In contrast, only 67 percent of smokers fell into this trap, suggesting that more of them must have looked at the figures more carefully and noticed that the percentage of nonsmokers who were depressed was much higher than the percentage of smokers who were depressed. For the researchers, these results were evidence that the smokers' vested interest in the topic motivated them to do more sophisticated reasoning. As this and other research has shown, the effect of motivated cognition is more subtle and perhaps more insidious than simply rejecting scientific evidence on an illogical or superficial basis; it can instead lead to the rational and systematic processing of evidence with the goal of reaching a conclusion that is more consistent with one's own beliefs or interests.

Another way to resolve the cognitive conflict between a vested interest and belief in human-caused climate change is not to doubt that it is actually happening (epistemic skepticism) but to believe that personal action regarding climate change is unnecessary or ineffectual (response skepticism). For example, whereas some academics concerned about climate change have called upon their peers to model a new social norm by limiting their carbon-intensive, work-related aviation to academic conferences, others avoid giving up the career and lifestyle benefits of frequent aviation by advocating the rather self-serving opinion that doing so is pointless because their contribution is so small relative to the number of people traveling worldwide.[46]

A third type of attitude root that can affect cognition is *social identity needs*, which refers to the human tendency to draw meaning and self-definition from belonging to social groups.[47] For example, self-identifying as a Republican tends to carry with it a bundle of expectations about national security, welfare, government regulation, and related issues, although such groups tend to allow some variation in beliefs. A member of that group who may disagree with some of its core aspects or beliefs, such as a Republican who favors emissions reduction, is faced with three basic options: try to reform the group, which is very challenging; leave the group, which could have far-reaching personal consequences for their own identity; or abandon their outlying belief and internalize the attitudes and behaviors considered normal for those holding that social identity.

In addition to meeting the expectations of other members of a group with whom one identifies, people sometimes meet their social identity needs by identifying themselves in opposition to a different group.[48] For example, when British participants in a UK study were asked to compare their own interest in energy conservation to that of the citizens of Sweden, who are known to be highly environmentally conscious, participants showed less interest in topics related to energy conservation; in contrast, when they were asked to compare themselves with people in the USA, whom they regarded as energy wasters, they expressed a higher zeal for green things, demonstrating the tendency of people in an in-group to move in the opposite direction of an out-group.[49] This phenomenon further helps explain the existing polarization around climate change: some current Republicans identify themselves as skeptics of climate change not only because their group takes that position but because they also want to oppose the out-group – Democrats – who do believe in climate change.

As sociologist Holly Buck points out, this polarization is also driven by "the political economy of [social media] platforms [which] depend on dividing the public in half and enraging one side with the other side's views, which increases time-on-site and supports the ad-driven business model of the platforms."[50] As a result, she says, "[p]eople live in completely different social realities, polarized by algorithms, and inhabiting different filter bubbles ... Twitter, Facebook, YouTube: their algorithms aren't built for nuanced, dialogic content that will allow people to gradually change their minds and question fossil fuels. They will show people content that is more extreme." Thus, motivated cognition is driven by a slew of attitude roots, including conspiracy thinking, vested interests, and social identity needs, which are further exacerbated by the influence of social identity and of social media that cater to and inflame such identities.

Cognitive Biases

Behavioral scientists have also demonstrated that people's ability to think clearly about the facts and threats of climate change is also affected by a number of inbuilt mental shortcuts called cognitive biases or heuristics. While these are adaptive in some scenarios, such as fast decision-making, they can also lead to systematic patterns of deviation from rationality, and they introduce errors of judgment.

In *Don't Even Think about It: Why Our Brains Are Wired to Ignore Climate Change*, George Marshall identifies a number of such cognitive biases that influence thinking regarding the climate crisis.[51] One of these is **confirmation bias**, which is the very human tendency to cherry-pick available evidence to support one's existing knowledge, attitudes, and beliefs. A person who considers climate change a myth, for instance, is more likely to see extreme weather events as additional proof that weather is naturally variable, whereas a person who accepts that climate change is real is more likely to regard such events as further evidence of global heating.

Another type of cognitive bias is **hyperbolic discounting**, or people's tendency to have a stronger preference for immediate payoffs than for later payoffs even when it will ultimately benefit them less, as has been demonstrated in laboratory-based observations in behavioral economics. According to Marshall, applied more broadly, this kind of bias might help explain why governments choose to defer paying about 1 percent of our annual global income to deal with climate change now rather than incur estimated annual costs of about 5 to 20 percent into the indefinite future. Yet another bias is the **sunk cost fallacy**, where we continue on our current trajectory – such as a high-carbon lifestyle, or for an institution, putting up new buildings, or investing in new fossil energy infrastructure – despite evidence suggesting that a course correction is necessary, simply because we have already invested so many resources in the status quo.

Yet as Marshall also points out, people's tendency to avoid costs and act only in their self-interest can sometimes be overruled by a sufficiently strong appeal to group identity and a visible social norm. Following the Fukushima Daiichi nuclear power plant accident in 2011, for example, many people in Japan responded to the public call to turn down their air conditioners despite sweltering temperatures and even though no one could see them, leading to a 20 percent reduction in Tokyo's energy demand. Such observations lead Marshall to conclude that "people will willingly shoulder a burden – even one that requires short-term sacrifice against uncertain long-term threats – provided they share a common purpose and are rewarded with a greater sense of social belonging."

Beyond these classic biases, researchers have identified a number of other factors that drive the formation of false beliefs, such as the previously noted belief that climate change is due to variation in the Sun's intensity rather than to the burning of fossil fuels. In an article on the psychological drivers of misinformation belief, Ullrich Ecker and colleagues note that one of the factors driving a false belief is that people trying to decide what is true often choose to "go with their gut" instead of taking time to deliberate.[52] They also point out that people are also more likely to regard a message as true if they have heard it many times before (familiarity), if the message is encoded or retrieved effortlessly from memory (processing fluency), and if the message has elements that are consistent with other information they have in memory (cohesion). As advertisers and propagandists have long recognized, the formation of a false belief is also strongly influenced by emotional content. Accordingly, misleading content that spreads virally on social media often appeals to emotion to be more persuasive. When the Green New Deal policy was floated in 2019 to deal with climate change, for instance, a speech that was widely distributed online attempted to stoke viewers' fears by claiming, "[t]hey want to take your pickup truck, they want to rebuild your home, they want to take away your hamburgers. This is what Stalin dreamed about but never achieved. You are on the frontlines of the war against communism coming back to America." [53] Thus, the psychological drivers of false belief include both socioaffective factors, including emotional appeals and social identity needs, and various forms of cognitive biases and failures.

Shifting Skepticism

Researchers who have looked at these issues and who advocate for improving public acceptance of climate science and reducing emissions have suggested a number of strategies for shifting beliefs. These include leveraging what is known about a skeptic's attitude roots to try to influence them to change views on climate change, using a messenger inside the skeptic's group to convey a climate change message, trying to get skeptics on climate change to support policy by pitching that policy at a different issue, and changing skeptics' minds by giving them information about the social norms around them.

The previous section described attitude roots, such as identity needs, and vested interests. Hornsey and Fielding have pointed out that awareness of a skeptic's underlying attitude roots creates leverage to make change, something akin to what they call **jiu jitsu persuasion** – the strategy of using an opponent's force against them.[54] This differs from the most common response to resistance to an evidence-based message, which is to continue to explain or add to that evidence. The problem with that common response strategy is that it assumes that resistance is based on ignorance or a failure to grasp evidence rather than on the goals and needs that lead the intended target of persuasion to maintain their beliefs. The climate change communication strategy of focusing on attitude roots rather than surface attitudes is similar to what happens in psychotherapy, where clients are encouraged to reflect on root issues rather than surface ones. A research study provides one possible example of how this could work for climate change.[55] The authors found that people who scored higher on the attitude root of system justification (i.e. that the current social/political system is the "right one") were more skeptical about environmental realities. However, in a follow-up they showed that a message that framed proenvironmental action as patriotic, as preserving the American way of life, turned things around – now those with stronger system justification beliefs reported stronger intentions to engage in proenvironmental action. It's possible that the jiu jitsu approach, when undertaken among family members and close friends, would be similar to the strategy taken by the same-sex marriage campaign of the late 1990s: engaging in slow work, family-by-family, conversation-by-conversation, until a transformation in attitudes made policy change possible.

A different communication strategy proposed by Fielding, Hornsey, and colleagues for changing attitudes regarding climate change is to take advantage of social identity links by having members of the targeted political group talk to others in that group about climate change, which is called **in-group messaging**. To test this idea, they asked more than 400 self-identified Democrats and an equivalent number of self-identified Republicans to indicate their level of support for ten socio-political issues, one of which was "the US will implement a tax on carbon emissions."[56] Next, each participant was asked to read a news media article outlining a carbon tax proposal that was identified as coming from a Republican or a Democrat source and to again rank their support for that proposal. The results showed that Republican participants tended to have a more positive attitude toward the carbon tax when it was presented to them by a Republican source than when it was came from a Democratic source, whereas the reverse held for Democratic participants (Figure 5.9). This approach has a real-world counterpart in the activities of the Climate Leadership Council, a conservative public policy think tank that

Figure 5.9

Messaging from a member of a political in-group is effective in shifting beliefs/policy support

In a research study, participants received a message about a carbon tax from either a Democrat or Republican source. The results showed Republican participants had more positive attitudes towards the carbon tax when it was presented to them by a Republican source than when it was from a Democratic source; whereas the reverse held for Democratic participants.

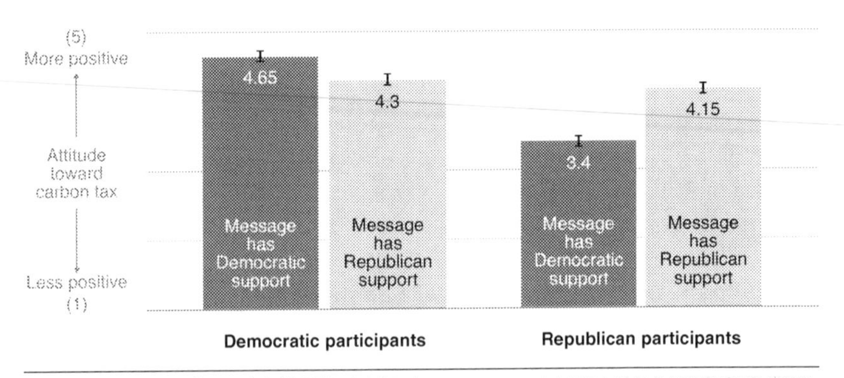

Adapted from Fielding, K.S., Hornsey, M.J., Thai, H.A. et al. Using ingroup messengers and ingroup values to promote climate change policy. Climatic Change 158, 181–199 (2020). https://doi.org/10.1007/s10584-019-02561-z

advocates for the introduction of a carbon tax and whose campaigns appear specifically targeted to influence conservatives.[57]

Another promising approach to gaining support for climate policy without directly challenging people's underlying attitudes and beliefs is to present climate policy in terms of **co-benefits** with which they are likely to agree, such as reducing air pollution, providing jobs and economic development through green industries, and improving health by reducing disease and promoting healthier lifestyles. That such framing could help advance climate policy is suggested by a 2016 study involving 6,000 participants from 24 countries that showed that participants who were unconvinced that climate change was human-caused but believed that actions to reduce emissions would improve economic development and scientific progress were more willing to engage in or contribute to pro-environment causes.[58] Indeed, as Chapters 9 and 10 show, a co-benefits framing around jobs is one aspect of the just transition that underpins policy platforms such as a Green New Deal – with sufficient organizing it may be possible to get some workers to support climate action policy even without changing their beliefs about climate change. Another kind of co-benefit that is likely to be particularly effective is to frame the need for emissions reduction in terms of reducing air pollution – something that is enormously concerning for citizens and industrial and agricultural regions and cities worldwide, given about ten million people per year die from air pollution and hundreds of millions are affected by it.[59] And co-benefits framing could be effective in conservatives too, for example by talking about national security. As Anatol Lieven points out in *Climate Change and the Nation State*, it might be quite impactful to some conservatives to point out that hard-nosed realists in the military are greatly concerned over the national security implications of climate change.[60]

A final suggestion for shifting skepticism is to harness **social norms**, or the shared standards that guide the behavior of members within a group, such as putting a card in hotel bathrooms saying "the majority of guests reuse their towel." One promising approach by Wes Schultz and colleagues has been to leverage social norms to reduce household energy use.[61] In one experiment, the results of which are shown in Figure 5.10, when households in California that had previously shown above-average energy use were provided with information about the positive social effects of their neighbors' reduced energy use, they incurred household energy savings of 1.5 to 2.5 percent.[62] This approach of providing social norm information could thus not only shift beliefs but also household action, a potentially large step given that about 40 percent of country-level greenhouse gas emissions are related to home energy use. Changes in the beliefs of household members could lead them to purchase

Figure 5.10

Social norms have the power to shape household energy consumption

In a California field experiment, households received information about how much energy they and their neighbors had consumed in recent weeks. Some households received *descriptive* norm information (what neighbors were doing) only, while others received descriptive plus *injunctive* norm information (what was socially desirable). The results show that a combination of descriptive and injunctive social norms is effective in reducing household energy use overall.

Change in daily energy consumption (kWh)

A typical household uses about 30 kWh/day. Error bars show the 95 percent confidence interval of the pair-wise difference between usage during the follow up and during the baseline.

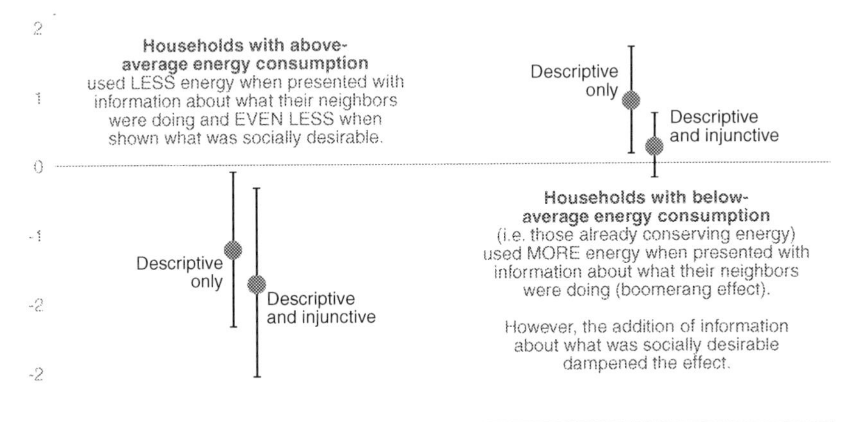

Adapted from Schultz et al. 2007, Psychological Science, Vol 18, No 5.

more efficient technologies for heating, lighting, and transportation; to install rooftop photovoltaics (PVs); to reduce energy consumption overall or when demand is high; and to shift patterns of consumption, such as charging EVs or running dishwashers, to time periods in which energy from the grid is less carbon-intense.[63]

Conclusion

An adequate response to the looming climate crisis will require understanding and shifting the sources of current skepticism about the problem. As this chapter has shown, accomplishing this vital goal will require advocates of climate action to find ways to counteract misinformation emanating from the fossil fuel industry and its supporters in the political arena and media, to seek out and share reliable sources of information, to demand fuller and more probing coverage from mainstream media, and to effectively undermine skepticism and shift beliefs among the people they know. Chapters 6 and 7 examine other ways to do that, including providing scientific information about climate change, such as the mechanisms underlying global heating and the consensus of scientists, the topic of Chapter 6.

6
Science Communication
Countering Skepticism and Delivering Information Clearly

> Errors and biases arising from the mental models and heuristics we use to evaluate risks and make decisions in complex systems. . . arise not only in the context of unfamiliar systems like the climate but also in familiar, everyday contexts such as compound interest or filling a bathtub. Therefore they cannot be remedied merely by providing more information about the climate, but require different kinds of communication.
>
> John Sterman

Even though preexisting worldviews and values play a large role in the level of epistemic skepticism toward the findings of climate science held by individuals and groups, facts – and how those facts are presented – do shape people's understanding and acceptance of those findings.[1] Based on that recognition, a number of researchers have investigated what specific modes of communicating those often complex scientific facts are likely to be most effective at promoting belief in the inconvenient and often frightening realities of climate change.[2] They have explored ways in which scientists, activists, journalists, policy makers, and others concerned about global heating can communicate climate science findings most effectively and persuasively to the general public. This chapter illustrates the value of science communication on climate change by focusing on five strategies: clarifying the mechanism behind global heating, emphasizing the consensus among climate scientists, counteracting misinformation about climate change, overcoming the challenge of uncertainty, and effectively communicating information visually.

Clarifying the Mechanism behind Global Heating

Although climate science is a vast field, only a subset of that information is probably vital to understanding global heating and changing skeptics' minds.

As suggested in Chapter 1, one piece of information that is especially important to this understanding is a **mechanistic explanation** of the greenhouse effect – an explanation of how human activity causes a rise in CO_2 levels that in turn causes global heating.

To test the hypothesis that teaching people about the chemical-physical mechanism of the greenhouse effect would make them more likely to believe in human-caused global heating, two psychologists conducted a series of experiments.[3] The results of their first experiment, a survey of a few hundred visitors to San Diego's parks, found that although 80 percent of the study participants accepted that global heating was occurring and 77 percent accepted that it was human-caused, only 3 percent could name the greenhouse effect and only 1 percent were able to articulate a key aspect of that theory: the difference between infrared energy and sunlight. They also found that those who demonstrated more knowledge about the mechanisms behind global heating were most likely to accept that it was happening (a weak correlation of r=0.22) and that it was human-caused (r=0.17).

In a second experiment, which included nearly a hundred students from Berkeley and a smaller sample from the University of Texas, participants were again asked about their acceptance of global heating and their level of knowledge about the greenhouse effect. Unlike the first experiment, however, they were then asked to read a 400-word explanation of the greenhouse effect and retested to see if their attitudes toward global heating and knowledge of the mechanism of heating had changed. The researchers again found that almost none of the participants had knowledge of the greenhouse effect before reading the explanation, but that after reading the explanation, 59 percent correctly stated that the Earth emits infrared light, and a good proportion now stated knowledge of other aspects of the mechanism of global heating, and their acceptance of global heating increased.

To test the durability of the impact of this brief intervention, a third experiment, also conducted with undergraduates, followed the same format but retested the participants much later, with an average delay of 18.5 days. The results showed that this brief learning experience increased participants' belief in global heating and that knowledge of the mechanism behind this effect lasted at least several weeks. Two following experiments showed similar effects with even longer delays.

Averaged across the experiments the researchers showed that these brief interventions increased mechanistic knowledge by 28 percent and increased acceptance that global heating is happening by 9 percent upon delayed retesting. Although these changes in belief were relatively modest and reflect self-reported beliefs and intentions rather than actual changes in behavior, these and several other published studies do suggest that teaching the mechanisms behind global heating can make a difference, even among politically conservative people.[4]

Another mechanism that is important to understanding global heating is the interaction between carbon emissions and carbon sink removals of atmospheric CO_2. As discussed earlier in this book, the level of global heating is determined by the concentration of CO_2 in the atmosphere, which can last for thousands of years from the time of emission and is increased by continued burning of fossil fuels. Although some of this CO_2 is taken up by the carbon sinks in the land and the ocean, those can absorb only about half the current concentration and they take time to work. As a result, slight reductions in CO_2 emissions will not immediately reduce atmospheric concentrations but simply slow the rate at which they are added to the atmosphere; only by ceasing or substantially reducing new CO_2 emissions will carbon sinks be able to remove enough existing CO_2 for concentrations to actually decrease. This is called the **stock-and-flow problem**, which can be explained with a bathtub analogy: the level of water in a bathtub will continue to rise as long as the flow from the faucet exceeds the flow out from the drain and will go down only when the flow from the faucet decreases enough for the drain to remove more water than is flowing in.[5] The more dramatic the decrease in flow from the faucet, the faster the water level will go down.

As simple as this principle may appear, it can be a difficult concept for people to understand. As demonstrated by the results of a study shown in Figure 6.1, even people with math and engineering training often incorrectly believe that we can keep CO_2 levels from increasing by simply stabilizing rather than drastically reducing our current emissions. That figure shows that when participants were asked to draw how emissions and removals would proceed after the year 2000 to match a stabilization of atmospheric CO_2 levels by about 2050, they correctly drew removals as a constant level, but they also drew emissions as stable, when emissions would actually need to dramatically decrease. This misunderstanding of the mechanisms that produce global heating is not a trivial matter, as it works against a wide recognition of the seriousness and size of the problem and leads many people to think that we can afford to wait to see how bad climate change actually gets before taking concerted action. (Incidentally, it explains why atmospheric CO_2 kept increasing during the pandemic in the year 2020 even though emissions were cut about 6 percent globally).

The work mentioned above on communicating the greenhouse effect and the stock-and-flow problem are both examples of providing mechanistic information in the climate science domain. Why such mechanistic explanations might be effective has been better explored in the domain of health

Figure 6.1

When thinking about reductions of atmospheric CO_2, people don't apply the basic stock-and-flow (bathtub) analogy, and so dramatically underestimate how much reduction is needed

(A) Scenarios with rising and falling CO_2

Subjects were asked to consider a scenario in which **atmospheric CO_2 gradually rises to 400 ppm**, then stabilizes by the year 2100:

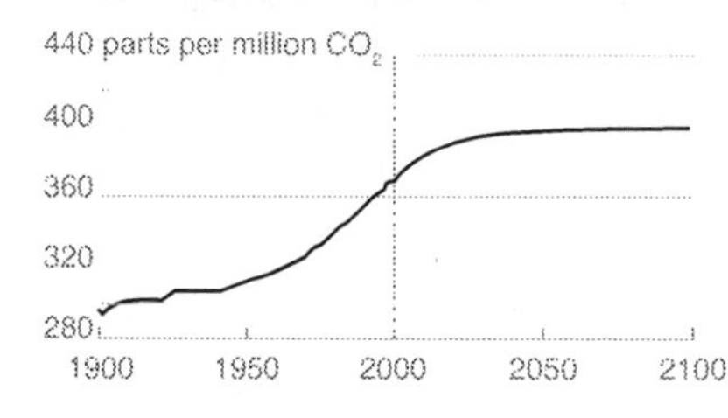

Alternatively, others were asked to consider a scenario where **atmospheric CO_2 gradually falls to 340 ppm**, then stabilizes by 2100:

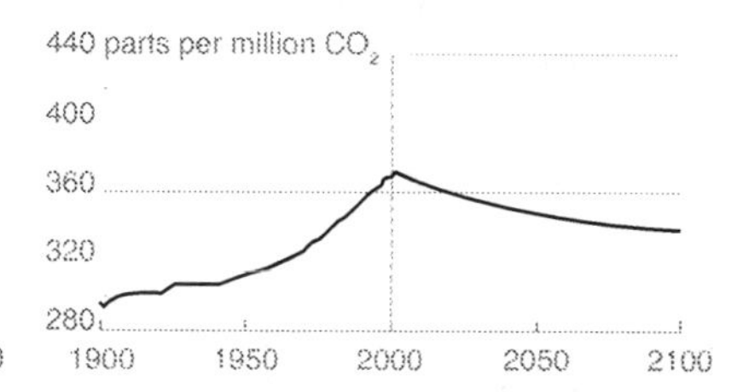

(B) Current emissions and removal levels

Given the scenarios above, both sets of subjects were shown the anthropogenic CO_2 emissions from 1900 to 2000 and current net removal of CO_2 from the atmosphere by natural processes were provided. They were then asked to sketch their estimate of (a) future anthropogenic CO_2 emissions and (b) future net CO_2 removal.

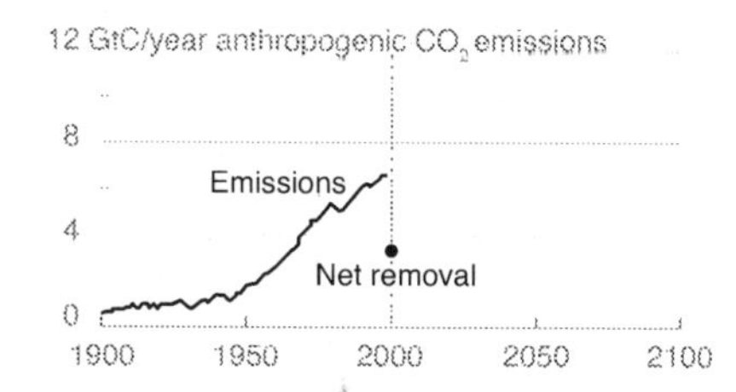

(C) Estimates

Although a scenario of unchanging emissions requires that emissions eventually equalize with removals (E = R), the subjects drew paths where they continued to outpace removals (E > R).

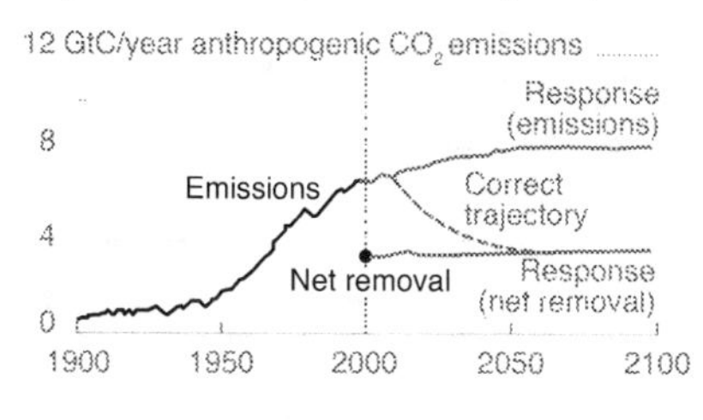

Although a scenario of declining emissions requires that emissions eventually dip below removals (E < R), the subjects drew paths where they continued to outpace removals (E > R).

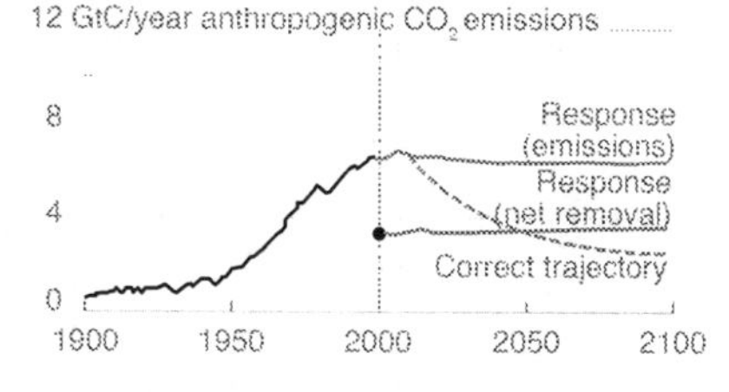

Adapted from Sterman JD and Sweeney LB. Climatic Change (2007) 80 213–238 DOI 10.1007/s10584-006-9107-5

interventions. One such study found that children who received biologically based mechanistic knowledge about the spread of viruses reasoned better about viral survival and were better at identifying risky and preventative behaviors than children who simply received information about the differences in symptoms between colds and flu and a list of behaviors to follow or avoid. Based on these and similar results regarding antibiotic resistance, vaccination, and nutrition, researchers Kara Weisman and Ellen Markman have theorized that mechanistic or theory-based explanations lead to more behavioral change because they focus on the underlying causal framework rather than superficial details and make people's understanding more robust to misinformation.[6] Such findings suggest that mechanistic explanations are special and that providing the public with more mechanistic knowledge about the greenhouse effect, the stock-and-flow relationship of atmospheric CO_2, and other key aspects of climate science and extreme weather attribution is likely to play a key role in counteracting skepticism or apathy about global heating and encouraging the public to demand more serious action to combat it.

Emphasizing the Consensus among Climate Scientists

That human activity is responsible for the global heating we are currently experiencing is no longer a source of scientific debate among climate scientists, as discussed in the previous chapters and reflected in the IPCC 2021 declaration that the evidence for human-caused global heating is *unequivocal*. Yet, years earlier Republican strategist Frank Luntz, recognizing that "[s]hould the public come to believe that the scientific issues are settled, their views about global warming will change accordingly," recommended that "[t]herefore, you need to continue to make the lack of scientific certainty a primary issue in the debate."[7] One manifestation of the vast organized disinformation campaign that followed is that, from 2007 to 2010, the most common argument in conservative op-eds on climate change was that there was no consensus among scientists that it was human-caused.[8]

This vast, organized disinformation effort has been stunningly successful: a 2014 survey of twenty nations, for example, found the USA had the lowest levels of belief in this expert consensus, and a 2016 study found that even many US science teachers were unaware of it.[9] As of 2020, only 55 percent of surveyed US adults believed that this consensus existed.[10] As shown in Figure 6.2, research has found that this **consensus gap** – the difference between the actual scientific consensus and people's belief in it – varies in degree between politically liberal and conservative respondents. The fact that it is greatest among the most politically conservative respondents suggests that it is influenced by cultural factors, but that it exists among even the most liberal respondents suggests that it is also influenced by lack of knowledge or misinformation.[11]

Figure 6.2

The gap between beliefs about scientific consensus and the actual consensus relates to both misinformation and culture

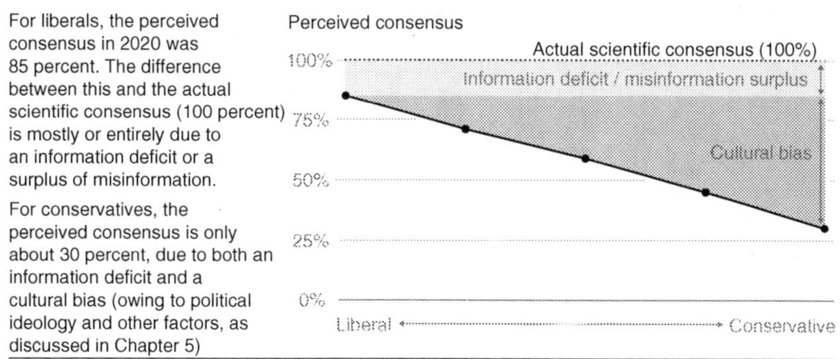

Adapted from a graphic in the consensus handbook. http://www.climatechangecommunication.org/all/consensus-handbook/. Updated with 2020 data from the Yale Program on Climate Change Communication. That set the consensus at 97%, however the IPCC 2021 report, backed by governments, calls anthropogenic global heating "unequivocal" which we take to mean 100% consensus.

The prevalence of misinformation on this point prompted a research group led by Sander van der Linden to investigate whether providing people with accurate information about the near-total scientific consensus on climate change could not only change their beliefs regarding that point but also serve as a gateway to develop accurate beliefs about climate change more generally, a theory they termed the **Gateway Belief Model** (Figure 6.3).[12] In a series of studies, including a 2016 large-scale replication study of more than 6,000 US adults nationwide, participants were asked to use a sliding scale to indicate their perception of the level of *scientific consensus* regarding global heating their level of belief in the reality of *global warming* and its *human causation*, and their level of *worry about global warming* and of *support for action on global warming*.[13] One group of participants was then asked to read a statement that "97 percent of climate scientists have concluded that human-caused global warming is happening," while other participants were either given no information or completed a neutral word-sorting task, after which all participants were asked to answer the same five survey questions again. The results demonstrated that those who were provided with information about the consensus among climate scientists reported an increase in belief in the level of consensus, and increased beliefs in global warming, human causation, worry about global warming, and support for action than participants who received no intervention. The same researchers later found that the impact of consensus information on the treatment group was even greater when video rather than

Figure 6.3

Belief in scientific consensus can be a "gateway" to other beliefs

Psychology experiments in large samples of participants measure people's perceived sense of the scientific agreement, their belief in climate change, human causation, and other variables. An information manipulation is then done to tell people about the consensus (i.e. consensus messaging). A statistical method called structural equation modeling suggests that such consensus messaging information then "affects" the other variables, leading to increased belief and support for public action.

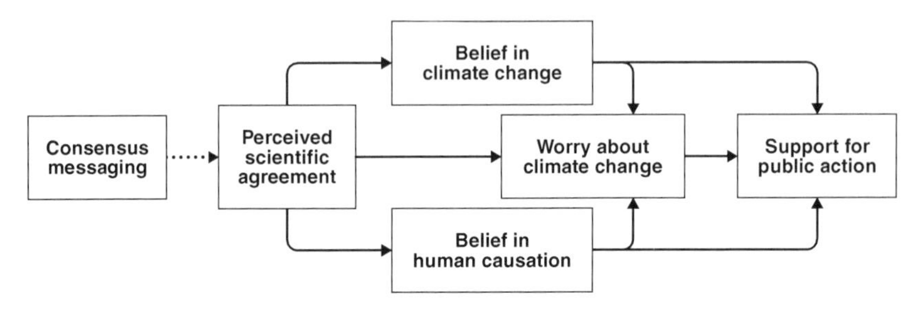

Adapted from van der Linden et al. Journal of Environmental Psychology, 62, (2019), 49-58.

text was used to highlight the consensus and that the effects of that perceived consensus were still evident six months later. [14]

Although other researchers have pointed to the limitations of such studies of the effectiveness of communications conducted in the lab rather than in the real world,[15] the results of these large-scale and replicated studies do suggest that consensus messaging can positively affect beliefs regarding climate change. In another example, researchers who showed participants a video clip in which comedian John Oliver filled his stage with a hundred people, three of whom were identified as contrarian scientists and the rest as mainstream scientists, found that this striking visual representation of how absurd it is to believe there is no consensus on climate change increased those participants' belief in that consensus.[16] Such evidence suggests that the consensus gap, which narrowed more than 20 percent between 2010 and 2020,[17] could be reduced even further by efforts among educators, journalists, political leaders, and activists to provide the public with more information about the overwhelming scientific consensus among climate scientists that global heating is the result of human action rather than natural variation.

Counteracting Misinformation on Climate Change

Another way in which science communication can help counter skepticism regarding climate change is to correct specific misinformation. One way to do this is to "inoculate" people against such information by refuting it in advance, which is sometimes also called **prebunking**. Just as vaccines employ weakened doses of pathogens to trigger the production of antibodies to protect people from later infection, inoculation against misinformation exposes people to weakened forms of misinformation so they can recognize and reject the real thing later. This inoculation typically involves two components: a forewarning to participants to expect a threat, and the pre-emptive provision of refutational information.

The way this works is illustrated by another study by the van der Linden group that attempted to use inoculation against the Global Warming Petition Project, one of the most potent misinformation campaigns against the scientific consensus discussed in the previous section.[18] The Global Warming Petition Project is an online petition started in 1998 that claims to have obtained the signatures of more than 31,000 US scientists in support of its declaration that human activity is not disrupting the climate.[19] As others have pointed out, 99.9 percent of the signatories were not climate scientists, many were not scientists at all, and many were not even actual people but had names such as "Spice Girls." And even if all 31,000 signers were actually scientists, that number would amount to only 0.3 percent of the ten million people in the USA who have a science degree. Nonetheless, a 2016 analysis found that the petition was frequently cited as evidence in social media posts regarding the supposed "climate myth."[20] The van der Linden group's study of more than 2,000 US participants found that providing participants with just a message conveying the scientific consensus message increased their perception of consensus by about 20 percent; that exposing participants to the Global Warming Petition neutralized the positive impact of subsequently presenting them with the consensus information; but that pre-emptively warning participants about politically motivated attempts to spread misinformation protected their perception of the existing consensus when they were later presented with the petition misinformation.[21]

Based on these results, the study authors recommended two main strategies that science communicators can employ to help people contend with the misinformation they are likely to encounter in real-world conditions. The first of these is to accompany communications about the scientific consensus on human-caused climate change with information forewarning readers or viewers that politically motivated actors seek to undermine belief in this established science. The second is to help the public build what they call a "cognitive repertoire" of information about disinformation campaigns in general, which could, for example, be built into a general school curriculum. (In 2019, for instance, Italy mandated that students of all ages receive thirty-three hours of climate change–related education per year.)[22] As these suggestions indicate, to effectively counter misinformation of various kinds, such education would need to not only communicate the physical facts of climate change but also address the political motivations of the fossil fuel industry and its allies and the history and practice of misinformation campaigns. For example, much of the current misinformation campaign has shifted its efforts from sowing doubt about the scientific consensus or facts of global heating to **greenwashing**, in which business-as-usual practices are represented as forms of climate action, such as the strategy of branding fracked methane gas as "natural" gas. Another strategy that is useful for prebunking, which could be part of a general education curriculum, is to teach people to do lateral reading – consulting other sources to examine the reliability of a piece of information or the credibility of the source.[23]

In addition to prebunking or inoculation, people can also be taught how to **debunk** misinformation – to counter specific misinformation by pulling apart its mistaken premises and conclusions, as in the specific example of the Global Warming Petition Project's central claim given in Figure 6.4.[24] Such efforts could be helped by equipping people with basic information on global heating

Figure 6.4

Example of how to debunk a climate myth

The structure of the claim that there is no global warming based on the Global Warming Petition Project. Premise 1 is seen to be false based on simple logic. Premise 2 is seen to be false by inspecting the names of signers.

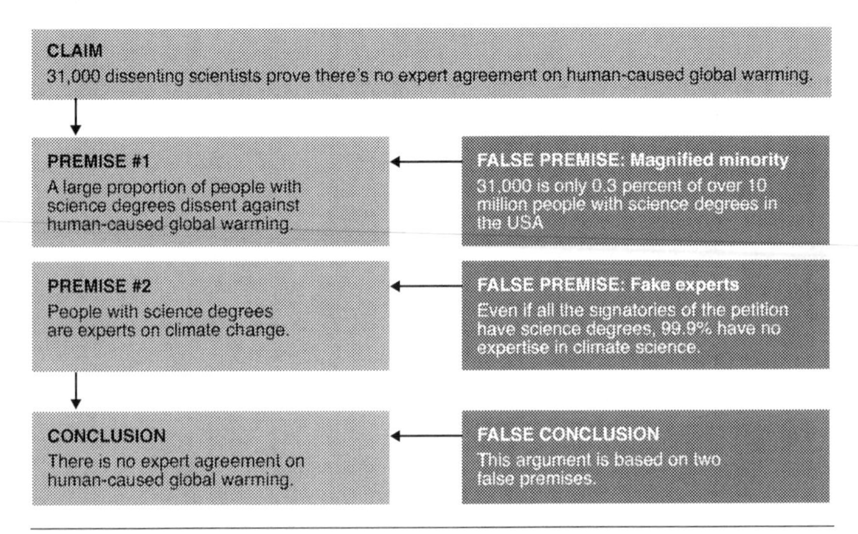

Adapted from https://www.climatechangecommunication.org/wp-content/uploads/2018/03/Consensus_Handbook-1.pdf

including not only the main mechanism of global heating as discussed earlier, but also the facts in Global Heating 1-2-3 we saw in Chapter 1, including specific facts such as the irradiance of the Sun has actually decreased somewhat in recent decades, which could be used to debunk claims that global heating is caused by sun spots. Even if someone being confronted by such misinformation does not currently have those facts within immediate recall, being aware of the wishful thinking and oversimplification that often accompanies such denials regarding global heating can at least help people think twice before accepting questionable claims.

Contending with misinformation is, of course, a skill that is important in many domains of life beyond climate change. Misinformation has now become a global problem of huge proportions, as demonstrated by recent misinformation campaigns targeted toward the Brexit campaign in the UK, 2020 election results in the USA, and Covid-19 protocols and vaccines. In recognition of this problem, several of the technology companies whose platforms have been used to spread such misinformation have begun to try to do something to prevent it, such as Facebook's collaborating with third-party fact-checking agencies to flag misleading posts and issue corrections and Twitter's using algorithms to label some tweets as misleading or disputed. Stirred into action by the health and political impacts of such misinformation efforts, the United Nations has launched a platform called "Verified," which is intended to build a global base of volunteers to debunk misinformation and to spread fact-checking content. Yet much remains to be done – as one recent research study on misinformation concluded, "the full potential of applying insights from psychology to tackle the spread of misinformation remains largely untapped."[25]

Misinformation can also be countered by making critical thinking and reasoned decision-making one of the primary objectives of education. As several researchers examining this problem have recommended, students can be taught how to directly address the arguments of climate change dissenters, be assigned texts that address these misconceptions explicitly, and be taught the process and methods of argumentation.[26]

Overcoming the Challenge of Uncertainty

One dilemma facing climate science communicators is that communicating accurately about global heating and its likely effects necessarily means talking in terms of probabilities and likelihoods rather than certainties. Scientific findings and predictions always contain some level of uncertainty and evolve over time as conditions change, new information is uncovered, and new techniques of investigation are developed. Scientists rightfully view this uncertainty as one of science's greatest strengths, as it discourages error and complacency and spurs them to continually ask new questions and engage in further investigation and research. Nonscientists, however, sometimes confuse this inherent uncertainty with unreliability, and climate scientists thus worry that accurately conveying limits on the certainty of their conclusions will sow confusion, undermine the public's confidence, and be exploited by climate deniers.

As Figure 6.5 illustrates, science communicators must find effective and accurate ways to communicate uncertainty from what already happened, and uncertainty about what will happen. Focusing on past uncertainty, it can be affected by a variety of factors, including *who* is communicating *what, in what form, to whom*, and *to what effect*. An article published in the National Academy of Sciences tested the "in what form" variety by communicating uncertainty about the magnitude of a number, and how this affected participants' trust in that number and in the source that was doing the communication.[27] In the first in a series of experiments, the authors asked more than 1,000 participants to read a short statement about global temperature: "An official report stated that between 1880 and 2012, the Earth's average global surface temperature has increased by an estimated 0.85°C." One group of

Figure 6.5

Science communication has to grapple with the problem of conveying uncertainty.

There is uncertainty about the past/present and also about the future. By being aware of the subcomponents of uncertainty (i.e. who, what, in what form, to whom and to what effect) science communications can more effectively tailor their messages.

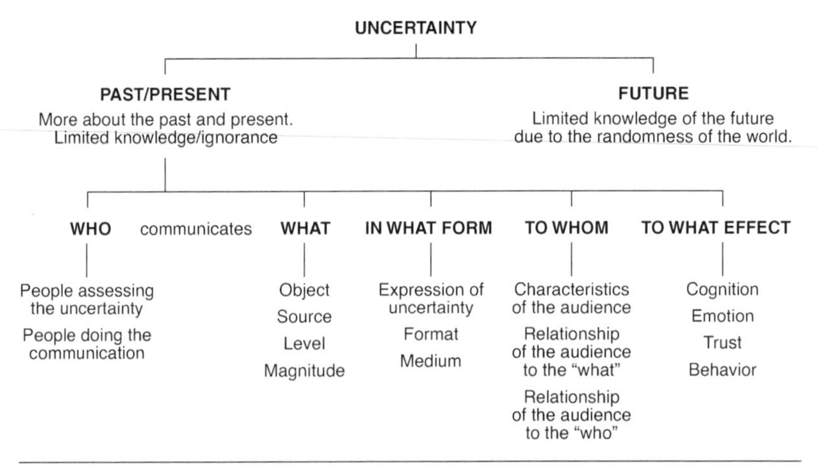

Adapted from van der Bles et al. 2019 Royal Society. Open Science. https://royalsocietypublishing.org/doi/10.1098/rsos.181870

participants received no further information about that statement, constituting the control condition with no uncertainty. A second group, which constituted the numeric uncertainty condition, was also given a numerical range appended to the estimated number: "with a minimum of 0.65°C and a maximum of 1.06°C." A third group, constituting the verbal uncertainty condition, were given a verbal statement appended to the estimated number: "the report states that there is some uncertainty around this estimate, it could be somewhat higher or lower." After reading the statement, participants were asked to recall the specific temperature and to answer questions about how reliable they perceived the number to be and whether they thought the writers of the report were trustworthy. As shown in Figure 6.6, this research found that, as one

Figure 6.6

Communicating uncertainty around data can be done in a way that preserves audience confidence in the data itself

Three different groups were presented with a short text (e.g. regarding the Earth's average global surface temperature). For one group the text did not present any uncertainty (just the estimate, Control), while for the other two it was expressed as a numerical range (Numerical) or a written statement (Verbal).

Example control statement: "Between 1880 and 2012, the Earth's average global surface temperature has increased by an estimated 0.85°C."

Example numerical uncertainty statement: "Between 1880 and 2012, the Earth's average global surface temperature has increased by an estimated 0.85°C, with a minimum of 0.65°C and a maximum of 1.06°C."

Example verbal uncertainty statement: "Between 1880 and 2012, the Earth's average global surface temperature has increased by an estimated 0.85°C. The report states that there is some uncertainty around this estimate, it could be somewhat higher or lower."

Participants were tasked to score how much uncertainty they perceived, how much they trusted the original estimate, and how much they trusted the source of information. Conveying numeric and verbal uncertainty *did* lead to significantly greater perceived uncertainty than control (A, dark bars). Decreases in trust in the estimate (trust in number, B) and in those communicating it (trust in source, C) were small and only significant when the uncertainty was communicated verbally.

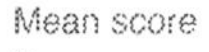

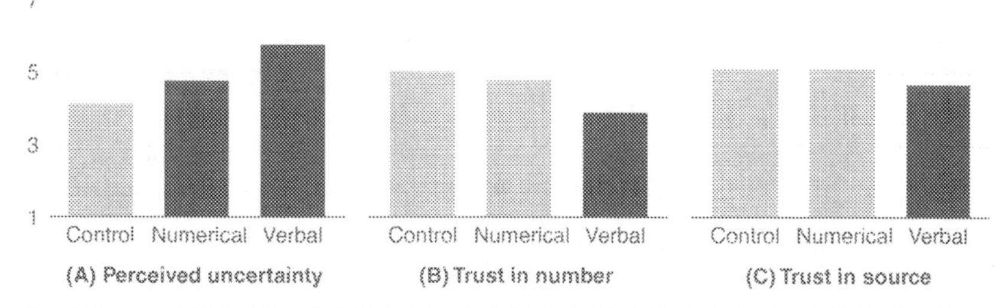

Adapted from Anne Marthe van der Bles, Sander van der Linden, Alexandra L. J. Freeman, David J. Spiegelhalter, The effects of communicating uncertainty on public trust in facts and numbers. *Proceedings of the National Academy of Sciences*. Apr 2020, 117 (14) 7672-7685, DOI: 10.1073/pnas.1913678117.

might expect, people did perceive the statement as being more uncertain when uncertainty was communicated, but the important finding was that their degree of trust in the actual number and in the source of information was not decremented when that uncertainty was expressed as a numerical range compared to when it was a verbal statement. Other experiments in the same study showed similar results, with the final one, Experiment 5, running a field experiment with the BBC website in which the researchers showed that the results could be generalized to some extent, beyond the online laboratory to the real world. These results led the authors to recommend communicating uncertainty in the form of an estimate followed by a numerical range, a practice that has been adopted in recent IPCC reports.

Another set of researchers conducted a similar experiment related to future uncertainty by asking a representative sample of US adults to read about predictions of the anticipated impact of global heating on sea-level rise.[28] One group of participants, serving as a control group, read that statement without any expression of uncertainty. For another set of participants, the **bounded uncertainty** group (akin to the numerical uncertainty group above), the uncertainty inherent in that statement was expressed as a range: "global warming will cause sea level to rise about four feet, but it could be as little as one foot or as much as seven feet." For a third set of participants, the **irreducible uncertainty** group (akin to the verbal uncertainty group above), the uncertainty was expressed verbally in terms of impacts, such as "storms induced by global warming could influence sea level in unpredictable ways." The results showed that, much as in the research described above, participants who were given the bounded uncertainty estimates reported increased trust in the scientists, which in turn increased their acceptance of the statement to an even higher level than that found among the control group, suggesting that participants found the science more trustworthy when it expressed uncertainty. That effect was reversed, however, among participants in the irreducible uncertainty group, who trusted the scientists less than those in the control group, which also reduced their acceptance of the message. These results indicate that differences in how future uncertainty is expressed can affect how scientists are perceived and the way their messages are received – in particular, audiences are willing to accept and even appreciate some uncertainty so long as its boundaries are made explicit.

The scientific community's success in promoting more widespread acceptance of the climate crisis and its likely impacts thus hinges on finding ways to convey the inherent uncertainty in such important questions as how soon we are likely to experience a sustained temperature of 1.5°C above preindustrial levels, how much the sea level is to rise in a given location by a given date, or what housing insurance might cost in 2030. Although more research into how to communicate such uncertainty effectively is definitely called for, these results do provide some basis for thinking that it is possible to express uncertainty in the climate domain without a serious loss of confidence in the science behind it.

Effectively Communicating Information Visually

Decades of psychological research has shown that visual representations can have a large influence on one's comprehension and thinking process, an insight that Jordan Harold and colleagues have shown also applies to climate and impact data.[29] As they argue in a 2016 article, scientific graphics such as those provided by the IPCC should be designed to be accessible for multiple stakeholders, including the lay public as well as other scientists and policy-makers.[30] In addition to pointing out ways in which such graphics have too often been sorely lacking, the authors have offered several concrete ways in which climate science communicators can adopt good principles of graphic design arising from insights in the research on visual attention, memory, and learning.

As illustrated in Figure 6.7, visual attention, or the process by which the mind selects or focuses on a subset of the information that is presented, is affected both by what Harold et al. refer to as bottom-up sources, such as color, shape, and size, which can stand out from other features, and by top-down processes of the viewer's expectation, which are driven by prior knowledge (their goal and reason for looking at the graphic and their earlier experiences). When a viewer looks at an image, both the bottom-up and top-down forms of visual attention operate to create a mental representation of the information that is stored in the viewer's memory and updated as the viewer further explores the graphic. According to the authors, the key to good comprehension is matching the perceived information to the viewer's expectations, which can be achieved by creating visual features in a graphic that match the likely prior knowledge of the audience and doing so as simply and clearly as possible.

Figure 6.8 provides an example of how several of the specific principles of good graphic design recommended by the authors can be used to improve the effectiveness of scientific graphics. Among these are reducing potentially distracting clutter by including only the visual information required to comprehend the intended information, using contrast or color to make important elements perceptually salient, putting text close to the graphic information it describes, and using arrows or text to guide viewers to important features of

Figure 6.7

How visual attention, memory, and learning work together

As both top-down and bottom-up visual attention processes do their work, a mental representation of the information is created in memory. The mental representation is updated cyclically as the viewer further explores the graphic.

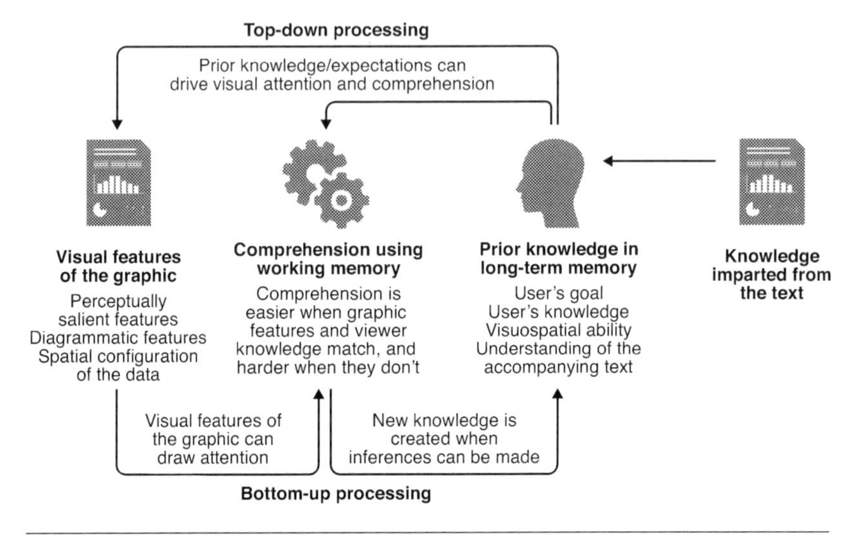

Adapted from Harold, J., Lorenzoni, I., Shipley, T. et al. Cognitive and psychological science insights to improve climate change data visualization. Nature Clim Change 6, 1060–1069 (2016). https://doi.org/10.1038/nclimate3162

the visual. Other principles mentioned in their article are breaking up the graphic into visual "chunks" that can be sequenced, choosing common graphic elements that viewers will be familiar with, and matching the visual data to metaphors (such as up and down) that aid comprehension.

Simulations are another form of data visualization that have been found to be very effective in aiding the public's comprehension of specific climate, impact, and energy information. An excellent example of this communication mode is the En-ROADS simulator developed by scientists and educators at MIT, which shows how much temperature will rise by 2100 on our current emissions trajectory and provides a set of dials that a user can turn to change that trajectory. As shown in Figure 6.9, a user can turn dials that will change the level of renewables, amount of economic growth, and amount of coal that will still be burned and immediately be able to visualize their impact on global heating via an underlying climate model. A research study that used this simulator to engage college students in competitive role-playing showed that the simulator improved students' knowledge of the many aspects underlying

Figure 6.8

Using graphic principles, science communicators can direct visual attention to improve comprehension

The original graphic output (right) is difficult to read and comprehend. Visual clutter like extraneous borders and backgrounds, extra decimal places, and default cryptic labels, as well as a lack of general contrast, obscure the data and make the viewer wonder where to look first. Perhaps more importantly, the overall lack of context forces the reader to search for meaning (what is the point of this graphic? What is it trying to tell me?).

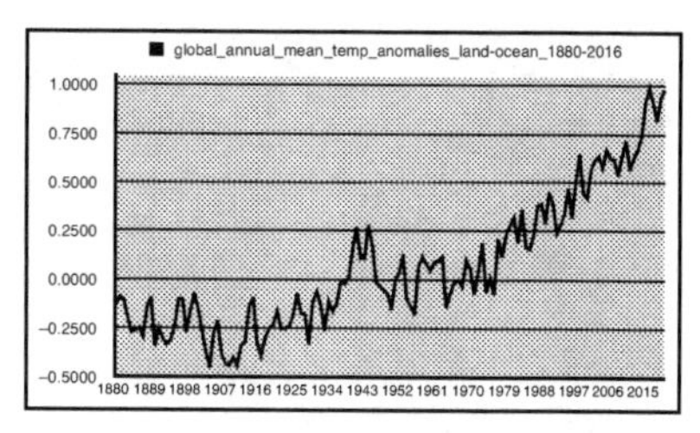

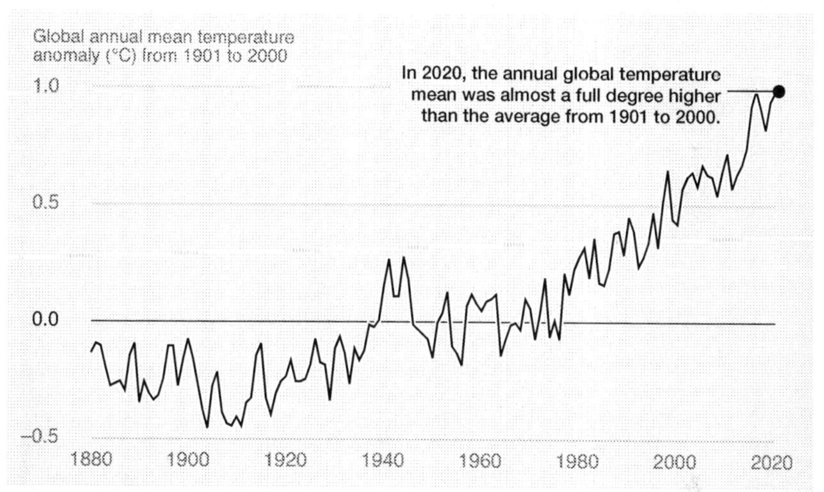

The improved graphic (above) uses graphic principles to focus attention to the salient information:

1. Only required information is presented to reduce clutter. Extraneous borders are gone, while axes have been simplified with fewer breaks and more intuitive intervals.

2. Important elements are highly perceptually salient to capture visual attention. Here, the data, zero anomaly reference point, and contextual explanatory text are darker to attract attention.

3. Graphic elements and text are close so the viewer's attention is not split. The y-axis text is directly above the axis, and explanatory text is close to the data it describes.

4. Arrows or text guide the viewer to important features. Here, explanatory text is linked to the relevant data point with a marker.

For more complex graphics, designers should also make sure that any graphic elements or symbols used are familiar to the intended audience, that data and visual metaphors "match" (e.g. "hot" is red/ "cool" is blue), and that the graphic is broken up into visual chunks that can be sequenced to lead the reader through the information.

Graphic principles from Harold, J., Lorenzoni, I., Shipley, T. et al. Cognitive and psychological science insights to improve climate change data visualization. Nature Clim Change 6, 1080–1089 (2016). https://doi.org/10.1038/nclimate3162. Data from NOAA National Centers for Environmental Information Global Surface Temperature Anomalies, available online at https://www.ncdc.noaa.gov/monitoring-references/faq/anomalies.php#anomalies

Figure 6.9

An internet simulator allows users to instantaneously evaluate the impact of changing sources of energy and other factors on the global heating trajectory

The En-ROADS simulator (www.climateinteractive.org) incorporates a climate model, as well as many assumptions about the dynamics of energy sources, economic growth, afforestation, and other variables.

(A) En-ROADS simulator default mode

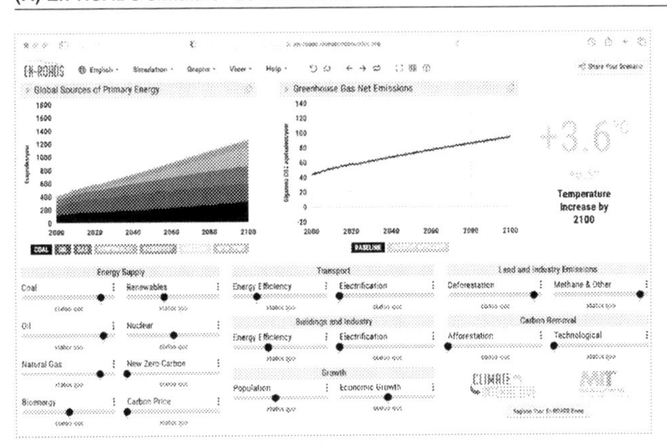

With the En-ROADS simulator controls in the default mode (roughly our current situation), global heating is predicted to rise approximately 3.6 °C by 2100.

(B) Adjusting the En-ROADS simulator

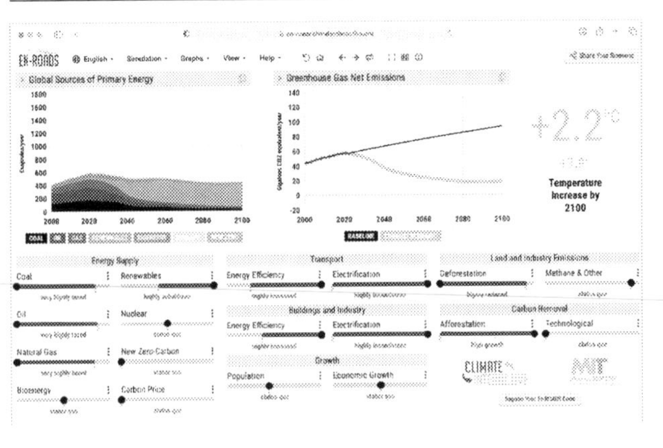

The simulator shows how specific changes lead to a dramatically reduced heating trajectory: here, all fossil fuels have been heavily taxed/banned, renewables have been maximized, efficiencies increased, deforestation decreased, and afforestation (new trees) increased. These changes limit the predicted heating to only 2.2°C by 2100.

Screenshots from https://en-roads.climateinteractive.org/

global temperature and increased their personal and emotional engagement with the topic of climate change.[31]

Another example of an effective simulator is a map of sea-level rise, which was shown to increase people's acceptance that climate change is already happening.[32] In another study in which people were given a textual description of sea-level rise and then some counter-information that expressed doubt about the source of that text, the presence of an animated map was found to help protect against the doubt.[33] Such simulations are likely to be particularly effective if they relate to one's personal location and could help make estimated effects of global heating such as wildfire damage or flooding much more vivid, one example being a virtual reality simulator of projected sea-level rise in Long Beach, California.[34] Already websites such as cal-adapt.org have been created to provide information about projections of future global heating, such as the predicted effect of the high-emissions pathway (SSP5-RCP8.5) on the number of days of extreme heat for any location on a California map. Such approaches could be made even richer and more consequential by providing more vivid visual displays and building in such information as predicted insurance costs, air pollution, and fire risk.

The visual arts are yet another way in which climate change information can be effectively conveyed. As the authors of one article on this topic have pointed out, the absence of verbal information in visual art can be an advantage in capturing and engaging people's attention, as it forces viewers to relate to climate change personally and create their own interpretations to make sense of the presented visual information.[35] One striking example is the sculpture *Unbearable* by artist Jens Galschiot, shown in Figure 6.10. This large outdoor

Figure 6.10

The sculpture *Unbearable* by Jens Galschiøt, 2015

The artwork shows the connection between the burning of fossil fuels (an oil pipeline), rising global temperatures, and species extinction.

The display of the polar bear resembles a public execution.

sculpture includes the figure of a bear impaled on an oil pipeline in the shape of the famous hockey stick temperature graph that, starting at 1700 and ending in the present, curves upward as a result of our use of fossil fuels. The work thus provides a powerful image of climate change, including its main cause, resulting rise in temperatures, and a major effect, the extinction of species. A research study on the impact of such artworks on climate beliefs presented at a UN climate summit showed that exposing hundreds of people to climate artworks increased their support for climate policy.[36]

Conclusion

Although people's views regarding climate change are substantially influenced by a range of factors that were discussed in Chapter 5, such as intuitive thinking and memory failures, elite sources cues, emotional state, and motivated cognition (reflecting values and worldviews), the fact that even liberals appear to have a knowledge deficit about the facts about global heating strongly suggests that, in addition to the approaches discussed in Chapter 5, effectively communicating the facts of climate science is a necessary strategy to increase belief in global heating and the steps that must be taken to combat it. Even though much of the extant research on this topic is limited to laboratory-based studies, in which participants are presented with specifically tailored information and asked about their beliefs without any observation of their subsequent actions, the results still provide valuable insights into how real-world communication can be done better. This may well include employing more integrated approaches, such as combining knowledge of the greenhouse effect, the facts of Global Heating 1-2-3, and the stock-and-flow problem and providing practice in confronting misinformation as part of a useful education curriculum in schools and colleges, an approach that is increasingly being considered and even deployed in some US states and countries. Furthermore, many other practices of science communication, such as communicating uncertainty as a point estimate and range (also known as bounded uncertainty) and employing graphics, simulations, and visual art, can help science communicators convey relevant scientific knowledge more clearly and powerfully.

Although science communication is not itself a panacea, as sufficiently overcoming skepticism about global heating will also require contending with people's worldviews and values and confronting the vested interests and power structures of the fossil, utility, and agriculture industries and their political allies, science communication is still an essential part of the toolkit that will be needed to shift people out of their epistemic skepticism, and to gain broader public support for meaningful climate change policies. Such science communication is likely to be more effective still if it also finds ways to convey the risk or threat of global heating, the topic of Chapter 7.

7

Elevating Risk Perceptions about Global Heating

> Perhaps it was because we were so sociopathically good at collating bad news into a sickening evolving sense of what constituted "normal," or because we looked outside and things seemed still okay. Because we were bored with writing, or reading, the same story again and again, because climate was so global and therefore nontribal it suggested only the corniest politics, because we didn't yet appreciate how fully it would ravage our lives, and because, selfishly, we didn't mind destroying the planet for others living elsewhere on it or those not yet born who would inherit it from us, outraged.
>
> David Wallace-Wells, *Uninhabitable Earth*

Whereas is might be possible to help people overcome the impact of misinformation and motivated cognition discussed earlier and shift them from being epistemic skeptics about the human cause and impacts, the time-delayed and abstract nature of the climate crisis often fails to adequately signal the actual level of risk it represents. The resulting low level of risk perception regarding global heating is another major factor contributing to skepticism, especially response skepticism. To better understand this underappreciation of risk, this chapter first looks at factors that have been shown to influence societal and individual risk perceptions regarding climate change and then at proposed strategies for breaking through the currently high level of denial or indifference without leading to a sense of hopelessness or despair.

Factors in Climate Risk Perception

In 2020, only 63 percent of polled US adults reported that they were worried about global heating and only 26 percent reported that they were alarmed.[1] That is a small proportion, given that, as we have seen, global heating threatens to widely disrupt organized existence within mere decades. Numerous factors help explain this perhaps surprising gap, including, as Chapter 5 showed, that people's perceptions of the reality of global heating are affected by misinformation and by motivated cognition shaped by values, affiliations, and self-interest. Yet research into the specific area of risk perception and decision-making has also uncovered several major social and personal factors that influence people's evaluation of global heating. As discussed in this chapter, these include both larger historical shifts in the view of humans' relationship to nature and individual factors such as personality, values, worldviews, and experience.

Historical Shifts in Attitudes toward Nature

As noted in Chapter 5, research has found that a concern for nature is one of the psychological factors that relates most strongly to belief in climate change, and as noted above, people with strong **biospheric values** (viewing one's and others' actions in terms of the advantages and disadvantages they pose to nature) also report having stronger climate risk perceptions.[2] Yet the percentage of the population with such values is currently too small to activate the kind of concerted change necessary to combat the real and looming dangers of global heating, which reflects not merely a matter of individual differences, such as those discussed later in this chapter, but a major historical shift in societal attitudes toward nature.[3]

Throughout most of human history people lived close to the natural world and were therefore deeply aware of their dependence upon nature for their very survival, which in turn led to them viewing nature quite differently than most of us do today. Such people could often describe the names and properties of hundreds of plants and animals around them in the same way most people today are able to identify celebrities and brands. That way of thinking tended to produce moral codes that prevented people from exploiting other living systems by taking more from nature than was sustainable or necessary to meet their own needs. Even today, many Indigenous peoples continue to fish, hunt, and gather out of a spirit not of extraction but of reciprocity.

Over the past several centuries, however, that intimate sense of connection with and responsibility for caring for nature changed profoundly with the rise of capitalism, modern science, urbanization, and industrialization, all of which resulted in removing more and more members of society from the land. As Jason Hickel and numerous other anthropologists and historians have argued, these developments, supported by ideas from world religions, philosophy, and modern science, together served to instill a dominant cultural sense of human

Figure 7.1

Capitalism, urbanization, and industrialization required a new story about Nature

If you think Nature is imbued with spirits behind every rock, tree, and stream, you'd be reluctant to extract from her. For large-scale agriculture, mining, and industry to accelerate, workers needed a new attitude. Over a period of hundreds of years, with the enclosures (that forced people off the land), there was a shift away from animist ideas (of spirits in nature) and toward a tendency for dualistic thinking (mind and body separate). There was also a discarding of the belief that the world is alive and worth respect, a distancing of ourselves from animals (which came to be seen as more like machines), and a greater focus on Nature as hard data that could be measured and counted to make a profit

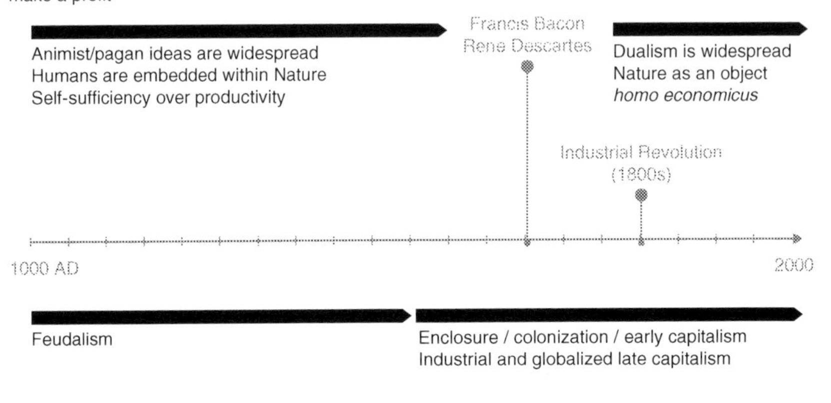

Concept by the author based on text by Hickel J (2021). Less is More. Penguin

separation from and dominion over nature, as opposed to a sense of human interconnection with and dependence upon nature.[4] As shown in Figure 7.1, the resulting alienation from nature was intimately connected with the rise of capitalism and colonialism discussed earlier in this book and with a resulting commodification and extractivist view of nature, or valuing nature in terms of what can be taken from it, now in a quest for profits rather than subsistence. But as such authors have argued, to exploit something, you need to first regard it as an object rather than as something infused with its own spirit and agency. That process of objectification was accelerated by early scientific and Enlightenment thinkers such as Francis Bacon, whose mechanistic view of nature further loosened ethical restraints against the possession and exploitation of nature, including animals, distant people, and natural resources such as land, water, and fossil fuels.

Although, to be sure, many modern urban dwellers continue to express affection and respect for nature, that interest often tends to be largely recreational or aesthetic in essence, which perhaps helps explain how it is possible for someone, such as a capital markets expert for instance, to visit a protected national park and marvel at its natural wonders and then return, without much reflection, to their job financing fossil fuel extraction operations. Ultimately, one of the reasons that perceptions of the risks that global heating poses for nature are not currently higher is that fewer people in the developed world have the same strong sense of nature and climate as a living organism affected by but independent of human life that was held by premodern and Indigenous societies and by many bioscientists and others who understand nature more intimately.[5]

Personality Traits

Beyond the societal level covered in the previous section, research has identified several psychological factors that influence individuals' views of nature and corresponding climate risk perceptions. One of these is **personality**, which is different from temporary mood states and refers to relatively enduring patterns of feelings and behaviors that distinguish people from one another.[6] Personality is typically studied using self-report scales that measure a number of main personality traits, such as the often used HEXACO scale, an acronym referring to six such traits: honesty-humility, emotionality, extraversion, agreeableness, conscientiousness, and openness (these overlap with the classic Big Five personality traits).[7] In an attempt to better understand how personality relates to the level of concern about climate change, one group of researchers examined how variation in HEXACO traits related to self-reported emissions reduction behavior and to the results of nature-connectedness scales in a study of more than 300 US adults.[8] The emissions reduction scale included fifteen behaviors, such as how often participants chose to walk, bicycle, carpool, or take public transportation instead of driving a vehicle by themselves. Their results showed that such emissions reduction behavior was most strongly related to the HEXACO personality factor of openness (r value of 0.28), which refers to abstract thinking and appreciation for variety and unusual experiences, and especially to a subcomponent referred to as aesthetic appreciation. As shown in Figure 7.2, it also found that the connection between openness and emissions reduction behavior was mediated by having a high sense of connection to ecology or nature, suggesting that the personality characteristic of openness leads people to be receptive to valuing the natural environment and in turn to engage in more emissions reduction behavior. A similar study conducted in 2021 further demonstrated that the personality characteristic of openness was also correlated with climate risk perceptions.[9]

Although personality has traditionally been thought to be unchangeable, evidence has shown that traits such as conscientiousness and emotionality can

Figure 7.2

Those who score high on "openness" and identify as nature lovers report more emissions reduction behaviors

Mediation analysis examines the impact of openness on emissions reduction behavior. The analysis of survey scale data from hundreds of participants suggests that the influence of openness is not direct, but rather is indirect and mediated by other variables (New Ecological Paradigm and Connectedness to Nature).

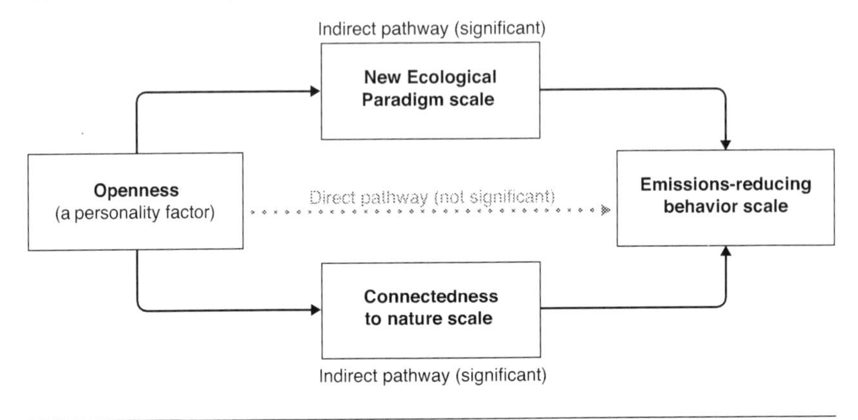

Adapted from Brick and Lewis (2014). Environment and Behavior. https://doi.org/10.1177/0013916514554695.

be modified through interventions and therapy, especially in early adulthood.[10] Such findings suggest that it might be possible to increase people's level of openness, especially of aesthetic appreciation, perhaps via early experiences in nature, in ways that could increase their level of concern over the effects of global heating.[11]

Personal Values and Socially Constructed Beliefs

A large body of psychological research has demonstrated that personal values are another key determinant in an individual's environment-related beliefs and behaviors. Values can be thought of as general goals or guides for how people live their lives that shape a wide range of specific beliefs. According to one review of the literature, values tend to be relatively stable and to serve as "deep-rooted personal criteria on which thoughts and actions are, often unconsciously, based and evaluated."[12] As that review notes, four types of value have been found to be most predictive of environmental beliefs and behaviors: in addition to *biospheric* values (valuing the well-being of the environment, as discussed above), these include *altruistic* (valuing the well-being of other people), *egotistic* (valuing one's personal resources), and *hedonistic* (valuing one's pleasure and comfort) values. Although all individuals hold all four types of value to some extent, the degree to which they prioritize some of those values over others tends to vary between particular situations, groups, and individuals. Research has shown that individuals who hold strong biospheric or altruistic values, or what are referred to as *self-transcendence values*, are typically more inclined to hold pro-environment beliefs and act in more proenvironmental ways than those who hold strong egotistic or hedonistic values, referred to as *self-enhancement values.* These relationships have been found to be consistent across a range of environmental beliefs and actions and across a number of countries and cultures.

Another factor that has been found to shape individuals' risk perceptions related to nature is how they view themselves in relation to others. The cultural theory of risk originally posited by anthropologist Mary Douglas and political scientist Aaron Wildavsky argues that individuals' sense of risk and their decision-making regarding those risks are largely socially constructed.[13] As discussed by Mike Hulme and schematized in Figure 7.3, this theory identifies four "ways of life" that vary along two dimensions: the extent to which individuals are group-oriented or individual-oriented, and the extent to which they believe rules are necessary to control behavior.[14] This schema places people into four major categories: fatalist, hierarchists, individualists, and egalitarians. On the left side of the social regulation axis, fatalists and individualists are both less group-oriented than the other two categories although individualists see little need for social structuring whereas fatalists accept their position as isolated people within a stratified society. On the right side, hierarchists and egalitarians are both strongly group-oriented, but hierarchists view social bonds as vertical, strongly ordered by rank or role, whereas egalitarians view those bonds as horizontal, linking fundamentally equal and voluntarily joined individuals. As illustrated in Figure 7.3 by a characteristic statement by persons falling in that category and also by an image of a ball on a landscape, each of these categories is likely to respond differently to climate change and the level of threat it poses to the learned values that underlie their way of life and where they perceive their interests to lie. As it shows, those on the left are less likely to engage in or demand action to respond to the threat than those on the right, and those who fall in the egalitarian sector are the most likely to see the climate system as precarious and to advocate action in response.[15] While this theoretical framework undoubtedly oversimplifies the fact that few of us fall into just one of these extremes in all areas of life or at all times, it can serve as a useful way to recognize the general patterns of socially informed response and archetypical voices in public discourse.

Figure 7.3

How one views risk, and how one thinks about Nature, is also determined by culture – are you a fatalist, hierarchist, individualist, or egalitarian?

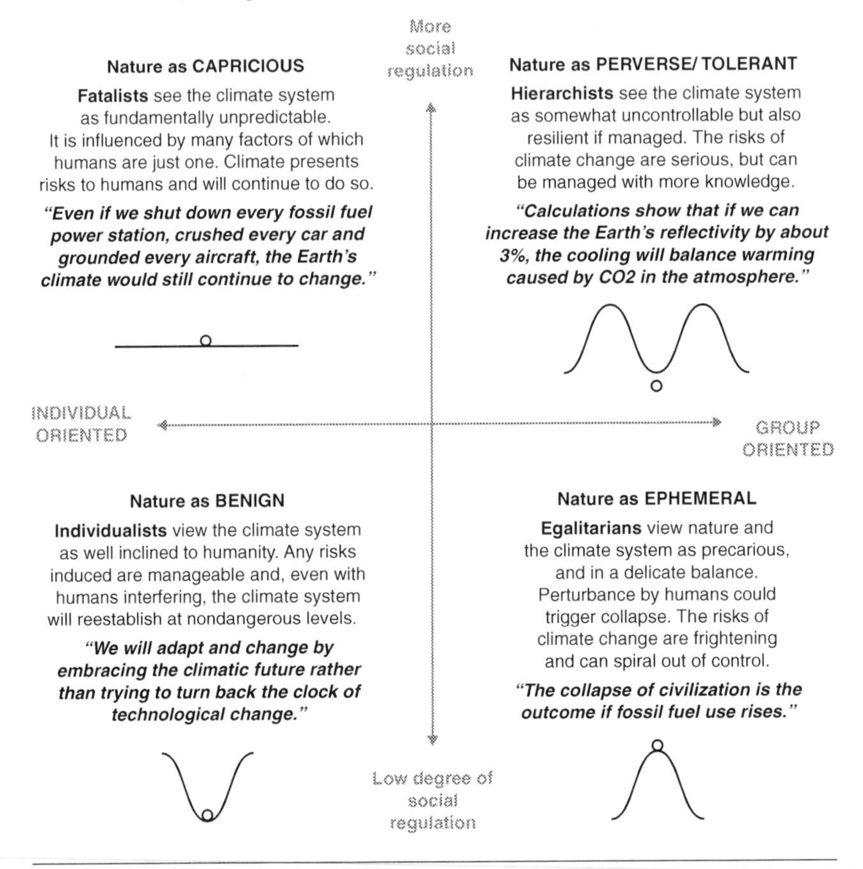

Studies that have applied this theory to the threat of global heating have found that egalitarianism does indeed appear to be related to a stronger sense of climate risks. One such study of nearly 700 US participants found a moderately strong correlation between egalitarian values and both an elevated sense of risk from global warming (r=0.45) and the endorsement of policies to reduce greenhouse gas reductions (r=0.48).[16] Further, a meta-analysis of sixty-seven studies found that individuals who scored higher on egalitarianism perceived more environmental risks (r=0.25).[17] Other research has hypothesized that higher egalitarianism translates into increased climate risks partly through a concern about the equitable distribution of environmental burdens and responsibilities among specific nations or people within nations, which might provide a social justice basis for climate risk perceptions perhaps even without concurrent high biospheric values.[18]

While research in children shows that egalitarian values are already evident in infancy and may be partly heritable, other research also shows that the degree to which they get expressed or acted upon also depends on social context and learning[19] – thus it may be possible to modify these values.

Experience

As the earlier discussion about human beings' historical sense of connection and reciprocity with nature would suggest, several research studies have found that individuals' personal experience with nature can play a key role in elevating their perceptions of risk to the environment. A 1993 study by Wendy Horwitz, which interviewed twenty prominent environmental activists from Pennsylvania and Delaware about the formative events or circumstances that guided their concern for nature, found that for most it began early in life as a result of academic and direct experience and of family and cultural or spiritual influences that led them to develop an environmental ethics in which their sense of connection to nature became part of their personal identity.[20] A perhaps iconic example of a person whose early life experiences conferred them with high biospheric values is Rachel Carson, the naturalist and author who first alerted the general public to the ecological dangers posed by pesticides. As Horwitz noted in her study, Carson grew up in rural Pennsylvania, where she was nourished by a spiritual, nature-loving mother and given access to books and opportunities to play in nature that led her to love the natural world from an early age. Her later career choices, mentors, and role models allowed her to develop her dual identity as naturalist and writer. Catalysts to her later activism included her scholarly study of the effects of pesticides on nature and personal experiences with the destruction of wildlife that generated moral outrage, which, as we will see in Chapter 10, is a key variable that drives people to action.

In his interviews with youth from fourteen countries who were committed to climate activism, Scott Fisher found they often expressed a concern for nature or for social justice, or both.[21] Dipendra from Nepal, for instance, told Fischer that "everything is inter-related. For me, I see it as protecting human and nature life and our futures. I know it might sound slightly dramatic and I am possibly not expressing myself the best way but everything is interconnected. The planet isn't compartmentalized." Other interviewees also suggested that growing up in communities in which they were aware of how everyday lives were directly connected with nature had contributed to their sense that it was impossible to separate a concern for nature from a concern for people.

Another major way in which people's climate risk perceptions are formed is through experience with extreme weather events related to global heating, such as storms, drought, and flooding, which (as noted earlier) have increased in frequency by around 400 percent since the 1980s. It would stand to reason that this dramatic increase in the observable physical evidence of the destructive effects of climate change would increase most people's perception of the risks it poses. Yet a considerable body of research into the psychological impact of extreme weather events has provided only mixed results concerning their short- and long-term impact. One study of Twitter messages emanating from areas experiencing such an event, for example, showed that messages referring to "climate change" or "global warming" did increase following an event – including floods, excessive heat or cold, wildfires, and tornadoes – but could not determine whether that increase was due to people's direct experience or reporting in the media, or even whether people were responding with sarcasm.[22] Another study that examined levels of post-event discussion and activism in fifteen areas across the USA found that differences among locales appeared to be related to a combination of such variables as political party affiliation, education levels, pre-event levels of newspaper discussion and local engagement with climate change, and the recent number of such events.[23]

A 2019 meta-analysis of seventy-three studies of the impact of extreme weather events found inconsistent results between studies, ultimately concluding that these studies provided only weak evidence that local extreme weather had an influence on climate opinions.[24] As the authors of that study and others have noted, there are probably several reasons for this surprising finding that extreme weather appears to be such a poor teacher, one of which is that some of the included studies were conducted decades before the dramatic impacts of extreme weather events became as evident as they are now. Another is that most of this research has focused on individual events or the same type of event, whereas anecdotal evidence suggests that the cumulative experience of multiple events over time may be more important in forming risk perceptions. For instance, when people in St. Louis, Missouri were interviewed about a 2012 heat wave that spanned ten consecutive days of triple-digit temperatures, some remarked that it was not that single event that drew their attention to climate change but successive extreme weather events that had occurred throughout Missouri – especially in 2011, when the state was hit with a blizzard, multiple floods, multiple tornadoes, and extreme heat and drought.[25]

A third reason that personal experience appears to be a poor teacher of the actual risks of global heating is that, as noted earlier, people's perceptions are highly influenced by their values and worldviews. As psychologist George Marshall speculated based on his interviews with survivors of extreme weather events in the US Midwest, people who already regarded climate change as a myth tended to interpret extreme weather as proof that weather is variable, while people who already considered climate change to be a real risk tended to regard extreme weather as further proof of climate breakdown. He also noted that after these extreme weather events, many of those survivors chose to just pick themselves up and start rebuilding, demonstrating what Marshall termed a "false sense of invulnerability" based on convincing themselves that because the event had already occurred, it would be less likely in the future, a version of the cognitive bias known as the gambler's fallacy.[26]

A fourth reason that extreme weather might be a poor teacher is that as extreme weather events become more common, they tend to become normalized in people's minds, leading to **shifting baselines syndrome**, or a gradual change in how people accept a changing situation as being normal.[27] In the realm of climate change, such shifting baselines are understood to represent a gradual change in accepted norms regarding the condition of the environment and to be due primarily to a lack of information about or experience of past conditions. Journalist David Roberts offered intergenerational fishing as an example of how such baselines can shift over generations.[28] Whereas one generation of fishers may be aware that the level of abundance of a given type of fish is lower when they retire than when they started, the reduced level becomes the new baseline for the next generation. As time goes by, the baseline for each generation of fishers shifts downward. Even though the fishers are operating in a degraded ecosystem, they tend not to notice it until the fish actually begin to go extinct. A similar phenomenon applies to global heating: few people are aware of how many hot summer days were normal for their grandparents' generation, even though extremely hot summers have become 200 times more likely than they were 50 years ago.[29]

Research evidence that shifting baselines affect people's perception of extreme weather includes another study of Twitter messages, which found that although the number of tweets about the weather did indeed increase when temperatures were unusually hot or cold for that particular place and that time of the year, the remarkability of temperature, as measured by the number of tweets, decayed rapidly with repeated exposure (Figure 7.4).[30] Furthermore, the researchers discovered that if people experienced extreme cold or heat at a

Figure 7.4

Extreme weather events become “psychologically normal” if they are experienced several years in a row

When people experience extreme cold or hot anomalies several years in a row in their local area, they stop finding them remarkable. The entire Twitter database for the USA was analyzed from 2014 to 2016 to find tweets related to extreme cold and hot events at the county level. This was matched to the meteorological record for temperature. (A) shows results for cold temperature extremes. Cold anomalies (–3 Celsius, 5.4 F) experienced between 2 and 8 years ago reduced the remarkability of a current temperature anomaly. An even more rapid decline in remarkability occurred for hot extremes (B). People shift their psychological baselines of what is “normal” after about five years on average of experiencing abnormal events. Error bars are 95 percent confidence intervals.

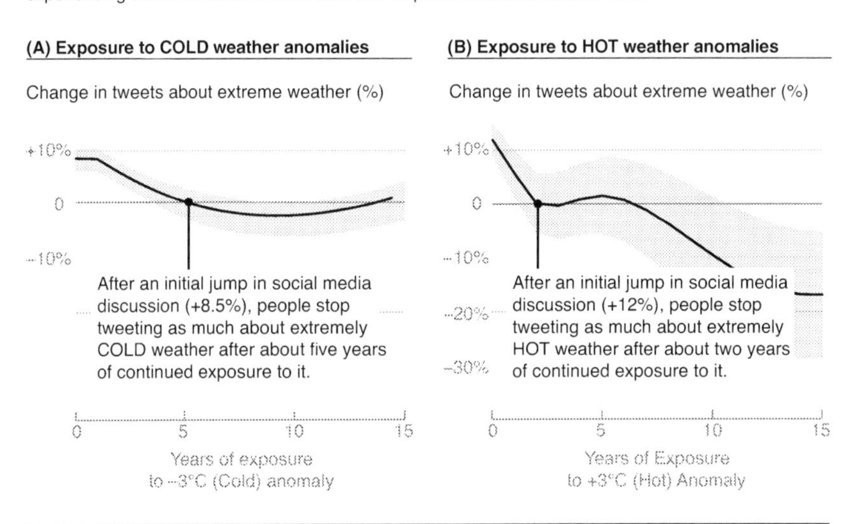

Adapted from Moore et al. (2019). PNAS. 116 (11). https://www.pnas.org/content/116/11/4905

particular time of the year for around five years in a row, it no longer became remarkable. Such evidence provides empirical support for a shifting baselines explanation for human experience of climate change, showing that the seeming remarkability of temperature is dependent not only on the absolute value but on one’s past experience and expectations, and particularly on one’s more recent experience.[31] More generally, it also suggests that the window of experience that many people find cognitively and emotionally remarkable may be too narrow to fully appreciate long-term changes in the biosphere and the level of risk they represent. (As Roberts points out, however, this does not apply to everyone, as demonstrated by Indigenous people who sometimes carry an enormous amount of accumulated knowledge that can help reveal what is lost.)

Although the research discussed above suggests that the impact of extreme weather cannot be counted on to serve as a clear teacher of the real risks of climate change, it is possible that future research looking at compound events occurring across time might reveal a different picture. It is also plausible that, as extreme weather becomes more common and more dramatic, its potential for raising the general level of climate risk perceptions may increase. For instance, a YouGovAmerica poll of late 2021 showed that while only 15 percent of Republicans who were not directly impacted by western state wildfires attributed them to climate change, this number rose to 26 percent for those Republicans who did experience these events.[32] Such perceptions might also be increased by simulations and time-lapse photos and videos of such climate-caused changes as melting glaciers and rising sea levels, including such efforts as the Baseline 2020 project, in which photographers will document changes in locations particularly prone to the impacts of global heating every five years until 2050.[33] While it is doubtful whether that particular approach could help shift attitudes fast enough to support the policies we need this decade, other related approaches using, for instance, interactive maps of sea-level rise might be useful, and will be discussed below.

To better understand which of the many different factors that appear to affect climate risk perceptions may have the strongest influence, a 2015 study of more than 800 participants in the UK measured survey participants' levels of risk perception by asking them such questions as how likely they thought that climate change would create risks to their own personal well-being and how serious a risk they believed it posed to the natural environment.[34] It then asked them to fill out surveys to measure a range of other variables, such as those discussed above, that the researcher hypothesized would influence their risk perception. The results, as shown in Figure 7.5, support the claims above that social norms and biospheric values were most strongly linked to risk perceptions (and that extreme weather events, abstract knowledge, and demographic variables were less so). The study also found that the single strongest explanatory variable was the participants' overall reported emotional response to the risks of global heating, the topic of the next section. Notably, however, this research was simply about self-reported risk perceptions rather than any measurement of behavior.[35]

Overall, while this section has identified a variety of psychological and sociological factors that relate to climate risk perceptions, the challenge is to leverage this knowledge to design and implement better communications and better interventions to elevate those risks and to translate them into action.

Figure 7.5

How someone perceives the risks of climate change is most impacted by their overall emotion and sociocultural influences

The model below depicts how ten independent psychological variables and sociodemographic variables (e.g. age) impact risk perceptions. This published model does not include personality, although it can also play an important role.

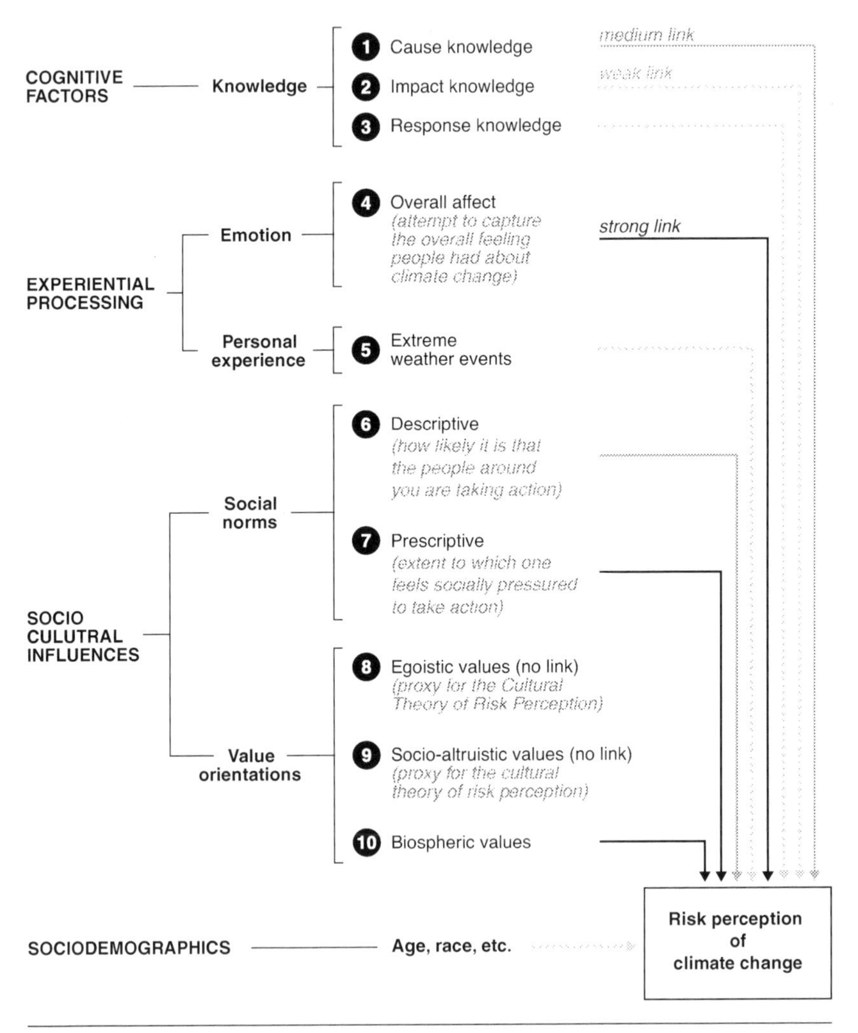

Adapted from Van Der Linden, S (2015). Journal of Experimental Psychology, Vol 41, Page 112-124.

Strategies for Elevating Risk Perception

Psychology has shown that humans recognize and evaluate risk through two different reasoning systems: an *affective* or *visceral* system that is fast, intuitive, and automatic, and works through processes of emotional association, and an *analytical system* that is slower, abstract, and learned, and operates through mental processes of reflection, deliberation, and judgment. One compelling explanation for why so few people currently feel driven to take action against the undeniable reality of climate change is that the abstract, delayed, and faraway nature of global heating may drive the visceral system too weakly to make many people pay attention, or at least too weakly to maintain the attention generated by extreme weather events in the face of factors, such as values and worldviews and other competing risks, that act to diminish their impact. Of course, inputs from the analytical system in the form of abstract or descriptive information about climate change also shape people's risk perceptions about the level of danger it represents. But unless people are earth scientists or have certain kinds of scientific training, such analytical information by itself typically does not create enough sense of urgency to motivate action (and even then, few earth scientists focused on climate are moved to action[36]). As Elke Weber concludes in a review of behavioral decision-making research, three decades of findings strongly suggest that interventions designed to capture the attention and engage the emotions of a wider public may be necessary to generate the level of public concern and action that will be necessary to accurately and effectively respond to the risks posed by global heating.[37] This section explores some suggested strategies for elevating people's risk perceptions regarding climate change without creating a counterproductive sense of hopelessness.

One way to elevate risk perceptions is to leverage the science communication approaches in Chapter 6, buttressed by scientifically accurate information about extreme weather or escalating impacts. So far, there appears to be little research of this kind. It could begin by conducting research with large groups of participants to determine what kinds of experimentally delivered descriptions and explanations are most effective in raising their climate risk assessments. Such descriptions and explanations could range from representations of RCP pathways to explanations of how scientists attribute extreme weather to climate change to depictions of likely damages and impacts on human migration and extinction rates. Such research might also provide more information about what kinds of communication are most effective for various groups of people, who obviously vary in their level of scientific and statistical literacy, biospheric values, and numerous other psychological and social factors discussed earlier. It may also be possible to tailor future risk communication interventions at the personal level, including providing such projected impacts as housing and insurance costs and flood damage for a person's own local zip code or for locations that are likely to be of value to them. One example of this approach is the National Oceanic and Atmospheric Administration's sea-level rise viewer; another is the *New York Times*' visualization of the impact of sea-level rise on Manhattan.[38]

Another approach that has already been shown to be effective in elevating climate risk perceptions is combining the effect of both emotional (vicarious and visceral experience) and analytical stimuli in novels and nonfiction books and in documentary films, videos, and simulations. One particularly notable example is journalist David Wallace-Wells' bestselling nonfiction book *The Uninhabitable Earth: Life after Warming* (and the viral *New York* article on which it was based) which, closely based on the latest science, painted a vivid and often horrifying picture of the present and future impact of global heating in multiple spheres, from extreme weather events and rising sea levels to species extinction, devastating fires and droughts, and increased geopolitical conflict and migration. Numerous reviewers, interviewers, and readers have described the book as terrifying and merciless in its compilation of worst-case scenarios. In a *New Yorker* article, for instance, science reporter Joshua Rothman reported reading the book soon after his son was born and growing "so alarmed that thinking about the problem became almost unbearable. I live on the North Shore of Long Island, close to the beach, in a village that already seems to be flooding. What did the future hold for my town, and my family? What would my son live through?"[39] Although some critics have accused Wallace-Wells of peddling despair or "doomism," many readers have also credited his work with first making them fully aware of how serious the problem of climate change really is and how they responded to the book as a stirring call to action.

Another work that has had a similar impact on many people's understanding of the risks posed by global heating is *Ministry for the Future*, a 2019 novel by Kim Stanley Robinson, a leading author in the rapidly growing genre of climate science fiction.[40] Like Wallace-Wells' book (and the viral article on which it was based), Robinson's novel offers readers riveting and sometimes horrifying images and scenes of the devastation wrought by extreme weather events. It opens, for instance, with a scene in which an international aid worker, Frank, is caught in a brutal heat wave that kills twenty million people in India in a single week and gives an excruciating first-person account of the suffering of townspeople unable to find any relief, even in a nearby lake, from the deadly heat and moldering corpses that cover the rooftops and sidewalks.

The book has been widely read and reviewed and was included in several major publications' lists of the most important books of the year. Although a work of fiction, it, too, is based on exhaustive research on the actual and predicted effects of global heating and current attempts to move the world's governments and institutions to address it. As several commentators have pointed out, however, Robinson attempts to balance his chilling vision of the near future with a sense of hope about the possibility of effective action, however piecemeal or toothless much of that real-world action proves to be. Hundreds of my own students who read the opening chapter of Robinson's novel as a class reading reported that his powerful description of the heat wave was truly frightening to them and made them want to better understand and do more about global heating.

The visual impact, accessibility, and wide reach of films and television programs have also shown themselves to be effective ways to inform and engage audiences regarding climate risks. *The Inconvenient Truth*, the 2006 documentary film of former US Vice President Al Gore's slide presentation about the reality and dangers of global heating, has been widely credited with raising international public awareness of the problem and reinvigorating the environmental movement. Although the film is in essence a lecture, its use of music and images of such phenomena as hurricanes and melting ice caps and visual aids such as the Keeling Curve showing the increasing rise in CO_2 levels were effective in capturing audiences' attention and engaging their emotional responses to the problem. Another popular documentary regarding global heating is *The Years of Living Dangerously*, a US television series that ran for two seasons (in 2014 and 2016) and whose executive producers included filmmaker James Cameron and former actor and politician Arnold Schwarzenegger. Each episode explored a different aspect of the problems caused by global heating and featured celebrity hosts and environmental activists such as Matt Damon, Harrison Ford, and Don Cheadle and well-known environmental journalists interviewing experts and ordinary people affected by the effects of global heating. The series, while conveying a wealth of information about the effects of and possible solutions (both technological and political) to the dangers of global heating, consciously attempted to connect those facts to human stories and to balance fear for the future with the possibility of effective action. Documentaries such as these have also been joined by movies that have dramatized the human impact of climate change. These include the 2017 movie *First Reformed*, which depicts a character so traumatized by the idea of climate breakdown that he ends up harming himself, something that several of my students have credited with being particularly impactful, and as the turning point at which they joined a climate movement.

In 2021, the movie *Don't Look Up* relied on the parable of a comet headed toward Earth to dramatize the threat of climate change and the kinds of denial that attend it. It was especially important at validating the feelings of climate activists, who often feel, like the scientists in the movie, that they carry very grave knowledge that people around them don't want to hear about.

Yet the impact and effectiveness of existing efforts to increase a sense of threat and risk has been a matter of dispute among experts and activists. For example, the climate scientist Michael Mann has argued that the deeply threatening picture of the future presented by works such as *The Uninhabitable Earth* might be dangerous in encouraging pessimism, or what he calls "doomism," so that people "look away" from the problem even more.[41] (This concern flags how the elevation of risk perceptions relates more tightly to action/inaction than does the communication of the facts of human-caused heating.) And some people, especially young people, may feel that their climate risk perceptions are, if anything, already too high. Many report a kind of global sense of dread as they become aware of climate change and begin to anticipate the impacts of a more hostile and unpredictable climate. Such dread has been termed **eco-anxiety**, a condition first defined by the American Psychological Association in 2017 as a "chronic fear of environmental doom" and reports of which exploded in 2019 as a surge of heat waves and other apparent climate-related disasters led to extensive international protests.[42] A 2021 study published in the *Lancet* of 10,000 participants between the ages of sixteen and twenty-five from forty-two countries found that three-quarters said they believed the future of the planet is frightening, 45 percent reported that their feelings about climate change had negatively affected their daily life and functioning, and 56 percent responded that humanity is doomed. Most reported feeling betrayed by their governments, although those who felt more confident about their government's response reported feeling less anxiety.[43] Whereas some people experiencing such feelings have turned to psychotherapy to seek relief for symptoms that have negatively interfered with their day-to-day functioning in other areas of their lives,[44] some in the profession have argued that at least some level of eco-anxiety is simply a sane response to a grave threat, and one that can prod many people to productive and satisfying action. Graham Lawton, a clinical psychologist who works with children as part of the UK's Climate Psychology Alliance, argued that eco-anxiety is a rational response to the enormity of a problem and thus the opposite of an illness. Instead, he preferred to call it "eco-awareness" to reflect that such feelings about the climate crisis reflect not a surge of mental illness but an overdue emergence of widespread sanity.[45]

The problem for activists and policy makers trying to elevate risk perceptions regarding global heating, therefore, is finding ways to do so without encouraging **climate fatalism or defeatism**, a form of response skepticism that leads people to feel that taking action against the threat is futile. Responding to works expressing this position by such noted authors as Paul Kingsnorth, Roy Scranton, and Jonathan Franzen, the climate writer and activist Andreas Malm has commented that "climate fatalism comes from those on top" and has attacked this position as a rejection of the political will to fight, one that risks becoming a self-fulfilling prophesy. As he puts it, paraphrasing the argument of the philosopher Catriona McKinnon, who has written on climate despair, "the more people who tell us that a radical reorientation is 'scarcely imaginable', the less imaginable it will be."[46]

The research picture on the role of emotion in climate change communication is complex, however, and does not consistently support the commonsense idea that making people feel threatened will cause them to avoid thinking about or taking action in response to the problem and that making them feel hopeful will increase their level of belief and engagement. For instance, increasing people's threat perception by making them aware of the horrific impacts that climate change has already had on many people in the Global South may make them more angry than depressed, and although anger is often considered a negative emotion, both history and research have also shown that anger can play a useful role in overcoming apathy and is the emotion most associated with motivating individuals to act against injustices.[47] On the other side of this equation, evidence from two large studies conducted with hundreds of people found that a hopeful message was more likely to foster complacency than gain support for emissions reductions.[48] In an even more recent study of the impact of three kinds of affective or emotional endings – optimistic, pessimistic, and fatalistic – on actual climate change appeals, another set of researchers found that appeals with a pessimistic ending increased both risk perception and outcome efficacy, or the belief that one's actions matter. Yet looking at other mediating factors in such beliefs, they also found that the impact of emotional arousal was more pronounced among political moderates and conservatives and people holding individualist or hierarchical worldviews.[49] After reviewing a number of such studies, psychologists Chapman, Lickel, and Markowitz concluded that we should not view emotions as a simple lever that can be pulled and instead counseled the climate change communication community to "adopt a more nuanced, evidence-based understanding of the multiple and sometimes counterintuitive ways that emotion, communication and issues engagement are intertwined."[50]

In a 2006 analysis of public discourse about global warming in the UK, Gill Ereaut and Nat Segnit identified three main linguistic repertoires, or systems of language routinely used to describe and evaluate events, actions, and persons, that appear in such messaging, each of which has a different emotional impact and likely effect on the actions audiences choose to take in response to the risks posed by climate change.[51] They described the first and most common of these discourses as alarmist and fundamentally pessimistic, employing dire imagery, an urgent tone, and a sense of acceleration and irreversibility. Despite such rhetoric's effectiveness at gaining attention and increasing people's perceptions of the risks posed by climate change, they point out that it also seems to negate the effectiveness of action or agency on the part of its audience and that its apocalyptic imagery can make the problem seem unreal or engender a feeling that it is simply too big for ordinary people to take on.

The other two repertoires that emerged from this analysis are more optimistic in tone, but to quite different effect. The first is an evasive or mocking discourse that reassures audiences that everything will be okay, but only by essentially denying or downplaying the real dangers posed by climate change. By rejecting alarmist rhetoric and invoking "common sense" to reject scientific evidence, it tends to be immune to argument and projects a falsely positive view of the future that discourages both concern and action. The second of these essentially optimistic rhetorical approaches tends to be more pragmatic in tone and to highlight small actions that people can take to help save the planet. Although this approach can encourage belief in both the reality of global heating and the efficacy of action, the authors point out that it can make the problem seem too minor and fixable to require extended engagement or to justify the kind of collective and concerted effort that is actually required to ensure the planet's future (this focus on individual consumption action is also, as Chapter 5 showed, one of the strategies the fossil industry has fostered).

Although this particular analysis is specific to the UK and does not reflect more recent trends in climate change discourse, it does point to the complexity of the issue of what kinds of appeals and messages are most likely to be effective in encouraging action. It also led the authors to make several general recommendations.

They argued that it is not enough to simply produce more messages about the reality of global heating and the urgency of acting, but it is instead important to harness tools from advertisers to make it desirable to act. They suggested targeting communications at groups bound by shared values and communications literacy instead of demographics; representing climate-friendly behaviors to large swaths of the population as natural, normal, and right and simply the kinds of things that "people like us" do; associating such behaviors with people's esteem-driven needs to feel special; and acknowledging that people increasingly seem to place more trust in individuals than in governments, institutions, and experts. These recommendations could also be supplemented by those in Chapter 5 that focused on co-benefits, for example, as social psychologist Kelly Fielding elaborates in Box 7.1 – by communicating about renewable energy jobs it may be possible to get many people to accept wider actions for climate policy even if they don't change their core beliefs about climate change. Further, as Chapter 10 lays out, particular kinds of communication strategies might engage people in collective, rather than individual, action as part of a broad social mobilization to leave fossil fuels in the ground and contest fossil fuel interests.

Even though the question of what kind of emotional content and tone is best to elevate climate risk perceptions remains open, there is a common perception among friends, colleagues, students, and activists that we must remain hopeful about success in dealing with the climate crisis. Yet it is also possible that the question of (hope of) success is not the main one we should be asking. How to overcome response skepticism and to engage in the struggle to confront fossil fuel interests for a more just world might be seen instead as more a question of human dignity. To quote the environmentalist Wendell Berry, "[w]e don't have a right to ask whether we are going to succeed or not. The only question we have a right to ask is what's the right thing to do? What does this Earth require of us if we want to continue to live on it?"

Ultimately, the question of whether and how to act in the face of likely disaster and even possible annihilation toward the end of the century may be as much a philosophical as a technological or political one. In *Radical Hope: Ethics in the Face of Cultural Devastation*, philosopher Jonathan Lear used the end of the Crow Nation in nineteenth-century North America as an example to explore these questions.[52] As the Crow people faced a cultural apocalypse with the destruction of the buffalo herds and their way of life, their last chief, known as Plenty Coups (Figure 7.6), had to decide whether to bow to the federal government's order to enter a reservation or to continue to fight to the end. Lear describes Plenty Coup's decision to lay down his arms and enter a reservation not as failure or resignation but as a courageous imaginative leap that Lear terms **radical hope**, which required a fearless consideration of what ethical values would be needed to live and find meaning within this new reality. According to Lear, "[w]hat makes this hope radical, is that it is directed toward a future goodness that transcends the current ability to understand what it is." Arguing for a similar stance of radical hope toward the climate crisis, psychotherapist and chair of the Climate Psychology Alliance, Paul Hoggart,

Box 7.1 Interview with social psychologist Kelly Fielding

Kelly Fielding is a social and environmental psychologist and professor in the School of Communication and Arts at the University of Queensland, Australia. Her research has included a focus on trying to understand climate change beliefs and identifying ways to address climate change skepticism and inaction.

AA: What are your thoughts and feelings about the climate crisis?

KF: Probably like most people who work in this space and really care about this issue, I experience a range of feelings. Those range from, yes, we will make it happen, to wanting to throw up my hands in the air. A lot of the research on emotions and climate change has focused on the role of hope and the role of fear and anger and despair and frustration. I feel all of these things in relation to climate change. You know, one of my PhD students did a meta-analysis of all the recent media articles on eco-anxiety. It was interesting that the media articles mainly focus on children and young people. I think that's because, as an adult, you're probably better at compartmentalizing. You say to yourself, there is a scary thing over here, climate change, but meanwhile I need to get on with these other things.

AA: You have done much to characterize the underlying attitudes, vested interests, and ideologies that relate to skepticism about human-caused climate change. Is that work complete, or is there more to understand?

KF: I think we should focus more on how we convince people to support action. We've got this window of time in which we need to make these serious and dramatic changes to our systems in order to prevent runaway climate change. The data are very clear in Australia – almost everybody believes that the climate is changing. So it comes down to do they think it's anthropogenic or is it natural? About one-third of Australians think it's natural fluctuations. So where do we focus our time? Try to persuade those specific people? Or spend our time focusing on getting support for mitigation and adaptation?

AA: But in Australia you still don't have federal climate policy, right? Coal mining continues rampantly. So isn't it key to get that one-third? If they believed it was human-caused, would that lead to a political shift or not?

KF: The thing is that we are a fossil fuel–dependent economy. Another thing is that you have conservative politicians who genuinely do not believe in anthropogenic climate change. We've studied them with surveys, where they were guaranteed anonymity. But more importantly, you've got these marginal seats that represent coal-dependent communities. So what you've got within the political parties in Australia is these factions of politicians who don't believe in climate change, whose political careers depend on being reelected by coal-dependent communities, and they are the ones who are really preventing progress on climate policy. And, about those rural coal-mining communities, I've been told anecdotally that if you go out and talk to people, then you shouldn't mention climate change. If you leave that out, they will talk to you about what's happening in their communities and what the future is, and that the writing is on the wall, coal is on the way out.

AA: So how do you propose to communicate solutions to people in the rural areas and the wider population?

KF: The data this year show that almost 91 percent of Australians support government subsidies for the development of renewable energy. That's huge. So, what some people would say is, don't talk about climate change, try to frame solutions toward energy independence, reducing air pollution. This way you don't get stuck back on the idea that everyone needs to agree that the climate is changing and humans are causing it. The problem with being stuck back at that stage is that you're not addressing the critical point, which is that we need policy right now that changes our energy systems.

AA: Most of the social psychology studies looking at belief and action rely on surveys of attitudes and intentions, what people say they will do, without verifying their behavior. Of course it would be hard to do field studies, but not impossible. Why so few?

KF: I think it's something about the disciplinary norms of our field. You can advance your career by *not* having to do those difficult field studies. They are expensive. You won't publish as fast. But it is important. And there are fields where this *is* being done – for example the social norm work, going out and looking at what people actually do. Among other things, I do water-related research. And we can link survey results with people's actual water use. We've worked with engineers to install water meters. You have to have the appetite to collaborate. You could think of experience-sampling studies, tracking where people go. Provided you can deal with the ethical issues.

AA: We're facing a disruption to organized human existence, probably in our lifetimes. But I look around the major psychology departments in the United States and I can count on two hands the people who are focused on the climate and ecological crisis. Even the social psychologists.

KF: Part of it is that the social psychologists who focus on climate change start labeling themselves environmental psychologists and go to different conferences. Many environmental psychologists are in Europe. There are far fewer in the USA and Australia. The other thing is that psychologists are like most people – they don't have a visceral feeling of climate change. It doesn't seem quite real to them.

AA: But it seems like a colossal failure from a systems point of view. Here the science-funding agencies are quite attuned to what is seen as human welfare. Researchers get multi-million dollar grants to look at the impact of exercise on aging and myriad other issues. Yet the system hasn't allocated resources to look at the behavioral implications underpinning a disruption to organized human existence!

KF: It might be because people don't think it's going to come to pass, or they think that technology will save us. There is a huge amount of techno-optimism.

AA: I want to ask you about the kind of person, say a colleague of mine, who knows climate change is happening, that the impacts are bad and are going to get much worse. And he has young children. He'll make minor changes, put solar panels on his roof, but he won't advocate, he won't spend the time with the grassroots. He'll work on his career, care for his family, fly them to Hawaii for a vacation for a week. How is he able to mentally separate his knowledge that something so enormous is coming at us, from carrying on with everyday life and not responding to it? He doesn't seem to experience eco-anxiety. What other psychological consequences are there? Doesn't it seem irrational?

KF: First, it sounds like he has low efficacy beliefs. He doesn't believe he can make a difference. I also think we human beings are really good at compartmentalizing. We're able to know something intellectually, but not feel it. It's not visceral for him, what's coming at him and his kids. You know what that feels like, when you embody something viscerally, you can't ignore it? Maybe for him, it's a set of risk perceptions just sitting cognitively in his head?

AA: How can we make more people embody it?

KF: People are social creatures. We use the people around us to tell us what we should be worried about. People around us are going on their international holidays, sending their kids across the country, living in big houses, driving their cars. They're not worried about it, so why should I be? The social cues are not there. We're not getting them from the elites. The social cues are coming from what might seem like to some lone voices. If almost everyone was talking from the same hymn sheets, and if everyone was saying this stuff, and if the descriptive norms showed us that people were changing their lives, then it would be very difficult for us to ignore it. Your friend would be more likely to embody his risk perceptions viscerally.

AA: Would you agree that another huge factor is that to really start taking this seriously, to be a serious advocate for collective action, requires change in your own life? You can't carry on with the high-intensity carbon lifestyle. You have to be willing to curtail. And that people's resistance to do this is one of the main reasons why they take on beliefs and views that don't require them to get there, such as their saying reducing flying doesn't make a difference, or their saying that it's pointless cutting emissions in the US when China now emits more overall?

KF: Absolutely.

AA: Is extreme weather going to help?

KF: The data don't show it. Including a major meta-analysis that we conducted. Think of the Australian Bush Fires of 2019. We were being called Climate Ground Zero for a while. It brought it to people's minds, but only temporarily. There's a part of me, the pessimistic part, that thinks we're not going to get it together until these things are coming at us left and right, heat waves, bush fires, floods. And people will be like, this is it. We can't ignore it. But it will be late in the day for mitigation.

AA: The proportion of people ringing the alarm bells is so miniscule. But this brings us back to the relevance of psychology. We need to understand how to get more people there, right?

KF: Yes. Despite the rhetoric about too many doom-and-gloom messages, I think people aren't worried enough. We probably don't talk about it enough with our family, friends, and colleagues. We don't want to depress people or get into a conflictual conversation. There is also research in Australia that people don't feel confident enough that they have the information to have those conversations. But, my experience is that when we are in a crisis – a flood, a pandemic – we rarely talk about anything else. It becomes the focus of our lives. Of course we need economic, political, and other system change, but I think we have to also believe that we, individuals, have a role to play, too, and to do whatever we can to address this looming climate crisis.

Figure 7.6

What can we learn from those who already faced cultural apocalypse?

Plenty Coups was the last traditional chief of the Crow people in North America. With the genocide and destruction of their way of life they faced collapse. Said Plenty Coup: "When the buffalo went away, the hearts of my people fell to the ground, and they could not lift them up again. *After this, nothing happened.*"

This profound statement reflects the inability to make sense of what was happening. Yet Plenty Coup was able to adapt with courage. This meant facing the collapse, and accepting the tragic destruction of his culture as a means to imagine a new future. This has been called *radical hope.*

Curtis, Edward S., photographer. *Plenty Coups A. Apsaroke*, ca. 1908. Photograph. https://www.loc.gov/item/[illegible]

points out that we should distinguish between *hopelessness*, which is an evaluation of probabilities, and *despair*, which is an emotional state, and recognize that the first does not necessarily lead to the second. In fact, he declares, giving oneself permission to imagine the very worst can be liberating and even energizing, as therapists or medical doctors sometimes see among patients accepting terminal diseases or observers sometimes see among soldiers facing incipient death in war.[53] When viewed from such a perspective, the process of struggling as part of a collective movement to confront fossil interests and stop emissions can also be seen as part of a larger process of creating a better and more just world, and an effort that is intrinsically worthwhile and rewarding in itself, including the positive feelings of comradeship and feeling of empowerment that come from having solidarity with others in a cause.[54]

Conclusion

Whereas some people can be shifted out of epistemic skepticism by coming to understand the human causes of global heating and some of its impacts, the process of elevating people's risk perceptions can also help them overcome response skepticism. Yet the time-delayed, sometimes abstract, and here-and-gone-again nature of the crisis posed by climate change tends to make it difficult for many people to perceive those risks with the appropriate amount of concern. As this chapter has shown, the factors that shape people's climate risk perceptions are various, including both individual differences and historical and social influences, and these factors interact with one another in complex ways. Climate activists, communicators, and policy makers can try to elevate climate risk perceptions with a wide range of strategies that engage emotional as well as analytical responses. But an appreciation of the risks may not be sufficient; getting over the last bump, to actually to do something, often appears to require having a clear vision of what kind of change is necessary and an understanding of how one can fit into the political and social engagement that is needed to make that happens. These are the topics that Part III now turns to.

PART III

From Belief to Action

8
Principles for Just and Effective Action

> The continuing disruption of the earth's climate system is not a technical problem to be "solved," but rather a systemic problem, rooted deeply in social and economic structures.
>
> Brian Tokar

Since the UN framework for tackling climate change was first established at the UN Earth Summit in Rio in 1992, scientists and climate advocates have proposed numerous legal, political, technological, and market approaches for reducing emissions around the world to prevent global heating, a number of which are examined in this chapter. But how can we know what kinds of actions are worth taking, capable of gaining wide support, and consistent with actually reducing emissions instead of continuing business as usual? This chapter also argues that any approach that meets those requirements must ultimately be a just one, meaning that it must fairly recognize and assign responsibility for historical and current levels of greenhouse gas emissions and must reject technical or market solutions that attempt to sidestep or delay that responsibility-taking.

In recent decades, this focus on just principles has been fostered and sharpened by the development of an international grassroots **environmental justice** and **climate justice** movement.[1] As Brian Tokar recounts in *Toward Climate Justice*, organizations such as the World Rainforest Movement and Friends of the Earth had begun to call attention to the environmental concerns of Indigenous peoples as early as the 1990s. A coordinated global movement focused specifically on climate justice first emerged at the 2007 Conference of the Parties (COP 13) in Bali, where representatives of land-based people's movements demanded agrarian reforms, an end to the conversion of tropical forests into biofuel plantations, and payments from the Global North for ecological damages suffered by the Global South.[2] This was followed in 2008 by the formation of Mobilization for Climate Justice in the USA, which included such organizations as Rising Tide North America and the Indigenous Environmental Network. At COP 15 in Copenhagen the following year, they and activists from Europe and the Global South joined to call for fossil fuels to be left in the ground, popular community control over production, Indigenous rights to be respected, and reparations to be made from the Global North.[3]

Since then, hundreds of international environmental and climate justice organizations have continued to make themselves heard right up to the COP 26 meeting in Glasgow in 2021. While COP 26 again failed to produce any binding commitments to leave fossil fuels in the ground,[4] climate justice advocates continued to play a critical role in exposing the falsity of carbon trading and technical fixes, and they lampooned the decision makers for their weakness and continued to insist on climate finance and reparations[5] and other principles, as summarized in Figure 8.1.

Figure 8.1

Demands of the international climate justice movement

- Developed countries must drastically increase domestic commitments and global targets to limit warming to 1.5°C, focus on real zero, not net zero. *Note: as of late 2021 only 4 percent of surveyed climate scientists still believe the 1.5°C target is possible.*
- Reject false solutions: say "no" to carbon markets and risky and unproven technologies.
- Developed countries must provide adequate money to scale up adaptation, ensure protection for climate migrants and those impacted by climate change, and commit to climate reparations for the loss and damage already happening in the Global South.
- Corporate interference in climate talks must end by adopting a conflict of interest policy that protects Paris Agreement implementation and global policy talks from obstruction by big polluters.

Principles of the international climate justice movement

- Developed countries must publicly recognize and act on their greater historical responsibility for greenhouse gases.
- Awareness of the growth imperative and concentration of wealth within modern capitalism and acknowledgement that a transition to a more inclusive and democratic economic system is needed to deal with the climate crisis.
- Adoption of a broadly intersectional perspective on the climate crisis that recognizes the common thread linking environmental injustice to discrimination on the basis of race, gender, and class.
- Awareness that policy changes must be driven by the agendas of grassroots campaigns, labor, and indigenous peoples rather than elite proposals that represent corporate interests and the wealthy.

Demands and Principles drawn from The People's Demands for Climate Justice, available online at https://www.peoplesdemands.org and COP26 Coalition: Our Demands, available online at https://cop26coalition.org/demands/. 1.5 target note from J. Tollefson. "Top climate scientists are sceptical that nations will rein in global warming." Nature 599, 22-24 (2021). DOI: https://doi.org/10.1038/d41586-021-02990-w

Determining Responsibility and Just Solutions

Viewing the imperative and challenges of drastically reducing greenhouse gas emissions through a justice perspective is not only a matter of morality, but it is practically and strategically essential to obtaining individual, collective, national, and international action and cooperation. As this section explains, fair and effective action will require differentiating between various countries' responsibility for reductions and damages in terms not simply of their current territory-based emissions but also their historical, per capita, and consumption-based emissions and finding more effective ways to acknowledge and enforce those responsibilities. It also discusses a just program of action to leave fossil fuels in the ground, which requires a commitment to avoiding the negative effects that the mindset and practices of mining and extraction have had on vulnerable peoples and places.

Acknowledging the Different Responsibilities of Nations

As discussed in Chapter 1 and demonstrated in Figure 1.3, the Rio Earth Summit of 1992 was alive to the historical inequities among nations in their disproportionate contributions to emissions, and this recognition was enshrined in the concept of common but differentiated responsibilities. Even though pressure from corporate interests and wealthy countries has prevented the UN process from producing binding commitments that reflect those responsibilities, the moral framework remains critical.

In 2020, according to the Global Carbon Project, the countries responsible for the largest total CO_2 emissions that year were China (10.7 gigatons), the USA (4.7 gigatons), India (2.4 gigatons), Russia (1.6 gigatons), and Japan (1.0 gigatons) – figures that are often cited to support the presumed pointlessness of a given country making reductions when the emissions of some others are so large. But issues of justice and fairness require that we look at these emissions not just as totals but on a per capita basis, that is, the amount of emissions divided by the populations of those countries. By that measure, the largest current emitters are Saudi Arabia (largely reflecting the vast emissions related to extracting fossil fuels), followed in order by Australia, the United States, Canada, South Korea, and Russia. Although India is currently the third-largest source of total emissions, it has nearly twice the population of the five largest per capita emitters combined; as a poor country in which hundreds of millions of people lack reliable access to electricity, its per capita emissions are less than 10 percent of all those other nations.

But as discussed earlier, an even more important distinction for the purposes of climate justice is determining the responsibility that various countries bear for the effect of their historical CO_2 emissions on rising temperatures. As the *New York Times* reports, just twenty-three wealthy, developed countries have been responsible for half of all historical CO_2 emissions, while more than 150 nations share responsibility for the other half.[6] Those twenty-three countries include the USA, which by itself is responsible for almost a quarter of all historical emissions,[7] followed in order by Germany, the UK, Japan, and France; the rest are other western European countries and Australia. Given that the world is already suffering the effects of the emissions of the past and that they will continue to raise temperatures for some time whatever steps we take today, countries that have benefited the most from those activities and resulting living conditions obviously have a debt to pay toward mitigating the effects of those emissions and making concerted efforts to cease them as soon as possible. (A related equity issue comes up when considering that wealthy people within countries emit far more than others; see Box 8.1.)

A just response to this historical inequity would require a country such as the USA to reduce a lot more than other countries. The theoretical emissions reductions a country commits to under the Paris Accord of the UN process is its **Nationally Determined Contribution** (NDC). While the USA's own NDC aimed for about a 50 percent reduction of emissions below 2005 levels by 2030,[8] this does *not* reflect the historical responsibility; moreover, the USA is patently not meeting this so far, and indeed is escalating fossil extraction.[9] A different calculation that factors in the historical responsibility, known as the Fair Shares NDC, estimates that the total US fair share contribution to emissions reductions would need to be equivalent to about a 195 percent decrease below 2005 levels by 2030.[10] One way to achieve this would be to reduce by 70 percent domestically by 2030 and achieve a further 125 percent reduction by committing to international climate finance contributions of about $800 billion between 2021 and 2030. Currently, such a decline is wishful thinking, but it highlights what justice requires.

Attributing responsibility for emissions in a fair and just manner will also require moving away from measuring and holding countries responsible only for emissions generated within their own borders or territories. To give just one example, although many materials and products consumed in the USA are manufactured, in whole or in part, in China, the emissions generated in producing or manufacturing those goods are considered by the Paris Accord (the UN monitoring process) part of China's emissions rather than the USA's, which allows the USA to enjoy the benefits of those products without having to take responsibility for their environmental impacts.

Box 8.1 Lifestyles of the wealthy are incompatible with climate justice

The climate justice perspective is also pertinent to the wealthy who lead high-carbon-intensive lifestyles. The world's top 1 percent wealthiest people, in North America, Singapore, Saudi Arabia, and elsewhere, have annual per capita emissions above 200 tons of CO_2 equivalent per year, while the people with the lowest income, in countries such as Honduras, Mozambique, and Malawi, have emissions that are 2,000 times lower, around 0.1 tons of CO_2 equivalent pear year.[11] Aggregated across individuals, the figure in this box shows data from 2020, where the wealthiest 1 percent of individuals were responsible for more than twice as much emissions as the 3.4 billion who make up the poorer half of humanity.[12]

These facts about the uneven contribution within nations thoroughly refutes the argument one sometimes hears that the emissions problem is due to "overpopulation." Of course, if the poorest people in the world keep increasing in number and start living an emissions-rich lifestyle then they *will* start accounting for a much larger proportion of emissions, but that is not the problem now, and it's not going to be for the next ten years when emissions cuts must happen.[13] Indeed, a recent study showed that lifting about one billion people out of poverty would increase emissions less than 2 percent.[14]

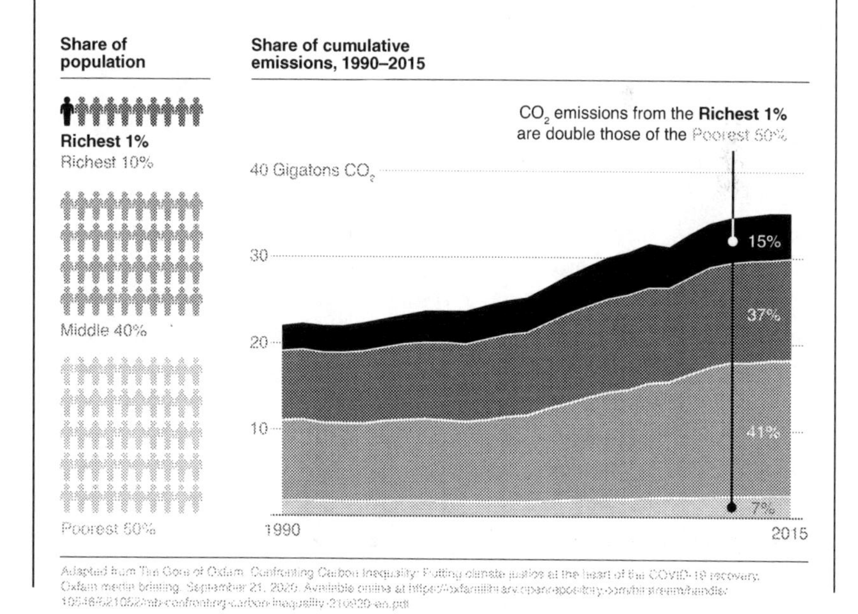

Such considerations of the unequal contribution of the wealthy, within countries, has prompted some to suggest that the burden of reducing emissions should rest mostly with affluent persons rather than affluent countries, a very different focus from the UN process.[15] More generally, these considerations point to the importance of consumption reduction for reducing emissions, covered in Chapter 9, and also considerations in how climate activism is targeted.[16] In late 2021, the climate activism group Extinction Rebellion blockaded the private Farnborough Airport near London, where elites come and go in private jets, with signs saying "1% of people cause 50% of aviation emissions."[17]

Assigning emissions based on consumption rather than production would require measuring **embodied emissions**, which refers to the emissions arising from goods, services, and commodities that are *imported* and then consumed by industry, commerce, the household, and the individual. The amount of such embodied CO_2 emissions from imports such as basic and fabricated metals, electronic and electrical equipment, mining, machinery and equipment, motor vehicles, and textiles is large and growing; one study estimated that in 2015, they amounted to 8.8 gigatons, or about 27 percent of global CO_2 emissions.[18]

The relative imbalances in who produces and who consumes these embodied emissions is demonstrated in Figure 8.2. As it shows, for China, India, and the Russian Federation, the emissions resulting from production are far greater than the emissions involved in goods and services consumed domestically, while in the USA, European Union, and Japan, the reverse holds, meaning that those latter countries are "net importers" of embodied carbon. Whereas some EU countries and the UK currently have some of the least energy-intensive economies in the developed world, owing to declines in their industrial sectors, their relative wealth also makes them some of the highest importers of emissions in the world. Yet under the UNFCCC process, countries must submit national emission inventories based only on emissions released within their territorial boundaries, which places a greater burden of emissions reduction on those countries producing energy-intensive goods than on those consuming them and once again benefits wealthier, developed countries more than poorer, developing ones.[19] As climate researcher Lucy Baker concluded, "wealthy countries have to a large extent exported or outsourced their climate and energy crisis to low and middle-income countries deliberately or otherwise," a phenomenon she has termed **carbon**

Figure 8.2

How CO_2 emissions are "embodied" in trade and how this creates a distorted and unjust picture of who is responsible

China, India, and the Russian Federation all generate more CO_2 emissions from production than they require from demand (or consumption). The reverse holds for the USA, the EU28 countries, and Japan. Since the international process of emissions-counting, via Nationally Determined Contributions, only counts "territorial emissions" this means that it looks, on paper, as if US, EU, and Japanese emissions are lower than they really are (since the emissions they are consuming are not counted), but do count in the production leger for other countries.

(A) Difference between CO_2 demand and production, 2015

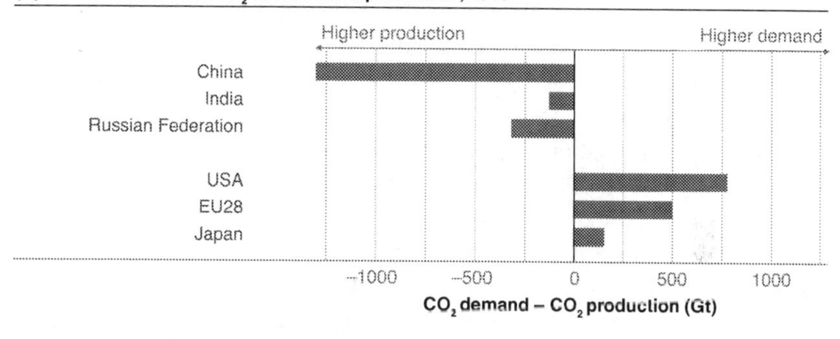

Adapted from OECD, Trade in Embodied CO2Database (TECO2)

colonialism.[20] Furthermore, emissions from the transportation of goods across borders – such as emissions from container ships that have increased over 100 percent over the last three decades,[21] and which are presently responsible for about 3 percent of yearly global emissions[22] – are not attributed to any one nation state, which also contributes to creating a distorted picture of who is responsible.

Revising Trade Agreements to Define and Enforce Responsibilities

The failure to take into account embodied emissions in the NDCs of individual countries is thus a further injustice in the Paris Accord and the wider UN process,[23] and one that is not widely acknowledged.[24] One way to reform that is to impose what are known as Border Carbon Adjustments, as the EU is currently considering.[25] This puts a carbon price on imports of specific products so that attempts to cut emissions in Europe really do lead to worldwide emissions reductions instead of allowing carbon-intensive production to leak outside Europe (known as carbon leakage) and then import it back in.

The issues around carbon leakage, neglecting embodied emissions, and carbon colonialism arise out of the system of global trade that was discussed in Chapter 4. As that chapter noted, global decision makers were determined to keep the imperative on climate change separate from the imperative for global "free trade" on the neoliberal model, an attitude that was already evident in the agreement at the Rio Earth Summit in 1992, and which endured through NAFTA and the WTO in the 1990s, right up to the Trans-Pacific Partnership that was signed by the Obama administration in 2016 (after the Paris Accord) with a dozen Pacific Rim countries. As Chapter 4 showed, these agreements are primarily about expanding trade, often in very extractive and energy-intensive sectors, protecting the rights of financial firms and corporations, and removing regulations that govern the public good, including environmental regulations. And the agreements under NAFTA, the WTO, and the Trans-Pacific Partnership granted multinational corporations special legal rights to dispute national policies, including challenging proposed bans on offshore drilling and fracking, and undermining attempts to install local renewable energy.

One obvious possibility to remedy these various injustices is to renegotiate the international trade deals, something that Senator Sanders was alive to in his 2020 presidential bid.[26] Other approaches could involve legal changes in WTO rules, implementing Border Carbon Adjustments (discussed earlier in this chapter) and improving transparency around and regulating fossil fuel subsidies.[27] But because even these incremental changes are likely to run up against the same powerful interests that have blocked international climate action for thirty years, insisting on change of this kind will require a much more powerful climate justice movement, and a much wider social mobilization, which is taken up in Chapter 10.

Confronting the Problem of Extractivism

As the previous section showed, inequity is at the heart of the climate crisis: the richest countries disproportionately created the problem, the richest people within countries continue to disproportionately drive it, and the poorest people everywhere are the most vulnerable to its impacts. Getting off fossil fuels presents an opportunity to redress these skewed power relations and to reduce global inequities. But many of the supposed solutions proposed by countries of the Global North and elite decision makers raise other issues of justice, as they would simply generate greater impacts in the Global South, not only through technofixes and offsetting mechanisms but through increased metal mining and extractive projects related to increasing the supply of renewable energy.

Already the emissions associated with primary mineral and metal mining are responsible for 10 percent of the total energy-related greenhouse gases, and the environmental and social impacts of such mining in the Global South have been severe.[28] Such mining is also slated to grow enormously under the renewable energy transition.[29] As will be discussed in Chapter 9, a near-total transition to renewables would require millions of wind turbines, tens of thousands of central solar plants and arrays of solar PV plants, and more than a billion rooftop PV installations; further, the needs for building retrofits and new electric transportation would also be huge. All of this raises profound ethical concerns related to extractivism, which refers to the high-intensity, export-oriented extraction of common ecological goods rooted in colonialism, and which creates large economic profits for the powerful few in the short term, but minimal benefits for the communities from whence the resources are taken.[30] Specifically, the worry is that the procurement of these critical metals will be left in the hands of the same mining companies that have been responsible for social and ecological crises all around the world, and it raises fears that territories in the Global South will thus become larger "sacrifice zones" (sites of ongoing colonialism, or neocolonialism).

An example of this neocolonialism is the operation of the Chevron oil company in Equatorial Guinea in Africa.[31] In *The Licit Life of Capitalism*, anthropologist Hannah Appel quotes former Chevron CEO Ed Chow's chilling explanation of how Chevron managed to control the political process in that country so as to extract its oil:

> You get the land but you don't provide a lot of jobs, you may be destroying the environment, and most of the profit goes to international capital. The companies . . . not only accept highly centralized government [in that nation] but . . . crave it. A strongman president [of that nation] can make all the necessary decisions. It's a lot easier to win support from the top than to build it from the bottom. As long as we want cheap gas, democracy can't exist.[32]

Nor is this exploitative and unjust history unique to Equatorial Guinea, as demonstrated by the profound pollution and fostering of violence engendered by Shell Oil, for example in Nigeria, or Chevron in the Amazon jungle[33] Environmental and international social justice advocates worry that the extraction of minerals and metals required for the renewable transition will entail similar problems.[34] Environmentally, mining has historically involved the rerouting, contamination, and depletion of water reserves, the destruction of habitats, the fragmentation of ecosystems, and a consequent loss of biodiversity. Politically, mining companies have frequently supported undemocratic regimes and engaged in explicit or implicit bribery. Socially, frontline communities have often been displaced and subject to armed force and threats against and even killings of dissenters and activists, as documented worldwide by the *Environmental Justice Atlas*, and have experienced eroded livelihoods, severe health impacts from contaminated air, soil, and water, and economic dependence on the mining operations.[35] Many such communities have also undergone a kind of spiritual loss or **solastalgia**, referring to the pain that occurs from the loss of a comforting place when a people's connection to the land is disrupted in the face of large-scale mining.

While the massive need for new mining for the renewable energy transition could easily end up recapitulating the same extractivist dynamic, such a fate is not inevitable. Numerous actions can be taken to minimize the impact of such mining, as shown in the examples specific to lithium mining in Box 8.2 and through some of the steps discussed in Chapter 9, such as reducing overall consumption, investing more in public transportation, recycling metals, and finding replacement materials.[36] But to avoid the unjust excesses of neoliberal extractivism, it will also be essential for partnerships between mining corporations and nations in the Global South to be governed and closely monitored by ethical oversight to prevent corruption, human rights abuses, and labor abuses. Nonetheless, the difficulties of prescribing ethical practices around mining and extraction are notorious. To cite just one such example, Apple corporation responded to an Amnesty International report on the use of child labor in cobalt mining in the Democratic Republic of Congo by declaring that it "believes every worker in our supply chain has a right to safe, ethical working conditions" and that "underage labor is never tolerated." [37] Yet a new report the following year showed that cobalt mined by children was still being used in the company's devices, leading Apple to suspend some purchases of hand-mined cobalt, although some have charged that the practice has continued via its subsidiaries.

Box 8.2 Case study of extractive mining: lithium mining in Bolivia

Lithium mining in Bolivia offers a case study of the costs and benefits of extractive mining. Lithium exists naturally as a dissolved mineral in the saline brines of the Bolivian Altiplano basins. To get it out, mining companies pump the brine into vast shallow pools at the surface, where evaporation over a period of eight months to three years concentrates the highly valued lithium salts. Global demand for lithium was projected to triple from 2015 to 2025, mostly driven by an eightfold increase in the demand for EVs.[38] A single EV battery requires 64 kilograms of lithium carbonate equivalent. To evaporate one ton of lithium carbonate requires two million liters of water – the equivalent of an Olympic size swimming pool.[39] Beguiled by the promise of economic growth, governments (including the current Bolivian government of the Movement towards Socialism party) have often chosen to partner with extractive companies to profit from their natural resources without a clear regard for the negative consequences of extraction.[40]

In the case of lithium mining in the Salar de Atacama basin in Bolivia, the mining companies SQM and Rockwood Lithium have claimed that the effects of groundwater extraction are localized to the mining areas and do not affect the local Indigenous communities, but other research has come to the opposite conclusion, showing that the groundwater system is connected throughout the basin.[41] The groundwater forms the basis of an ecosystem, a complex web of flamingos, grasses, and aquatic microbes and of Indigenous communities that think of themselves as integral to this system and view the threat to groundwater as a threat to all the interconnected life. Environmental justice advocates charge that current plans in Bolivia to massively upscale lithium for battery production, at least the way as such mining is currently being done, threaten to deplete one of the driest regions in the world of its ancient water reserves, irrevocably destroy the unique desert environment, and perpetuate the colonial cultural violence against the Indigenous peoples of the South American Altiplano.

Yet there are also numerous steps that could be taken to mitigate these negative effects while still providing the lithium necessary for a transition to renewables. Lithium has also been found elsewhere, including Argentina, China, the USA, and Australia (overall amounting to ten billion kg of known land reserves[42]), and lithium mining has not been equally devastating at all these sites. Hard rock lithium mining in Australia, for example, has higher capital outlays but does not contaminate groundwater, and huge deposits of lithium could be extracted using renewable geothermal energy from beneath the artificial Salton sea in California.[43] Technical improvements in lithium extraction could also speed up and improve the process, a considerable amount of lithium could be recovered through recycling,[44] and substitutes for lithium are possible, such as iron-nickel batteries for transportation. Importantly, even as governments and regulators explore all of these alternatives, they must also keep in mind that the current alternative to lithium for batteries for transportation is using fossil fuels for combustion, which has already been proven to poison the land and kill people through air pollution on a massive scale.

It is also important to compare the damage that is likely to emerge from the new mining that will be required to transition to renewable sources of energy against the enormous social and environmental damage from current fossil fuel extraction. In the USA alone, for instance, there are more than a million active drilling sites and more than three million abandoned ones, many of which are leaking and poisoning neighborhoods and schools.[45] Worldwide, air pollution from fossil fuels is estimated to cause about nine million deaths per year, far more than the Covid-19 pandemic of 2020–1.[46]

Evaluating Technical and Market Solutions

Since the international community began to recognize climate change as an existential issue several decades ago, numerous technical and market approaches have been proposed to help solve it. Yet to successfully avoid the most threatening climate scenarios, governments, organizations, and individuals will have to invest their efforts and resources in actions that will be both effective and just. That will require being clear-eyed and properly suspicious about proposed solutions that appear to promote the interests of the rich and powerful over those less so or which offer ways to relocate or put off necessary changes to a more distant future. This section examines a number of such proposals that, despite their seeming promise, have been deemed false solutions by many climate justice advocates and climate specialists, beginning with a number of widely promoted technical fixes and then several proposed market mechanisms.

Technical Fixes

Among the many potential technical fixes to slowing global heating proposed by governments and technological experts, several have raised concerns among climate specialists and activists who have questioned their feasibility, unpredictable consequences, and unjust impacts on vulnerable populations around the world. This section examines both the promises and large questions raised by new large hydropower projects, the use of bioenergy, new nuclear plants, geoengineering approaches, and technical carbon capture methods.

Large Hydropower

Hydropower is typically generated by damming rivers to hold back the potential energy of the water, which is then released to fall through turbines that spin to generate electricity, which at first glance seems a fantastic solution to generating energy from non-fossil sources. And indeed, although many people may be thinking of wind and solar power when they speak of "renewable" sources of energy, as of 2018, 71 percent of such energy was provided by hydropower, and governments, in their quest to increase the amount of energy they can capture from renewable sources, have proposed the building of 3,700 new large dams worldwide.[47] But unlike small hydropower projects, such as run-of-the-river turbines that sit in the river flow, building enormous dams often has large negative impacts on the environment and on the people they displace or who live near them,[48] and these dams can also generate substantial emissions.

The proposed new dam construction projects are expected to displace tens of millions of people, adding to the impacts that have already been felt by as many as 472 million who live downstream from existing dams.[49] The most extensive of the thousands of new projects proposed are located on the Mekong, Amazon, and Congo river basins, which are among the most biodiverse freshwater river systems in the world and home to many Indigenous peoples. In the Mekong alone, it is estimated that 60 million people who rely on those rich fisheries will lose their livelihoods. In Canada, many of the new large hydro projects that have been proposed are expected to expose Indigenous populations to unacceptable levels of methylmercury as an effect of flooding the soil.[50] The construction of new dams in Canada has also led to substantial deforestation for the reservoirs and transmission corridors; for example, the Muskrat Falls project required 700 miles of transmission lines, involving clear-cuts of intact boreal forest at a width of two four-lane interstate highways side by side.[51] This not only disrupted the affected forests' function as carbon sinks but has opened virgin landscapes to yet more mining and forestry that could further reduce that function.

And before the industrialized world decides to gamble away its last wild rivers in its attempt to reduce greenhouse gases, it may want to consider that the reservoirs created by some of the dams are themselves a significant source of greenhouse gases released into the atmosphere, such as methane from the decomposition of organic matter.[52] While there are huge differences between such facilities in their greenhouse gas emissions,[53] cumulatively the impact is quite enormous, estimated at about 2 percent of the global average.[54] In the words of the director of the environmental group International Rivers, "it would be a grave mistake to continue to finance those [large dams] with the impression that they were part of the solution to the climate crisis."[55]

Bioenergy, Biomass, and Biofuels

Another proposed technical solution to getting off fossil fuels is **bioenergy**, or generating energy from **biomass**, which refers to any material derived from living or recently living material. Converting biomass into energy includes burning wood to generate electricity or deriving liquid **biofuel** from corn or rapeseed crops to replace standard diesel fuel. Although some forms of bioenergy could be helpful for our response to the climate crisis, analysts have pointed to several problems with dedicating large amounts of land specifically for this purpose.

Writing in the Guardian, two authors pointed out that one problem is that diverting arable or currently uncultivated land to growing biomass for bioenergy would mean sacrificing land for much-needed food, timber, and natural carbon sinks.[56] Another problem they mentioned is that bioenergy emits CO_2, just as burning fossil fuels does, and thus does not generally cut greenhouse gases. But most calculations of the potential effects of bioenergy do not count the CO_2 released from burning on the assumption that it is offset by the CO_2 absorbed by the growing plants that are harvested for the biomass. But if those plants were going to grow anyway, turning them into bioenergy does not remove any additional carbon from the atmosphere as a climate solution, and if they were grown specifically to become biomass, they would displace natural carbon sinks or farming. (In response to this problem, some bioenergy advocates have proposed the use of BECCS, which is deeply problematic in its own right, as discussed later in this chapter.) And finally, a book-length study pointed to the danger that using arable land to grow crops for biofuels production may also simply fuel the growth imperative and profits of the global agribusiness industry while many in the Global South remain hungry.[57] In this way, bioenergy can be seen less as a market response to climate change than as an example of how concern about the climate can be used to legitimize a market-led distortion of agriculture.

At the same time, using some forms of bioenergy that do not increase competition with food or land instead of fossil fuels could help reduce greenhouse gases to some extent, such as biomass grown as winter cover crops to improve soil fertility or converted from timber processing or urban wood wastes, modest amounts of agriculture residues, and landfill methane (also called biomethane) that arises from the decomposition of food or farming waste. Capturing and burning biomethane to generate electricity has been posed as an attractive option because it would convert methane into the less potent greenhouse gas CO_2, potentially leading to less overall release of greenhouse gases and replacing some fossil fuel–generated electricity.

Skeptics point out, however, that this process could provide only 1 percent of current fossil gas needs in the USA and escalate the growth of industrial agriculture and feedlots (itself a major source of potent greenhouse gases); it is also less likely to be used to generate electricity on-site than to be injected into methane gas lines, thus bolstering the use of standard fossil fuel infrastructure and prolonging the economic and political power of the industry.[58]

This analysis suggests, therefore, that while small-scale bioenergy has some merit, large-scale bioenergy is not only an unlikely practical solution to the problem of global heating but an unjust one. It threatens to compete with the use of land to grow much-needed food (especially in the Global South), fudges the calculations on the amount of greenhouse gases it is actually likely to save, and helps legitimize a market-led distortion of agriculture that increases profits for industrial agriculture without helping to feed the world's growing population.

New Nuclear

Many stalwart advocates for climate action and leaving fossil fuels in the ground, such as James Hansen and the environmentalist George Monbiot,[59] strongly support the expansion of nuclear power, especially next-generation reactors. These proponents view nuclear power as an essential component in a shift toward non-fossil fuel energy, because once nuclear plants are up and running, they can produce huge amounts of electricity with seemingly no emissions. Despite safety concerns prompted by the atomic bombs dropped on Hiroshima and Nagasaki and the nuclear energy accidents at Three Mile Island, Chernobyl,[60] and Fukushima, US submarines and aircraft carriers have apparently operated for decades powered by small onboard nuclear plants without a single reported accident, and countries such as France currently derive over 70 percent of their electricity needs from nuclear power.[61]

Despite the appeal of nuclear power, the notion that it produces "clean" energy overlooks the reality that at every stage except the actual production of electricity, nuclear power produces significant amounts of greenhouse gases and other forms of pollution. In addition to the emissions involved in mining and transporting uranium, for instance, building a nuclear facility requires vast amounts of fuel-burning equipment and concrete, which is itself responsible for considerable greenhouse gases. Even once a plant is in operation, the issue of storage is deeply problematic; already 300,000 tons of spent nuclear fuel has been accumulated worldwide, which will remain dangerous for thousands of years, and finding sites willing to accept such waste is difficult. According to a 2018 US Government Accountability Office report, "[a]fter spending decades and billions of dollars to research potential sites for a permanent disposal site, the future prospects for disposal remain unclear."[62] Transporting such waste is also a risky and fuel-intensive process as is decommissioning the reactors. For instance, the transportation of the reactor pressure vessel from Unit 1 of the San Onofre nuclear plant in southern California to Utah for storage in 2020 involved a rail carriage with thirty-six axles, the largest in the world, and a trailer with 384 tires moved by six large trucks for 400 miles with a maximum speed of ten miles an hour over a ten-day period. The potential pollution danger and huge CO_2 emissions from all the other stages of the process of producing nuclear power certainly must be considered in any honest accounting of how clean it actually is.

Furthermore, the cost and construction time involved in adding new nuclear plants, especially in comparison to those of other renewable power sources, also makes them look less attractive as options for meeting the urgent IPCC 2018 goal of reducing emissions by 45 percent from 2010 levels by 2030. For example, the decommissioning of the San Onofre nuclear plant referred to above is estimated to take about ten years and cost over $4 billion,[63] or enough money to cover the installation of solar power for 400,000 homes in southern California, something that could be done much more quickly and cleanly and would generate energy for decades.[64]

As of 2020, twenty-seven of the planned forty-six new nuclear units under construction around the globe were already behind schedule, some delayed by a decade or more. In 2018, the total operating capacity of nuclear plants already operating was 370 gigawatts, while 165 gigawatts of power from renewable sources were added to the world's power grids just that year alone, and the share of global electricity produced by nuclear power declined to just 10.15 percent from a peak of 17.5 percent in 1996.[65] As one energy analyst has pointed out, "before accounting for meltdown damage and waste storage, a new nuclear power plant costs 2.3 to 7.4 times that of a same-capacity onshore wind farm (or utility photovoltaic farm), takes 5 to 17 years longer between planning and operation, and produces 9 to 37 times the CO_2 emissions per unit of electricity generated."[66] For those reasons, non-nuclear options save more carbon per dollar, emit less pollution during their life spans, and pose fewer safety and environmental risks, including the risk of nuclear weapons proliferation.

Nonetheless, engineers in the USA have continued to develop designs for reactors they say would be safer and more efficient, and in May 2020 the Department of Energy proposed prototypes that might be ready within seven years.[67] Yet even proponents of nuclear energy doubt that these new designs are likely to spur the construction of new commercial reactors while other sources remain so cheap. According to Robert Posner, a physicist at the

University of Chicago, "[n]ew builds can't compete with renewables. Certainly not now."[68] Meanwhile, China is currently developing new nuclear plants, but it is apparently doing so largely to cut deadly air pollution in its cities rather than to provide cheaper energy or to reduce greenhouse gases.[69]

Although the relatively small health consequences of the Fukushima disaster of 2011 have led some environmentalists, such as George Monbiot, to be even more supportive of nuclear power,[70] the costs of that disaster have been estimated at close to $1 trillion and the recovery period to take as long as forty years.[71] In April 2021, when the Japanese government proposed to dispose of one million tons of contaminated water into the sea,[72] there was an international outcry, as this would disperse radionuclides throughout the food chain, as far away as California.[73] In the meantime, Fukushima is generating 140 tons of radioactive water per day, and nothing has been done to handle the radioactive mess in the two reactors.[74] In 2011, the German government, responding to the Fukushima disaster, which magnified fears in some of its population who had been exposed to radiation fallout from Chernobyl, committed to phasing out existing nuclear power plants. Ironically, this forced Germany to increase its reliance on lignite coal (the dirtiest kind),[75] since it had stalled its initially fast rollout of renewable energy.[76]

All told, there is probably a good argument for keeping existing nuclear plants running for as long as possible, but the evidence suggests that building new ones is not economical (relative to building wind and solar), takes way too long in the context of the need to cut emissions by at least 50 percent in the next ten years, and incurs huge emissions in construction, supply, and storage. Furthermore, nuclear technology poses myriad safety issues that future generations will be forced to deal with, raising issues of intergenerational justice. Shifting investment from a capital-intensive power-generating technology to cleaner and cheaper renewable technologies could also help bridge current inequities between rich and poor nations.

Geoengineering

A newer and more experimental set of technofixes that some climate advocates have proposed is geoengineering. This refers to two major types of supposed solutions: decreasing solar radiation by blocking the Sun's rays, and CO_2 removal, which seeks to capture and/or remove carbon from the atmosphere. Those who have enthusiastically pushed geoengineering as one solution to the emissions problem include several members of today's billionaire class, such as Bill Gates and Richard Branson; environmental organizations such as the Natural Resources Defense Council; think tanks such as the Breakthrough

Institute; and fossil fuel corporations like Exxon Mobil. Geoengineering in the form of decreasing solar radiation is currently being explored by several governments including the USA,[77] and as discussed earlier, the IPCC has incorporated negative emissions strategies based on geoengineering into nearly all of its climate models.

The solar radiation management approach was largely inspired by the 1991 eruption of Mount Pinatubo in the Philippines, which injected millions of tons of sulfur dioxide and particulate matter into the atmosphere that led to a global temperature drop of 0.5°C from 1991 to 1993. Accordingly, one proposed geoengineering approach is stratospheric aerosol injection, which uses hoses, balloons, or planes to inject sulfur particles into the atmosphere to block the Sun's rays. Another is marine cloud brightening, in which a special fleet of 1,500 unmanned, satellite-controlled ships would roam the ocean spraying tiny drops of seawater into the air that would evaporate and leave salty residues that would reflect the incoming solar radiation.

Yet as former IPCC lead author Raymond Pierrehumbert has pointed out, these approaches are still largely theoretical and pose enormous potential hazards.[78] Stratospheric aerosol injection, for instance, could have massive, unpredictable effects on the global water cycle, such as shutting down the Indian monsoon system and disrupting agriculture for two billion people, and it risks affecting photosynthesis across much of the world and depleting the ozone layer. The approach would also have to be repeated over and over, and stopping the process could lead all the built-up carbon in the atmosphere to produce a rapid temperature increase by as much as 2–3°C in a decade. He and others have expressed similar concerns about the process of marine cloud brightening.

The second major bioengineering approach is CO_2 removal, comprising several methods. One of these is fertilizing the ocean with iron to boost phytoplankton and the uptake of CO_2. Yet this provokes widespread concern about the impact on the ecological cycles of creatures, ranging from phytoplankton to whales and questions about practicability, given the colossal areas that would need to be fertilized in order to have a measurable effect.[79]

As mentioned earlier in the book, the form of CO_2 removal that has attracted the most attention is BECCS (Figure 8.3).[80] The aim of BECCS is to pull CO_2 out of the air by first cultivating crops to absorb CO_2, then burning the resulting biomass to generate electricity, and finally capturing the released CO_2 and burying it underground, in the process presumably removing more carbon than the process emits, amounting to negative emissions. Many countries see BECCS as an attractive option because subtracting these negative

Figure 8.3

The two main technical proposals for pulling CO_2 out of the air are both colossally expensive and have myriad other problems

The minimal estimated costs for BECCS and direct air capture are currently $89 trillion to $535 trillion. They are also associated with other major problems.

BIOENERGY WITH CARBON CAPTURE AND STORAGE (BECCS)

Fast-growing plants are harvested and then burned to generate electricity. The CO_2 from burning is captured and buried.

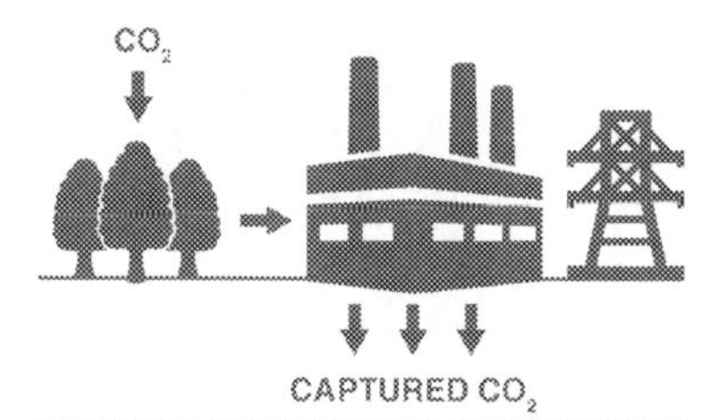

Enormous cost, land, and water requirements

Carbon capture unproven

Moral hazard: postpones cutting fossil fuels now

Competes with food production

Fosters huge agribusiness at the expense of local farmers

DIRECT AIR CAPTURE

Vast arrays of filters are built and powered. CO_2 in the air "sticks" to chemicals in the filters, and once captured it is buried.

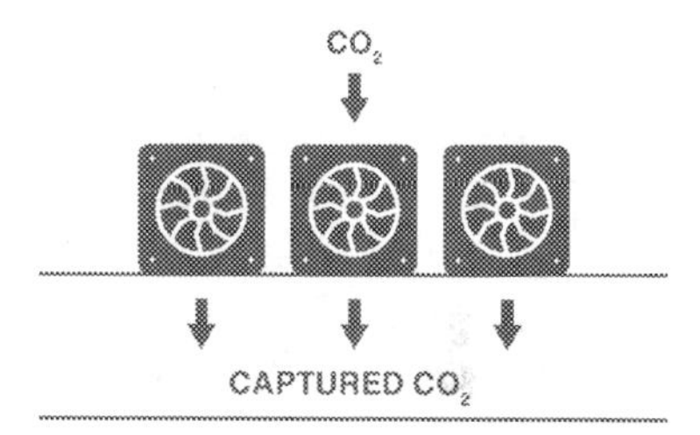

Enormous cost and material extraction requirements

Carbon capture unproven

Moral hazard: postpones cutting fossil fuels now

Cost estimates from Hansen, J. M. et al. (2017). Earth Syst. Dynam., 8, 577-616, doi:10.5194/esd-8-577-2017.

emissions from their total emissions offers them a way to promise to reach net zero levels by such and such a date without committing to eliminating CO_2 emissions now. Many climate justice advocates accordingly dismiss net zero as a greenwashing strategy (see Box 8.3 for definitions of greenwashing, and for net zero and carbon neutrality) because they appear to allow governments and institutions to look good while they keep extracting, subsidizing, and supporting fossil fuels now.

BECCS also poses enormous practical problems, including its colossal land and water needs. Using BECCS to meet even the IPCC's medium-emissions pathway of RCP4.5 would require removing 630 gigatons of CO_2 from the atmosphere, or around two-thirds of the total emitted between the Industrial

Box 8.3 Greenwashing: natural gas, carbon neutrality, and net zero

An institution or government is said to engage in greenwashing when it provides misleading information to make others think its practices are more environmentally sound than they really are.

In one example, in 2002 the huge energy extraction company British Petroleum infamously branded itself "Beyond Petroleum," unveiling a new logo that included an image of the Sun, reminiscent of solar energy.[81] But when a 2010 explosion at its Deepwater Horizon oil rig unleashed one of the greatest marine disasters in human history – a massive oil spill of 4.9 million barrels into the Gulf of Mexico – BP quietly dropped the new branding.

Another example is the buses in many parts of the USA that proudly carry logos claiming they are "Powered by Natural Gas – The Cleaner, Greener Way to Go." Yet there is nothing "clean" about natural gas, itself a greenwashing name for methane which, in the USA, is usually acquired by fracking, a process so toxic to local groundwater and air that it has been implicated in cancers and birth defects.[82] Although many utilities and politicians have described natural gas as a "bridge fuel" that can help us get us from coal to renewables, methane has a much greater global heating potential than CO_2, and leakages at wellheads and pipelines make it hardly cleaner than coal.[83]

Carbon neutrality is another term frequently used as a form of greenwashing by companies and institutions that proclaim their intention to achieve this state by a given date. But what it typically means is that the organization maintains its business-as-usual emissions while attempting to "neutralize" those with carbon offsets (the current plan, for example, of the ten-campus University of California).[84]

Another term often used in a greenwashing fashion is **net zero**, or the notion that the emissions one generates can be balanced by negative emissions, such as pulling CO_2 out of the atmosphere using geoengineering such as BECCS. Yet, as this chapter shows, this relies in practice on a mostly unproven and colossally expensive approach that means little or no actual cuts in emissions now on the shaky presumption that the problem will be solved later.[85]

Revolution and 2011. To accomplish this would require between 0.4 and 1.2 billion hectares of land, or an astonishing 25–80 percent of all land currently under cultivation.[86] Advocates have failed to explain how that would be possible while trying to feed 8–10 billion people by the middle of the century, preserve native vegetation (including carbon-fixing forests), and preserve biodiversity, all while BECCS requires the mass planting of alien species that can be quickly grown and cut down.

Some of these problems are already clear in the one major BECCS demonstration project in Europe. That BECCS facility, which was converted from a former coal plant in the UK, is now being fed biomass in the form of wood transported from virgin forests in Estonia.[87] Meanwhile, the carbon capture aspect of the technology has yet to advance beyond preliminary demonstration projects. In 2021, energy giant Chevron admitted that the world's largest demonstration project of carbon capture and storage, situated in Western Australia, and costing three billion dollars, had failed to meet its targets;[88] it is also the case that 80 percent of other attempts have failed.[89] A second major problem with BECCS is its exorbitant cost: several hundred trillion dollars this century.[90]

A different approach, as also noted in Chapter 2, is known as direct air capture, which involves pulling CO_2 out of the air with machines and then sequestering the gas underground. The largest direct air capture demonstration project to date was recently built in Iceland at a purported cost of $15 million. Yet even if this plant runs all day, every day, for a full year, it is expected to sequester only as much CO_2 as three seconds of global emissions.[91] Like BECCS, direct air capture also relies on the unproven carbon capture method, and it would also likely cost several hundred trillion dollars this century.[92]

Overall, although geoengineering has its boosters and technical innovations are always possible, the various approaches are riddled with problems. Chief among these are substantive injustices: from unanticipated effects on weather cycles for vulnerable farmers from atmospheric aerosol injection, to damage to ocean food chains for ocean fertilization, to competition with food production and the control of vast areas of land in the Global South for BECCS crops. Relatedly, the promise of technofixes appeals strongly to some members of the wealthy elite and decision-making class, since it seems to promise a climate "solution" without entailing any change to their lifestyles or the wider socioeconomic-political structures, which the writer David Wallace-Wells has called **Worshiping at the Church of Technology**. The reliance on technofixes also represents a profound injustice because it postpones confronting the need to stop using fossil fuels now, even while air pollution (typically experienced in poor parts of the world), toxic local environmental damage from extraction, and greenhouse gases keep building up.[93]

Market Solutions

Another set of proposed responses to the problem of global heating depend on market solutions, which, unlike classic command-and-control government regulations, are premised on allowing market forces to find the most efficient, low-cost ways to reduce greenhouse gas emissions. Underlying all of these proposed market solutions is the assumption that increasing the price of burning fossil fuels to cover its social costs in the form of damages due to global heating will lead consumers (governments, industries, and individuals) to find less expensive ways to meet their energy needs and in so doing to also reduce greenhouse gas emissions. Despite the appeal of such logic for democratic capitalist governments and institutions, some critics and climate justice advocates view such approaches as yet another set of false solutions. As discussed in this section, these market strategies include cap-and-trade, carbon offsets, and carbon taxes.

Cap-and-Trade

Cap-and-trade refers to a system where governments use regulations to limit or 'cap' carbon emissions from the economy and issue permits to emit carbon that match the cap. For example, if the cap is 100,000 tons of carbon, there are 100,000 one ton permits. Now every polluting entity, such as a power plant or refinery, is required to hold a number of permits equal to the emissions they produce. The polluting entities are supposed to buy the permits by bidding for them in an auction (in reality, as is shown below, they are often assigned the permits for free). Once the polluting industries have the permits, they are able to buy and sell them among themselves and other market agents. The overarching idea is that the number of permits is limited and therefore valuable, and so those polluting industries subject to the cap will have to cut their emissions to reduce the number of permits they need to purchase. An additional aspect is that, over time, the regulations require the cap to decrease, which is supposed to reduce emissions.

But practical lessons from how this scheme has been implemented point to myriad problems. As discussed in Chapter 1, one example is the European Union's Emissions Trading System, the largest carbon-trading market in the world. In the first phase, way too many permits to pollute were issued, which resulted in huge profits for polluting companies and a failure to actually cap emissions; in Germany electricity companies ended up being allocated 3 percent more permits than they needed – a windfall worth about $374 billion.[94] Further, the Emissions Trading System was subject to fraud and scams, which reinforced a carbon economy by allowing polluting industries to avoid cutting emissions.

Another large example is the cap-and-trade system of California. While it was encouraging to see that overall emissions in the state did drop by about 5 percent in 2019 from when the program started in 2013,[95] it seemed very unlikely that this was due to cap-and-trade itself; indeed, carbon emissions from California's oil and gas industry *rose* 3.5 percent in that period.[96] Problems with implementation also plagued this program; for example it set initial caps too high so that, by one analysis, by 2018 companies had banked about 200 million permits, enough for as many tons of CO_2 as the expected total reductions from cap-and-trade by 2030.[97] California also gave away many free permits to companies it was worried would face competition from out of state, and perversely, many oil and gas companies got free permits until as late as 2020.[98] This reflected continuing attempts by the oil and gas industry to weaken the cap-and-trade program: from as early as 2006 it spent $88 million in lobbying.[99] Another problem is that the program allows many companies to cancel out some of their CO_2 by buying carbon offsets, which are often deeply problematic, as discussed in the next section.

While advocates for cap-and-trade respond to these problems by arguing that cap-and-trade *could* yield better outcomes if it were done correctly,[100] there is little reason to think that it *will* be done correctly, and globally, with strong enforcement mechanisms in the short time that remains to actually cut emissions.

The cap-and-trade approach is also undercut by the fossil fuel subsidies that are continually provided by governments, which also serve to maintain the political power of those industries. The subsidies are massive, ranging from 500 billion per year for the combined G20 countries[101] (if a subsidy is calculated as the difference in consumer prices paid by fuel users that are below the actual costs of fuel supply) to nearly six trillion for all countries per year, as estimated by the International Monetary Fund[102] (if the subsidy adds the environmental costs of global heating, deaths from air pollution and taxes applied to consumer goods in general).[103] However, there is also a deeper, more fundamental problem with cap-and-trade – it does not directly challenge society's reliance on fossil fuels.

Carbon Offsets

Carbon offsets are voluntary payments that governments, businesses, or institutions choose to make to support CO_2-reduction schemes elsewhere, such as providing cookstoves and wood pellets for people in East Africa on the premise that doing so will allow those Africans to cut down and burn fewer trees and produce fewer emissions, or paying coal mine owners to capture the methane from mines instead of letting it leak.[104] By making these payments, the buyer tries to "make good," that is, offset, their current emissions.

There are numerous problems with this approach, above all uncertainty about whether such offsets actually confer benefits. Notably, offsets schemes can only truly neutralize the impact of consumption if they meet the **additionality requirement**, meaning that the resulting emissions reductions would not have occurred without the payments, which is notoriously hard to establish. For instance, if a local African government or a philanthropic organization might have paid for those cookstoves anyway, an offset payment from a university or business in the Global North would not be additional. According to a 2016 analysis of one major offsets scheme, the CDM under the Kyoto Protocol, only 2 percent of the sponsored projects spanning over a decade actually met the additionality criteria.[105]

Another complicating factor is that some offset schemes may actually harm the environment by creating perverse incentives. For example, methane capture schemes funded from offsets have been found to increase the profitability of coal mines and thus potentially prolong their life span.[106]

Furthermore, most offset payments are so small as to raise the question of whether they could possibly be effective. For example, some companies sell CO_2 offsets to airline passengers for around $15 per ton for a long-haul flight that emits about two tons of CO_2 per person, yet the approximate cost of a truly additional direct air capture scheme is about $600 to $800 per ton.[107] And indeed, the consequence of paying the substantial fee for this near certain and additional benefit would be to actually reduce flying itself since it would mostly be too expensive. Meanwhile, it would be much better to simply fly less and substantially reduce one's emissions, than to fly more and pretend one can "make good" with small offsets payments.[108]

A good example of the multiple problems undermining the supposed intentions of offset programs is the Reducing Emissions from Deforestation and Forest Degradation program, known as REDD, which was launched with $5.4 billion in funding by COP 13 in 2007.[109] REDD is an international payment for ecosystem services scheme that proposes to protect forests and the rights of local communities. Although those are both potentially good things, the program is also premised on the dubious morality of paying some of the world's poorest countries to absorb carbon pollution from some of the richest, effectively allowing the latter to continue their enormous emissions and thereby risk the health and welfare of the entire globe. Like other offsets,

REDD is also beset by significant technical issues, not only additionality but **permanence**, or whether the forests will continue to stay protected after rich countries have paid for it to be so, and carbon leakage, or whether putting one particular forest out of bounds will simply prompt loggers and hunters to move to another forest.[110] In 2014, FIFA, the world governing body of soccer, bought 250,000 credits intended to reduce deforestation in an area of Brazil to satisfy a sustainability pledge made before holding the World Cup in that country. Yet the large amount of that money that went to international groups for managing the project made some tribal members so upset that they colluded with loggers to sabotage it, resulting in more than 300 trucks per day leaving their territory filled with wood.[111]

Another issue with offsets is that they often scream of a glaring injustice that is neocolonial because they are often implemented in the developing world, where they tend to be much cheaper, subject to less scrutiny, and can even end up controlling people's lands and forests. Many climate justice advocates find it perverse that people in low-income nations have their forests "sequestered" in order to allow the rich world to continue to pollute, which some have described as analogous to the Catholic Church's practice in the Middle Ages of selling indulgences to the rich to absolve them of their sins.

Even if offsets payments worked as desired to "neutralize" local emissions (never mind actually cutting them in half by 2030), there is another problem which is they let us off the hook with regard to contesting the fossil fuel industry. Political scientist Matto Mildenberger makes the point that one ton of greenhouse gases reduced "over there" does not really balance one ton of greenhouse gases emitted here within the overall picture. This is because generating emissions here arises from making payments to domestic fossil fuel companies and in so doing increasing their profitability and power to contest local climate action policy.[112]

To be sure, good projects such as preventing logging, supporting reforestation, and providing cookstoves in Africa all have their benefits and definitely deserve funding, but linking that funding to offsets has the dubious effect of serving as a dodge that allows polluting countries, industries, and institutions to avoid taking full responsibility for their emissions and undertaking cuts. The fundamental problem is that those who purchase offsets are not changing their own planet-heating behavior. Both simple fairness and the need for massive and immediate reductions in greenhouse gases demand that high-emissions countries and communities cut their own emissions, rather than paying someone else to find ways to do that for them. To be sure, wealthy entities can and should give financial support to create better health and living conditions in the

Global South – and which climate reparations demands – but this should be done as a *good thing in itself along with the wealthy entities cutting their own emissions too*, not as an offset.

Carbon Pricing or Taxing

A final kind of technical fix that employs market mechanisms is a carbon price, which is a strategy in which governments put a tax on the social or external costs of CO_2 emissions that would be paid widely across society, whether levied directly on the consumption of fossil fuels (such as a fuel tax) or on the greenhouse gas emissions produced by particular goods or services (which would be paid by producers and passed on to consumers). On its face, this seems to be a fairer system than some of the others proposed, as the cost is paid by those directly benefiting from the emissions. The World Bank and many economists consider a carbon tax to be the most flexible, cheapest, and potentially effective market mechanism for reducing greenhouse gas emissions, and a number of countries, states or provinces, and cities have already adopted a carbon price, and as of 2021, five carbon-pricing proposals were under consideration by the US Congress. Nonetheless, critics have raised several problems with this approach, leading many climate justice advocates to consider it another false solution, and one that could even create more harm than good.

One obvious problem is the difficulty of determining the actual social price of the harm caused by emissions and global heating itself. As we have seen, economists' attempts to calculate the social cost of carbon are merely guesses that are often seemingly inadequate. As Chapter 3 showed, this results from problems inherent to the neoclassical tradition, but it is also difficult because we cannot predict the future. Meanwhile, the price that is actually charged is inevitably driven by what is politically acceptable. As a result, most carbon-pricing schemes have set this price or tax lower than the actual estimated social cost of carbon, typically at less than $10 per ton. Yet, according to IPCC estimates, carbon prices or taxes would need to range anywhere from $135 to $5,500 per ton by 2030 to keep heating below 1.5°C.[113] And while some studies have found that a carbon price *can* lead to a reduction of emissions, such as a 6 percent reduction of transportation emissions in Sweden when a tax eventually rose to more than $100 per ton, other studies have shown little effect.[114] But beyond the specific results of particular programs, and despite carbon pricing being introduced in dozens of countries, emissions have mostly continued to grow in those countries, undoubtedly because the very thing that tends to make a carbon price politically palatable – its modest price – also appears to makes it ineffective at actually reducing emissions.[115] Thus, to function as intended, a carbon price would need to be higher than most governments are likely to have the political will and stomach to impose and many citizens are likely to accept.

As this fact reveals, another major problem with a carbon price or tax imposed directly on consumers is that it is regressive, meaning that its effects are likely to be felt more intensely by those who can afford it the least. Further, in some countries the imposition of a carbon tax has led to polarization around the climate issue by stoking class divisions and reinforcing the myth that climate policy necessarily penalizes the poor and the working class.[116] In France, for instance, a small increase in the tax on gasoline sparked what became known as the Yellow Vest protests, in which thousands of workers who worried the tax would affect their livelihoods were enraged by the way the elites seemed to shrug off their concerns.[117] In Australia, conservatives managed to use the tax as a wedge issue to replace the progressive Julia Gillard as prime minister by the right-wing climate-denying Tony Abbot, perhaps delaying wider climate policy for ten years.[118] Indeed, any carbon price that is likely to have any real effect on decreasing emissions is going to force most residents to face higher fuel and energy costs and threaten the job security of workers in emissions-intensive industries, which can also undermine broad support for climate action.[119] One way to avoid such unintended negative impacts of a carbon price is to require a measure that would rebate some portion of that price to affected individuals, such as a credit for low-income citizens or the plan known as carbon fee and dividend.[120] This would rebate a substantial part of any carbon price as equal per-person dividends, thereby making the overall impact of the carbon price progressive, meaning it would help decrease inequality. Figure 8.4 shows how this would work for a $200 per ton price when it is disbursed as dividends to US households, resulting in the lowest quintile receiving a rebate equal to 20 percent of their income.[121] While this seems like a very defensible approach, the worry is that there is nothing to stop these lower-income households from spending this money on more carbon-intensive products and services – the problem of rebound effects – unless this is also done in conjunction with a range of other policies to transition our energy system away from fossil fuels, and to also reduce consumption.

Furthermore, the likely impact of governments' efforts to tax carbon is currently undercut by their subsidizing of fossil fuels, which, as noted earlier, was estimated in 2017 by the International Monetary Fund to be as high as $5.2 trillion worldwide, or about 6.5 percent of global GDP.[122] Although such subsidies work directly against the intent of carbon pricing by artificially reducing rather than raising the price of fossil fuels, there is currently little

Figure 8.4

Not all carbon taxes are unjust: a carbon fee plus dividend program can return more money to poorer people

The figure shows the predicted impact of a $200 per ton carbon price when it is disbursed as dividends to US households. The households are divided into quintiles (fifths) based on their expenditures. The simulation uses 2017 numbers.

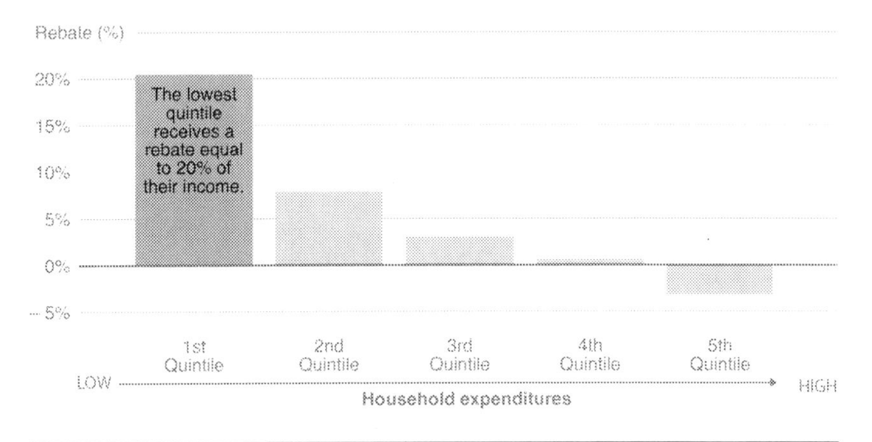

Adapted from from Boyce JK (2018). Ecological Economics. 150: 52-61. https://doi.org/10.1016/j.ecolecon.2018.03.030

political appetite for revisiting these huge fossil fuel subsidies in the USA, given the untold millions of dollars that have flowed to politicians from fossil fuel interests.[123] Unless some way is found to end those subsidies or the political impact of the fossil fuel industry and its supporters, the effectiveness of carbon pricing as a solution to our current emission problems remains questionable.

Carbon pricing appears a tantalizing option for reducing greenhouse gas emissions in large part because it is one of the few strategies that conservatives and free market ideologues can get behind.[124] As the analysis in this section makes clear, however, if it is to meet its objectives for tackling climate change, a properly implemented carbon price must be adopted as part of a suite of policy measures to make it more equitable and just (such as carbon fee and dividend) and which challenge the political influence of fossil fuel interests.

Assessing the overall picture on market mechanisms, the climate justice advocate Brian Tokar remarked: “The idea that we need to use the mechanisms of the market as an alternative to public policy is an approach that’s failed everywhere it’s been tried,” which is elaborated in Box 8.4.

Box 8.4 Interview with Brian Tokar

Brian Tokar is an activist and author, lecturer in Environmental Studies at the University of Vermont, and a board member of 350Vermont and the Institute for Social Ecology.

AA: Can you distill the concept of climate justice down to some essence?

BT: Sure, on the most basic level climate justice focuses on the disproportionate impact of the climate crisis on the most vulnerable people in the world who are also the people who contribute the least to the problem of excess emissions of CO_2 and all the other greenhouse gases. So it's bringing a social justice lens to our understanding of the climate crisis. It brings together a number of distinct currents. Probably the predominant one is the movements of Indigenous and other land-based peoples who most embody those disproportionate impacts and whose responses to climate disruptions has been incredibly important and inspiring around the world. Here in North America, many of the most articulate voices come out of the long-term environmental justice movement, which is of course mainly a movement of people from communities of color that have been disproportionately impacted by environmental threats, starting in the late 1970s and early 1980s with issues of toxic chemical contamination and exposure to toxic waste facilities. But that movement has continued to broaden over the years and has given us an understanding of the central role that race relations play in people's disproportionate exposure to all environmental threats, including climate-related ones.

AA: Is the climate justice framing a strategically useful way to build a movement of movements?

BT: Yes. I would focus on two reasons. One is that I think it brings the climate crisis home much more. We're able to focus on the ways in which it directly impacts people's lives. When I first got involved in climate politics in the mid-2000s the effects that most people were talking about were very remote and ones that not everybody could really identify with. The polar bear was the symbol of the climate movement fifteen years ago and people pretty much like polar bears, and for good reasons, but in terms of a sense of urgency, in terms of people's priorities, the direct impact on people's lives, the polar bear images are pretty remote. And focusing on the social justice dimension really brings it home. The second reason I think it's an asset to the movement is that it focuses people's attention on underlying systemic issues. The ways in which the entire social and economic system that we live under perpetuates all the worst manifestations and accelerates the disruptions to the climate system. I think movements that focus on systemic causes are inherently more effective and have the potential to bring the kind of fundamental change that is absolutely necessary if we're going to not only reduce emissions, but craft a way of life that is conducive to living well under conditions of less pollution and less consumption and certainly much lower fossil fuel use.

AA: What's the problem with market mechanisms and why is this a climate justice issue?

BT: For proponents of this approach, the market is the only way to modify human behavior. The reality is that the capitalist market has always been very highly manipulated by the most powerful players, namely the largest corporations and financial institutions. They hold all the cards when it comes to managing the effects of changes that may be attempted within the constraints of the financial system. And we have to think outside the box. The idea that we need to use the mechanisms of the market as an alternative to public policy is an approach that's failed everywhere it's been tried. And that's true in climate policy as it is more broadly. There have been attempts since the Kyoto Protocol was crafted in 1997. There were various attempts to use the market to mediate climate policy and it's been an absolute failure. It has created loopholes and has made it possible for the most powerful players to continue business as usual, while putting up a smokescreen of responsibility. That's not to say that, under some carefully designed and controlled conditions, putting a price on carbon might not help. There are maybe some practices in the right direction. There's a lot of debate about that within the climate justice movement. Some say, maybe it can work in a limited way, sometimes. Others want to reject it. They even reject carbon pricing outright. But certainly the buying and selling of pollution permits is problematic. It actually predates carbon trading, it goes back to policies established by the EPA as far back as the late 1970s. And it really became enshrined in federal policy with the Clean Air Act amendments under the George Bush Senior administration in the early 1990s, where they allowed utilities to buy and sell permits to emit sulfur dioxide as a way of controlling acid rain. And when you compare the US acid rain program with measures against acid rain, for example in Europe, where they didn't use emissions trading, where it was done strictly through regulation and other forms of public policy, they got rid of their sulfur emissions much, much faster. And there were fewer side effects and distortions resulting from the inherent manipulations of the trading system as well.

AA: What would you say is the proper role of technology in the climate crisis?

BT: Over the forty-odd years that I've been following energy policy there have been tremendous changes in technology. Back in the late 1970s solar energy was primarily the purview of hobbyists and inventors. With the first public push for renewable energy, under the Carter administration, there were small solar companies. Then large concerns like General Electric just ran them into the ground because they weren't profitable. It has all changed dramatically over really just the last ten years. Now the marginal cost of new electric generation from solar and wind is cheaper than any other source and we're approaching the point where it's cheaper to build new solar and wind facilities than even to keep operating existing fossil fuel–generating capacity. Battery technology is advancing at a phenomenal pace, the electric vehicles and all the rest, and all those technologies are part of the solution. But they're not the whole solution. I think we're seeing the development of renewable energy being skewed in a way that favors gigantism, that favors the interests of large investors and large technology companies. And that's a problem. Yes, we're seeing more solar and wind energy being developed, but it is being merely added to existing fossil fuel energy production. That's no good. So in order to address those problems that again are built into the economic system, we need a much wider critique. We need to discuss how technologies are developed and controlled, and who owns them, and in whose interests are they developed? We need to democratize the whole process of technological decision-making, which is currently in the hands of huge corporations for the most part. They will only continue to develop renewable energy to the extent that it's profitable and that's the wrong criterion for a decision that has such profound implications for really the whole future of life on Earth.

Conclusion

As the first chapter in this final part of this book, which addresses how to engage people to take action on the climate crisis, this material has tried to narrow the range of possible actions by using the climate justice lens to ask which ones are worth taking, both to gain wide support and to effectively and quickly reduce emissions rather than encouraging business as usual. The principles provided by the grassroots social movements remind us to consider the interests of vulnerable and Indigenous people when evaluating proposals that too often represent the interests of corporations and the wealthy at the expense of the broader community and globe. They also affirm that those countries who have historically benefited the most from the emissions that are now threatening the entire world must take greater responsibility for providing solutions and financial and practical support for dealing with the damages that global heating has already produced and threatens to make worse in the future.

The existential importance of this issue also means that we must be both clear-eyed and fair about the solutions we embrace going forward, including rejecting proposed technical and market fixes that threaten to perpetuate the same inequities, corporate agendas, and extractivist mentality that got us into this climate and ecological crisis in the first place. As the analysis of these supposed solutions helps makes clear, the only sure way to prevent more global heating is to leave remaining fossil fuels in the ground and invest in a fast and massive build-up of renewable energy sources. Chapter 9 examines what steps must be taken to make such a renewable energy transition feasible from a technical, social, and justice point of view.

9
A Technical and Social Framework to Guide Climate Action

> We could ignore ecological necessity and drive on past the exit ramp. No one can predict exactly how that would end, but I don't think we want to run that experiment and find out. If on the other hand we do take the exit . . . the necessity will be . . . to free ourselves from fossil fuels as soon as we can, to establish ecological stability and to ensure fair shares for all. We don't know exactly how that will turn out either. But maybe if we do it, the Earth that our generations leave behind, though gravely wounded, will remain livable.
>
> Stan Cox, *The Green New Deal and beyond*

Taking purposeful and collective action to enact local, national, and international climate policy will require a clear vision of a technical framework that can actually cut emissions – both immediately and in the long term – and a socio-political framework for doing so in an effective and just way. To that end, this chapter first examines the feasibility of achieving a more rapid electrification of end-user devices and a near-total reliance on renewable forms of energy without waiting for or depending upon unrealistic technical or policy solutions. It then outlines a political program of substantial economic support, regulations, and social programs and a set of strategies for reducing consumption to sustainable levels that, all together, will be necessary to accomplish the climate goals set by the IPCC.

Technical Feasibility of the Transition to Renewable Energy

The actions necessary to make the shift from dependence on fossil fuels to renewable sources of energy in the USA obviously vary for each of the current major sources of greenhouse gas emissions, as shown in Figure 9.1A. Although the emissions levels in all of those sectors can and must be addressed immediately and simultaneously, some are more urgent and currently possible,

Figure 9.1

US electricity production and greenhouse gases by sector

(A) The shift to renewables/electric-end-use devices will at first impact emissions for electric power (25 percent), transportation (29 percent), and commercial/residential (13 percent).
(B) Renewables are about 13 percent of US electricity generation, more if hydropower is added.

(A) Greenhouse gas emissions by sector, 2019

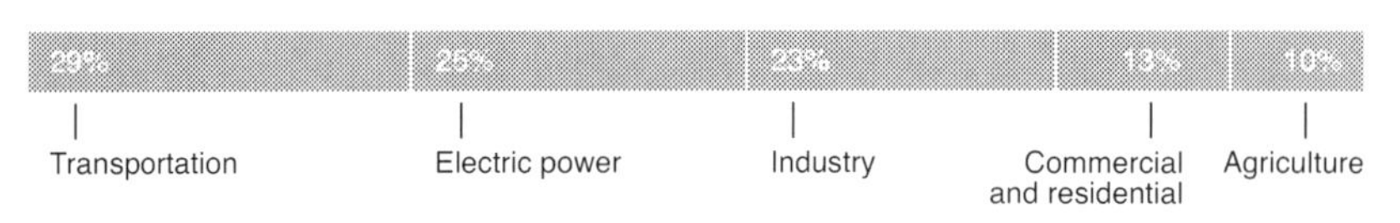

(B) US electricity generation by fuel source, 2011–2022 (forecast)

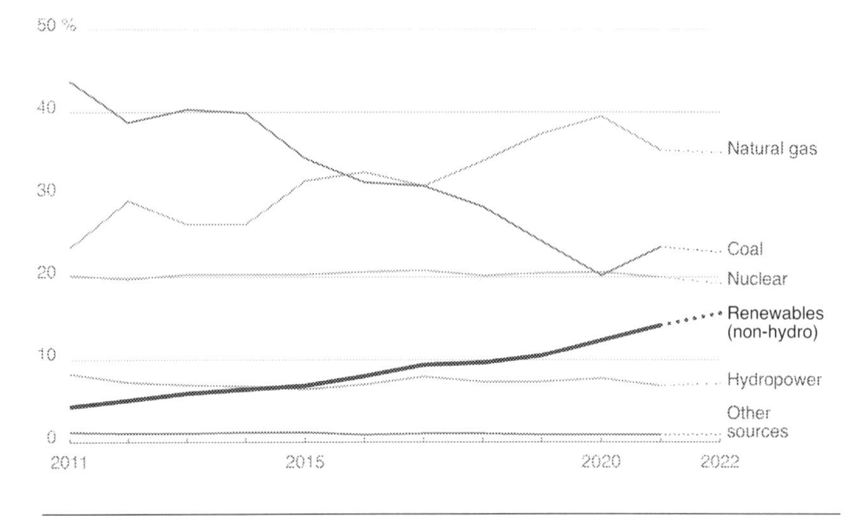

Electricity generation from US Energy Information Administration (2021). Short-term Energy Outlook. Online at https://www.eia.gov/outlooks/steo/report/elecricity.php. Greenhouse gas emissions data from US Environmental Protection Agency (2021). Inventory of US Greenhouse Gas Emissions and Sinks: 1990 - 2019. Online at https://www.epa.gov/ghgemissions/sources-greenhouse-gas-emissions

while others are likely to pose more of a challenge and perhaps take longer. The first and most important step in this process will be to accelerate the use of renewable sources of energy to generate electricity, as power generation currently accounts for about a quarter of all greenhouse gas emissions and achieving emissions reductions in most of the other sectors will also require a shift to electricity. That transition has already begun, but in the USA only about 13 percent of the energy used to produce electricity currently comes from such renewable sources as solar and wind, and about another 8 percent from hydropower (Figure 9.1B).[1]

Another major component of this transition will be to reduce the 30 percent of emissions currently produced by the transportation sector by shifting vehicles from fossil fuels to electricity. The increasing number of personal vehicles, and in some countries buses, being electrified each year clearly demonstrates that this transition is technically feasible, although making this shift within the shipping and aviation sectors (which each currently account for about 3 percent of global greenhouse emissions) is a trickier problem. One possibility for reducing emissions from shipping is using **green hydrogen**, which involves using renewable electricity to create hydrogen by electrolysis – a very inefficient process but one that may become feasible when there is an excess of renewable energy; and for aviation, research on synthetic fuels and electrification is currently underway.[2]

Moving to the remaining major sources of greenhouse gas emissions, the 13 percent produced by the residential and commercial sectors is attributable primarily to the use of "natural gas" (methane) for heating and cooking, which can be reduced by shifting those functions to existing electric methods.[3] Likewise, the 23 percent of emissions produced by the industrial sector can be reduced by such methods as using electric resistance furnaces instead of methane burners. Although finding ways to electrify mining operations will be more difficult, given the challenging environments in which machinery must work, replacing diesel-powered equipment with electric alternatives is not only cheaper in the long run but also reduces noise, heat, and air pollution that requires abatement or ventilation in tight spaces.[4] The remaining major source of greenhouse gas is agriculture, where emissions reductions are likely to involve such measures as livestock and manure management to reduce methane and nitrous oxide emissions, soil conservation practices to promote carbon sequestration, and replacing carbon-intensive synthetic fertilizers with organic sources of fertility such as nitrogen-fixing crops.[5]

Beyond the specific changes outlined above, the transition to electric power will lead to reductions in emissions simply because, in many cases, it is much more efficient than energy produced by burning fossil fuels.[6] For example, EVs are much more energy efficient than internal combustion vehicles: whereas 70 to 88 percent of the fuel used by EVs is converted to mechanical energy, that is true for only 17 to 20 percent of the fuel used by gasoline-burning vehicles. Furthermore, EVs also use less fuel because their regenerative braking systems are able to recapture energy lost in conventional braking and use it to operate the vehicle.

Although we already have ample evidence that making the kinds of changes described above are capable of reducing greenhouse emissions considerably, several research groups and institutes have developed specific proposals for converting almost all of our power needs entirely to renewable sources of energy, or what will be referred to hereafter as a near-total reliance on renewables.[7] Among several plans,[8] a particularly prominent and especially well-worked out example, which has also formed the basis for Green New Deal proposals, is the 100% Wind Water and Solar proposal by the Jacobson research group at Stanford University. This argues that such a transition is possible by electrifying or providing direct heat for all energy uses, obtaining that electricity or heat entirely from renewable sources, storing that energy, and reducing energy use.[9] (See Box 9.1 for specific information regarding each of those four major elements of their plan.)

To determine what such a transition in the USA would require and whether it is in fact feasible, the Jacobson group first made state-by-state projections of business-as-usual power demands for the electricity, transportation, heating/cooling, and industrial sectors between 2015 and 2050.[10] They then estimated the technical requirements and economic costs of electrifying each sector, assuming the use of some renewably generated hydrogen in transportation and industry and improvements in end-use energy efficiency. Based on their findings, they concluded not only that a shift to a near-total reliance on renewables in the USA is both technically and economically feasible but that it also would create 3.1 million more jobs, reduce US power demand by about 40 percent by 2040, and save 63,000 lives from air pollution annually. Since then, the group has aggregated their results across most of the countries in the world, as shown in Figure 9.2.[11]

Although some other experts and critics have questioned particular methods, assumptions, or aspects of the Jacobson Group's proposal,[12] there is a large and growing consensus that the kind of shift to renewables they have outlined is both feasible and the only possible way to meet the IPCC goal of cutting emissions by 50 percent by 2030 from 2010 levels.[13] Indeed, the IPCC 2022 Working Group 3 report on mitigation also laid out the feasibility of decarbonizing energy system, transportation, and industrial processes, flagging in particular the huge reduction in the price of renewables and the fact that most required technology already exists, even if it has not been scaled up as quickly and widely as needed.[14]

Generating enough support and public will to make such a transition happen will require rejecting both overly optimistic boosterism for certain technologies and naysayers who insist that such a transition will never be possible. When considering such criticisms and claims, it is important to evaluate the financial interests of those advancing them, many of whom are supported by the fossil fuel industry or by some electric utility companies that are directly threatened by the declining costs of renewables.[15] For instance, in some

Box 9.1 Core components of the 100% Wind Water and Solar proposal

Electricity-Generating Technologies

Onshore and offshore wind turbines have capacities of 1 kW–5 MW (the latter of which are half the height of the Eiffel Tower); each MW could furnish power to about 500 homes, running at 33 percent of maximum turbine capacity.

Wave and tidal refers to free-floating devices that bob up and down with a wave, creating mechanical energy that is converted to electricity and sent through an underwater transmission cable to shore.

Geothermal energy is extracted from hot water or steam that comes from beneath Earth's surface. The energy can be captured via pipes for heating homes but also by spinning turbines in steam plants or heating organic fluids that evaporate and spin turbines.

Hydroelectric power can range from small run-of-the-river turbines to very large reservoirs with turbines, such as the Three Gorges Dam in China, which produces a massive 22.5 GW.

Solar PVs are arrays of cells containing a material that converts solar radiation into direct current electricity and can be placed on roofs, walls, car parks. In India, a 3.6 km stretch of a canal covered with 33,000 solar PV yields 10 MW of peak power, saves cropland, and protects water from evaporation.

Concentrated solar relies on mirrors or lenses to focus sunlight onto a collector that heats a fluid to a very high temperature, which then flows to a heat engine where a portion of the heat is converted to electricity. The fluid (pressurized steam or molten salt) allows the heat to be stored for many hours to generate electricity overnight or when skies are cloudy.

End-Use Electrification and Resulting Energy Reductions

Improved efficiency of electricity makes a battery electric vehicle 70–88 percent efficient, needing only about a quarter of the energy to drive the same distance. Hydrogen fuel cell cars are about 30 percent efficient. For internal combustion engines, only 17–20 percent of the end-use energy embodied in the fossil fuel is used to move the vehicle; the rest is waste heat.

Improved efficiency of electricity for high-temperature heating means that an electric resistance furnace requires about 82 percent of the raw energy input compared to a methane gas furnace to obtain the same work output.

Improved efficiencies from moving heat with electric heat pumps means that heat pumps require 75 to 85 percent less energy than methane gas boilers. Heat pumps are devices that extract heat or cold from the air, ground, water, or a waste stream of heat/cold and move it to where needed. A heat pump operates in reverse to provide cooling when needed, whereas air conditioners only provide cooling. In a ductless mini-split heat pump system, for instance, each room or zone of a building has its own heat pump, with the indoor part connected to an outdoor compressor.

Storage

Concentrated solar power with storage relies on molten salt to be heated by solar power so that it can then be stored and its heat discharged for about fifteen hours before being depleted, depending on the capacity of the turbines. A plant in Seville, Spain, has provided electricity continuously, twenty-four hours a day, for stretches of up to thirty-six days. This method can ramp power up and down more quickly than coal or nuclear plants.

Hydroelectric power dam storage acts as a huge battery: the flow of water through turbines can be turned on or off as needed. Usually the water only flows one way, but a pumped hydropower storage system uses excess electricity in the system to pump water from a low reservoir to a high reservoir, which is then opened when electricity is needed.

Lithium ion batteries can be charged and discharged 500 or many more times before they must be replaced. It is estimated that there is enough lithium available worldwide to run 5.08 billion EVs with a Tesla-size battery pack. Although lithium mining is environmentally damaging, it must be weighed against the damage of unmitigated use of fossil fuels (see Chapter 8). Many other kinds of batteries exist and are being developed.

Fourth-generation district heating/cooling is much more efficient than heating or cooling individual buildings and a suitable choice for university campuses, corporations, and districts within dense cities. The heating and cooling is done by heat pumps and excess heat and cold is captured in central chillers or boilers and distributed to individual buildings as needed through insulated piping networks.

Underground thermal storage relies on excess heat generated by solar during the summer that can be used to heat water and stored in a borehole field or large underground heat exchanger drilled into the soil, and returned to heat homes and domestic water in the winter. In the Drake Landing Solar Community in Canada, such a system stores enough solar-produced heat to satisfy 100 percent of the entire winter heat needs of the fifty-two homes involved.

Figure 9.2

Timeline for 143 countries to transition from conventional fuels (business as usual), BAU to 100 percent wind, water, and solar (WWS)

The energy sectors that are transitioned include the electricity, transportation, building heating/cooling, industrial, agriculture/forestry/fishing, and military sectors. The percentages next to each WWS energy source (dark fills) are the 2050 estimated percentage supply of end-use power by the source. By 2050, 100 percent of end-use power in the annual average will be provided by WWS, while the 80 percent transition is proposed to occur by 2030. End-use power demand reductions occur for five reasons: (1) the efficiency of moving low-temperature building heat with heat pumps instead of creating heat with combustion; (2) the efficiency of electricity over combustion for high-temperature industrial heat; (3) the efficiency of electricity in battery-electric (BE) and hydrogen fuel cell (HFC) vehicles over combustion vehicles for transportation; (4) eliminating the energy to mine, transport, and process fossil fuels, biofuels, bioenergy, and uranium; and (5) improving end-use energy efficiency and reducing energy use beyond in the BAU case. The total demand reduction due to these factors is 57.1 percent.

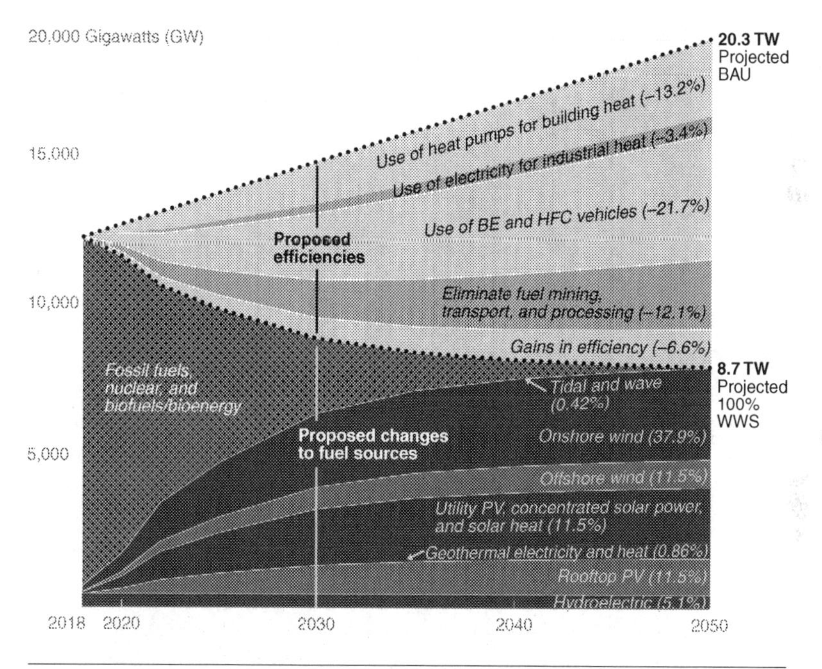

Provided by Mark Jacobson, and updated based on Jacobson et al. 2017 Joule, (1) 1: 108–121.

countries, the growth of rooftop solar has reduced growth in demand for grid electricity, and elsewhere the low operating costs of wind and solar are reducing electricity prices and the viability of coal-fired and nuclear power, in some cases driving their closure.[16] And as the authors of a supportive analysis of a near-total reliance on renewables have noted, the debate about the feasibility of this shift is often skewed because proponents of other fuels do not properly factor in the benefits of a near-total reliance on renewables beyond its elimination of emissions. These include reductions in respiratory diseases from air pollution (which is mostly from fossil fuels and kills as many as ten million people per year, more than the total from two years of the Covid-19 pandemic[17]), water pollution, water use, land degradation, cancers from pollution (from oil and gas extraction[18]), nuclear accidents and nuclear proliferation, and also the creation of more local jobs per unit of electricity than those generated by other sources.[19]

Technical Challenges for a Near-Total Reliance on Renewables

The transition to a near-total reliance on renewables faces other challenges beyond natural inertia and the resistance of the fossil fuel and electric utility industries. As this section discusses, some of these challenges are technical, while others involve cost and available resources.

The Intermittency of Renewables

Because the Sun does not always shine and the wind does not always blow, one of the biggest challenges for a shift to renewables is their intermittency. This is less a problem for hydropower, which already accounts for close to 100 percent of the renewable electricity in the countries of Iceland, Norway, New Zealand, and Bhutan.[20] Meanwhile, other countries and regions, such as Denmark and the northern German state of Schleswig-Holstein, are able to achieve nearly 100 percent of their power needs from wind on some days but not on others, and solar power obviously is not available at night. Sometimes, both wind and solar can fail substantially, leading to what is sometimes referred to in German as *dunkelflaute* (dark doldrums).

At a theoretical level, computer simulations have shown that a near-total dependence on renewables could be reliable, given an electric grid system that is capable of matching flexible demand with wind, water, and solar supply and storage.[21] Practically, such a system would need an integrated grid that incorporates geographically dispersed and variable wind, wave, solar, and other sources to provide a more constant supply (Figure 9.3).[22] It would also require the correct sizing of wind, water, and solar electricity systems; heat generation that meets the needs in each region and sector; the use of electricity to fill storage reservoirs (such as pumped hydropower systems); and the provisioning of other backup systems that are flexible and quickly dispatchable (such as hydroelectricity with dams, concentrated solar thermal with thermal storage, batteries, and geothermal).

Figure 9.3

Solar and wind resources in the USA

There are ample solar (Southwest, A) and wind resources (coasts and Midwest corridor, B) in the continental United States.

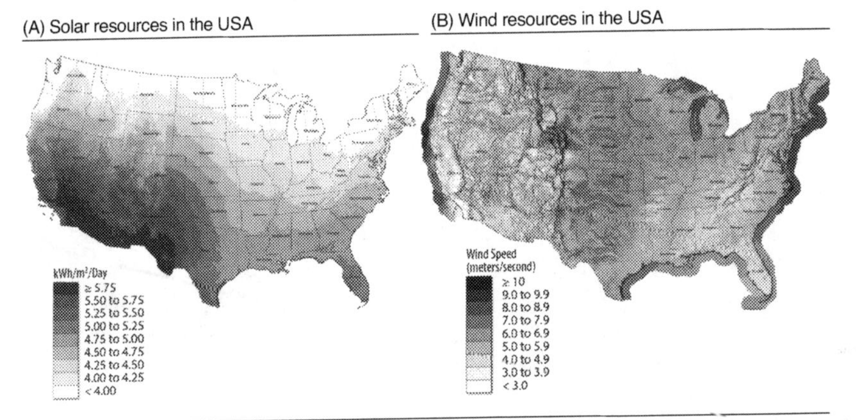

Solar map courtesy of Sengupta, M., Y. Xie, A. Lopez, A. Habte, G. Maclaurin, and J. Shelby. 2018. "The National Solar Radiation Data Base (NSRDB)." Renewable and Sustainable Energy Reviews 89 (June): 51-60. Wind map courtesy of Draxl, C., B.M. Hodge, A. Clifton, and J. McCaa. 2015. "The Wind Integration National Dataset (WIND) Toolkit." Applied Energy 151: 355366.

Europe currently provides an example of a large, integrated electric grid that works well between countries. If the supply of wind and solar power on a particular day is low in Germany, for instance, some of the excess solar capacity in Spain can be "bumped over" to France and some of France's nuclear-generated capacity similarly shifted to Germany. This system has been shown to work effectively even with a total power failure in one sector, and it could work even better with more pumped hydro storage and over-building of wind and solar capacity to cope with days with less wind and sun.[23]

Cost

Another frequent criticism leveled at proposals for a near-total reliance on renewables is the high cost of implementing these new systems and infrastructure. The Jacobson group estimated that the cost of installing more solar and wind, increasing electrification, adding more storage, and constructing new transmission lines for nearly 150 countries around the world would be $6.8 trillion a year between 2019 and 2050.[24] Although this is a huge expenditure, it is considerably lower than the current cost of business as usual, including all the subsidies now provided for fossil fuels. The cost of this transition in the

USA alone between now and 2050 is estimated to be about $7.8 trillion, a cost that seems relatively small compared to the predicted cost of increased climate damage if emissions are not cut, which the report by the insurer Swiss Re estimated would reach $23 trillion worldwide by 2050.[25] In sharp distinction to the rosy projections of mainstream economists, this would amount to 18 percent of GDP.

And as the Jacobson group also pointed out, much of the estimated investment that would be needed to kick-start the transition would eventually be recouped by profits from electricity sales over the lifetime of the new energy, storage, and transmission equipment. A similar analysis by economist Robert Pollin estimated that the level of necessary investment globally would average about $4.5 trillion per year between 2024 and 2050, spending that he expected would be equally split between public and private investment and that would largely pay for itself over time with huge cost savings in lower energy costs for consumers.[26] The IPCC 2022 mitigation report estimated that $1.7 trillion per year was needed globally for a two-in-three chance of meeting the 2°C limit, but this was for the electricity sector alone.

Helpfully, as the International Energy Agency noted in late 2021, the world's best solar plants already offer the "cheapest energy in history," already cheaper than gas in most major countries.[27] As Figure 9.4 shows, whereas five years ago, fossil gas was the cheapest form of baseload energy, or energy to meet the minimum demands of the electrical grid, energy from renewable sources is already cheaper in many countries. Given those costs and that depending on dispatchable renewables with battery storage for intermittent periods could soon be cheaper than building new fossil gas plants to meet increased demand, most countries will probably face strong incentives to make the transition.[28] Just as more than a hundred coal-fired plants in the USA have been shut down or replaced since 2011 because the fracking boom made natural gas cheaper than coal, in recent years some utility regulators and states, such as California, have begun to turn away from natural gas because of falling renewable costs.[29] Renewable energy installations are therefore increasing in size and number and are anticipated to amount to approximately 30 percent of the US energy market as early as 2026 (Figure 9.5).[30] In February 2022, renewables supplied nearly a quarter of electricity[31] and in April 2022, the renewables supply into the California briefly reached over 100 percent.[32]

Land, Raw Materials, and Energy Requirements

Critics of moving rapidly to a near-total reliance on renewables have also noted that the land use requirements for wind and solar generation will be very large. Yet analyses by the Jacobson group show that these land requirements

Figure 9.4

Cost tipping points for renewable energy

The chart shows the cost per MWh and four main tipping points for renewable energy. Note that LCOE is the levelized cost of electricity. This represents the average revenue per unit of electricity generated that would be required to recover the costs of building and operating a generating plant over its life. These are the tipping points:

(1) New renewables are cheaper than building new fossil fuel plants; this has already happened.

(2) New renewables are cheaper than the operating cost of existing fossil plants, which is happening in some places is predicted to be a feature of the 2020s.

(3) New dispatchable renewables (those with batteries to deal with intermittency) are cheaper than building new fossil fuel plants.

(4) New dispatchable renewables are cheaper than the operating cost of existing fossil plan.

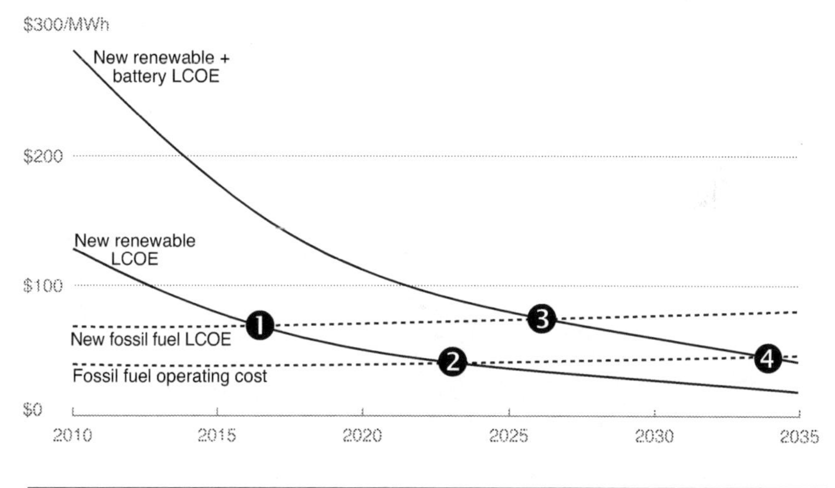

Adapted from Carbon Tracker: https://carbontracker.org/reports/the-trillion-dollar-energy-windfall/

would amount to less than 0.5 percent of the total US land area (Figure 9.6), and a roughly similar estimate was provided by renewable engineer Saul Griffith.[33] On a global level, an independent calculation by Bloomberg suggested that 1 percent of land around the globe would be needed to supply the whole world with renewables.[34] These estimates reflect the fact that solar panels can be installed on rooftops and sites such as parking lots, and that wind turbines can be erected on agricultural land with only minor losses to productivity. The amount of land required by onshore wind turbines also goes down as the blades increase in size, and leases have been issued for very large offshore wind such as 25 GW off the coast of Scotland. To put these area requirements in context, about 1.3 percent of US land area is already taken up

Figure 9.5

Surging generation from utility-scale solar and wind projects are on track to push US renewable market share to 30 percent by 2026

The IEEFA estimates that in just five years, wind and utility-scale solar will generate roughly 850 million MWh of electricity annually, equivalent to powering just under 80 million households per year, or all residential and commercial use in California, New York, and Texas combined. Assuming total US electricity demand remains essentially flat, as it has since 2010, this would equal to more than 21 percent of the total demand. When rooftop solar and hydropower are included, the total renewable generation is expected to generate approximately 1,180 million MWh, nearly 30 percent of total US demand.

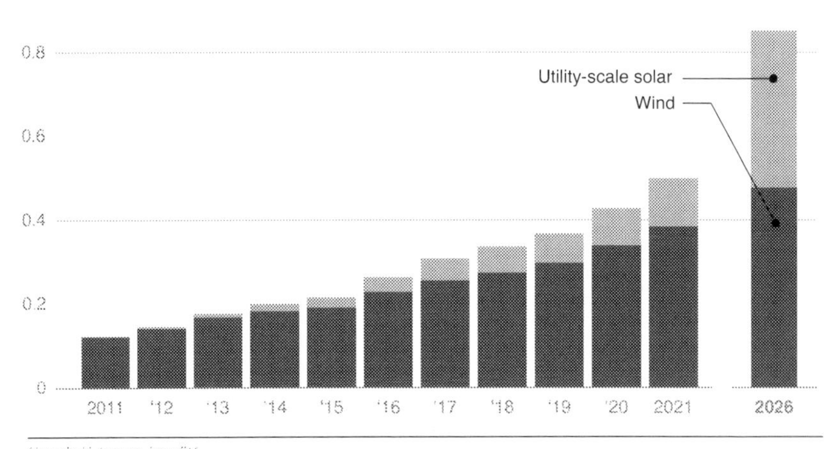

Household data use from EIA.
Adapted from Institute for Energy Economics and Financial Analysis (IEEFA), October 4, 2021. IEEFA U.S.: Surging generation from solar, wind on track to push renewable market share to 30 percent by 2026. Online at https://ieefa.org/ieefa-u-s-surging-generation-from-solar-wind-on-track-to-push-renewable-market-share-to-30-percent-by-2026/

by fossil fuel infrastructure, some of which could be repurposed for wind and solar plants or for other uses.[35]

Another issue that raises questions about the feasibility of a near-total transition to renewables is the scale of build-out and raw materials that will be needed. Worldwide, it has been estimated that the transition to near-total reliance on renewables will require the production and installation of about 3.8 million 5 megawatt wind turbines; 40,000 300 megawatt central solar plants; 40,000 300 megawatt solar PV plants; and 1.7 billion 3 kilowatt rooftop PV installations.[36] The material requirements of this infrastructure will be enormous. Solar panels and wind turbines require large amounts of mined materials, including copper, aluminum, iron (steel), and harder-to-extract metals

Figure 9.6

The land use footprint requirements for 100 percent wind, water, and solar (WWS) is less than 1 percent of the US surface

Footprint is the physical area needed for each individual energy device (e.g. roof photovoltaics (PV)). The spacing need is greater. Spacing is the area between some devices to minimize interference (e.g. for onshore and offshore wind turbines). However, spacing area can also be used for other things, e.g. to grow crops.

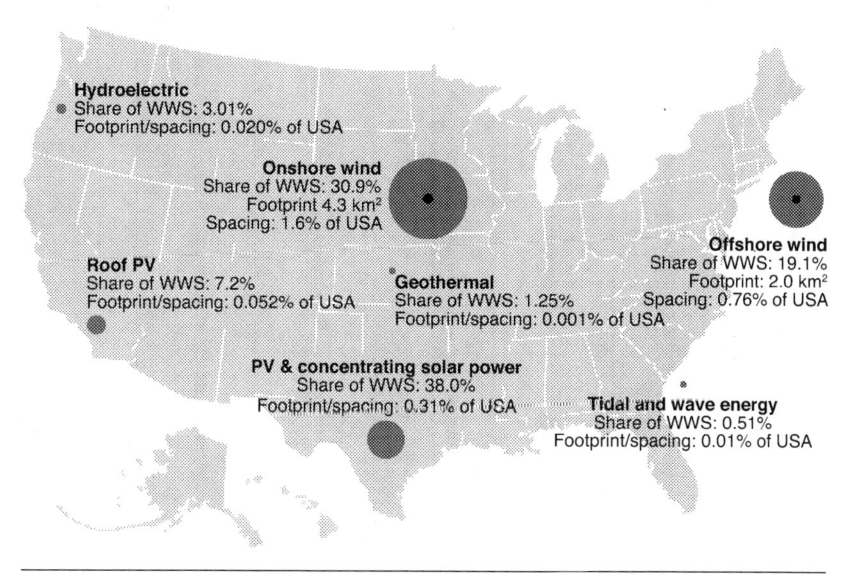

Adapted from Jacobsen et al. 2015. Energy Environ. Sci, 8:2093.

such as neodymium for magnets, lithium for batteries, and indium for solar panels. One analysis of the availability of such materials for the time period 2016–50 showed that bottlenecks were likely for thirteen elements – cadmium, chromium, cobalt, copper, gallium, indium, lithium, manganese, nickel, silver, tellurium, tin, and zinc.[37] Others, however, have pointed out that such bottlenecks could be overcome through a heavy increase in recycling (including a new plant being built in Georgia in the USA that could recycle 30,000 metric tons of lithium per year) and the development of substitute materials, as for the nine elements shown on Figure 9.7.[38] These measures, as well as such others as reducing consumption, investing more in public transportation, and increasing oversight of human rights abuses, could go some way toward ameliorating the most devastating and unjust consequences of extractivism that were discussed in Chapter 8.[39]

Figure 9.7

How can we substitute or recycle the metals that are key to renewable energy?

	Importance to renewable energy	Materials efficiency or ease of substitution	Current ability to be recycled
Aluminium	**Very important** Used for wind, batteries, and photovoltaics.	With some loss of performance (steel, plastic).	~70–80%
Cadmium	Low importance. Small share of PV market.	Efficiency increasing. Can shift to other photovoltaic types.	~77%
Cobalt	Medium importance. Li-ion dominant battery technology.	Efficiency increasing. Can shift to other battery types	90%
Copper	**Very important** Used for wind, batteries, and photovoltaics.	**Difficult to substitute** in most applications.	~34–95%
Dysprosium	**Very important** Used for wind, batteries, and photovoltaics.	Can shift to other magnet or motor types	**Not currently recycled**
Indium	Least important. Small share of photovoltaic market.	Efficiency increasing. Can shift to other options.	**Not currently recycled**
Lithium	Medium importance. Li-ion dominant battery technology.	Efficiency increasing. **Used for all Li-ion and Li-S.**	**~10%**
Manganese	Medium importance. Li-ion dominant battery technology.	Efficiency increasing. Can shift to other battery types.	**Very limited recycling**
Neodymium	**Very important** Used for wind, batteries, and photovoltaics.	Can shift to other magnet or motor types, or non-PMG wind.	**Not currently recycled**

Most important | Difficult to substitute | Non-recyclable

Medium importance | Some substitutions | Some recyclability

Least importance | Easiest to substitute | Fewest challenges

Adapted from Dominish, E., Florin, N. and Teske, S., 2019, Responsible Minerals Sourcing for Renewable Energy. Report prepared for Earthworks by the Institute for Sustainable Futures, University of Technology Sydney.

Any discussion of the feasibility of the transition to renewable energy must also acknowledge the considerable amounts of energy it will take to get those minerals out of the ground, to turn them into batteries and solar panels and giant rotors for windmills, and to dispose of them when they wear out. In the short term, at least, those mining operations will be worked primarily by fossil fuel–burning vehicles and carried across the seas by fossil fuel–burning container ships. Additional energy will also be needed to rebuild the electric grid to make it more efficient, to upgrade buildings to the highest standards of insulation efficiency, and to develop low-carbon transportation networks based on EVs and high-speed rail. While all of this is likely to actually increase greenhouse gas emissions in the short term, once the new large-scale renewable infrastructure is in place, it can be expected to lead to rapid reductions in emissions and to serve its purpose for decades. (See Box 9.2 for an interview with a leading renewable energy expert on the feasibility of the transition). The bigger picture of the scale of emissions arising from the transition to renewables requires what is known as **life-cycle assessment**. This approach takes into account the emissions from the manufacture of, for example, wind turbines, the years of use (twenty-five or more), the amount of energy generated, the amount of emissions saved, as well as the recycling of materials to make new components. On a life-cycle basis, onshore wind energy emits eleven grams of CO_2 equivalent per kWh of electricity produced compared to rooftop solar at 41 grams, fossil gas at 490 grams, and coal at 820 grams.[40] The carbon payback time for a wind turbine that operates for thirty years is about nine months.[41]

Box 9.2 Interview with Mark Jacobson

Mark Jacobson, the lead researcher for the 100% Wind Water and Solar proposal, is a professor at Stanford and an internationally recognized expert on the problems of air pollution, global heating, and developing large-scale renewable energy approaches.

AA: I'm interested in what you have to say about demand reduction. There's been a rapid rise of renewable energy and it's going to accelerate even more, but total energy use for the world accelerates much faster. Even with your compelling technical vision of 100% WWS (Wind Water and Solar), is it going to be enough if the wider system is having to grow at 3 percent GDP per year?

MJ: Our plans account for growth to 2050, so it's not a problem. There's still plenty of renewables. There's enough solar to power the world more than 300 times over, and wind seven times over. You're going to use a combination in the end. This is just a distraction.

AA: Yes, sure, but if you keep growing the economy and you need more and more energy, even from 100 percent renewables, you're going to have to do a lot more mining to keep pace with

that, and is that really going to make the emissions go down? All that neodymium and lithium and aluminum?

MJ: First of all, when you transition to electrifying the energy sector, you have a 57 percent reduction of demand just by electrifying. And if you reduce that much, you're not going to grow energy demand worldwide that much. People are also going to be using energy-efficient technologies. I use a heat pump in my home [in California]. The home is all electric. It hardly uses any energy. Consider, Texas uses two and a quarter times the electricity per capita as California. Why is that? It's just because they're inefficient. There are no regulations in Texas to make their homes and buildings more efficient. Sure, you could have a completely inefficient world in the future and just waste energy. But if you actually use energy-efficient appliances and have energy-efficiency standards and electrify everything, then your demand will go down a lot.

Now, in terms of mining, I mean that's just a red herring. In the USA alone there are 50,000 new oil and gas wells drilled every year. That's what you need to compare with. And there are 1.3 million active oil and gas wells and 3.1 million inactive ones. You know the entire gas and oil industry and fossil fuel industry in the USA takes up 1.3 percent of the land. So the renewable shift requires orders of magnitudes less mining, no matter how much lithium you mine.

And there are many batteries that aren't lithium. Yes, you need lithium for transportation, because it's light. But for stationary you can use sodium, sulfur, basalt. Neodymium is needed for permanent magnets. And even if you wanted to power everything by wind by 2050, that's only one-seventh of the neodymium we know exists. Plus you don't have to use permanent magnet generators [which have neodymium]. You can use induction generators [without neodymium]. For lithium, there's enough for five billion cars. We now have 1.1 billion. I certainly hope we don't have more than that, but there's certainly enough resources for that.

AA: **What about Jevon's paradox, or Jevon's effects – as you get more efficient, you just use more of the resource? Are you concerned about that?**

MJ: A little. I'm thinking about it in my own home context. I haven't paid an electric bill or a gas bill or a gasoline bill for four years. But I also get paid for the extra solar electricity I produce. I get paid about $1,000 per year. And even though I'm using way less electricity, the more electricity I save, the more money I get paid at the end. So that's one thing. [He explained that this is because he's going through Community Choice Aggregation and not the local utility.] I think there are other reasons. I'm not going to drive around more just because my car is more efficient.

AA: Ok, so we need to get off the fossil fuels as soon as possible. Now even if 100% WWS happened very soon, it's an enormous need for new materials, hundreds of thousands of turbines, and so on. We'll have ships coming from across the oceans. This will *increase* our emissions, right?

MJ: Yes, you're going to get an increase in emissions in the short term, before it decreases. It also speaks to the strategy – what should you electrify first? In California, they wanted to use all this energy to build the high-speed rail – to dig and drill with fossil fuels. Why not electrify everything first, and then do that at the end and build the railway with renewable energy? So there's a strategy to minimize emissions by transitioning industry. In fact, in Texas there's a plan to make a lithium mine 100 percent renewable so they're not emitting from the mining. You make sure the electric input to the mine is renewable. Ideally, you do it for the machines, too. I'm not sure they're going that far, but they should.

AA: What about the justice issues around extraction of rare earth metals from the Global South? Even if one can replace neodymium with other materials to some extent, there needs to be a huge up-scaling of mining, and that will have damaging effects there.

MJ: First of all, rare earths are not rare, they're everywhere. They're not rare elements, but they are dispersed and not found in many concentrated deposits that are economically exploitable. And for something like lithium, every time they look, they find more; it's a question of looking. Sure, it's partly related to where these things are mined now, and things don't change overnight. But there are a lot of choices of materials for batteries. Lithium is being recycled already for car batteries. The former CEO of Tesla started his own battery recycling program. He made a statement that the largest lithium mine may be in the drawers of America.

AA: You've articulated a technical vision very well for how we can respond to the climate crisis. We already have the technology we need. Here's the puzzle. So many people around me, STEM students, other professors, keep harping on about how technologies that don't yet exist are going to save us. Why?

MJ: I think there are just enough people pushing those ideas. Especially in a university like my own, I can see where those ideas come from. They come from the energy resources engineering department, which used to be called petroleum engineering. They come from a few economists who, for a lack of a better word, like fossil fuels. They might have been climate deniers, and now they believe but they resist rapid change, they like the status quo. There was a Princeton study, it did a largely renewables scenario, but it was funded by ExxonMobil and British Petroleum (BP). MIT has large funding from fossil fuel interests, and Stanford does, too. At our School of Earth Sciences at Stanford, every faculty member gets two paid-for students. That comes from royalties from the oil and gas industry, since the 1950s. That's not to say it's influencing the students directly. But a lot come there to study oil and gas. So there are enough people who graduate and go into the field and keep pushing these ideas. So they are asking, how can oil and gas fit in? So they want to look into carbon capture and sequestration, it keeps oil and gas going. Look at mechanical engineering departments – a lot of the work they do is on combustion. Making better engines for cars and airplanes. They want new ways to get funding. It's a way to keep themselves going. And there are huge institutes getting fossil fuel research dollars. At Stanford, there is one called the Global Climate and Energy project, a greenwashing name. It was a huge amount of money, like 225 million over 10 years. It funded a lot of students. And Berkeley has a program like that funded by BP.

AA: Do you sometimes get discouraged and daunted, and how do you sustain yourself through those periods?

MJ: I'm actually very positive because I know we can do it. I've seen a lot of progress. Costs have come down, renewables are growing substantially. I know there's a huge way to go, so I don't want to be under an illusion that we're there, but I feel more and more confident. In 2009, I did the first paper on going 100 percent renewable. And people just joked about it and said that's pie in the sky and the utilities wouldn't even talk about more than 20 percent renewables on the grid. And now the discussion has changed. People talk about whether you can have 80 to 100 percent vs. 100 percent renewables on the electric grid. There are eleven countries that are 100 percent renewable in the world. There are states that are really far along, Iowa is 60 percent from wind. And California is close to 50 percent renewable electricity, on average. And there are laws in sixty-one countries to go to 100 percent renewable electricity. We're making progress on electric power, but the other sectors are slower. We need really aggressive laws, especially in industry, and also buildings and transportation. So there's a little less opposition than before. But the nature of the opposition has changed. Now it comes from the "all of the above" crowd. They want to try everything that doesn't work, nuclear and carbon capture and direct air capture and geoengineering. There's no excuse for this now. In March this year, the growth of solar and wind hit a new record, growing 34 percent more than a year ago and now accounting for 13 percent of electricity generation. It's there. But we need a lot more.

A Political and Socioeconomic Plan for Action

In 2021, the seemingly cheery and convincing picture regarding the feasibility and costs of renewable sources of energy and the imperative to cut emissions led the International Energy Agency (IEA), an intergovernmental body established in 1974 to foster energy cooperation and that has historically advocated fossil fuels, to recommend that governments should no longer grant permission for the development of new oil and gas fields or new coal mines and mine extensions.[42] Yet the IEA's accompanying call to reduce the burning of oil by only 75 percent and of gas by only 55 percent by 2050, a decline consistent with projections by other energy analysts, is simply not great or fast enough to meet the IPCC goal of a 50 percent cut in 2010-level emissions by 2030. As shown in Figure 9.8, that currently projected course appears to view renewable energy as an *addition* to rather than a *replacement* for fossil fuels and to anticipate that fossil fuel extraction will continue for a long time, a view also

Figure 9.8

Projections show that global energy consumption will still be mostly fossil fuels even in 2050

World energy consumption is projected to increase by nearly 50 percent by 2050: renewable use grows fastest but it's still a fraction of the overall consumption.

World total energy consumption by fuel (quadrillion Btu)

1000

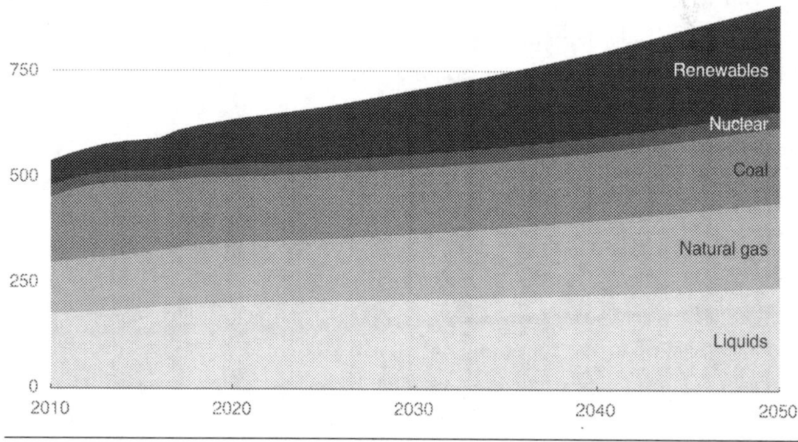

Adapted from U.S. Energy Information Administration, https://www.eia.gov/todayinenergy/detail.php?id=41433

consistent with the major energy company DNV-GL's prediction of the ongoing demand for oil and natural gas to 2050.[43]

Such projections – and anticipated resistance from current beneficiaries and allies of the fossil fuel industry – make clear that accomplishing the transition to a near-total dependence on renewable sources of energy in time to meet IPCC goals to avoid ever more devastating effects of global heating will require more than technical approaches. As shown in Figure 9.9 and discussed in the rest of this section, this transition will also require two additional shifts: to a much more active role by governments, and to reduced consumption at a societal and individual level.

A Framework for Political Action

As noted, the transition to renewables is already underway, and much of that is through the private sector, such as individuals installing solar panels on their homes, airports installing solar arrays with third-party private providers, Tesla delivering more than 600,000 EVs in the first three-quarters of 2021, and private equity investments in renewables hitting nearly $22 billion in the USA in 2021.[44] Yet there is ample reason to believe that accomplishing this transition at the necessary speed and scale and elevating the public good over special interests will require governments to take a comprehensive, far-reaching, and active role. Many of these necessary decisions and initiatives, in the USA and elsewhere, are already being and will continue to be made at the local, state, and regional levels (such as Ithaca becoming the first city in the USA to try to electrify all buildings, and the state of California's mandating 60 percent clean electricity by 2030). Ultimately, however, national governments are the only actors with enough power and resources to facilitate and require many of the large-scale and integrated changes.[45]

Meeting current IPCC goals is likely to require a comprehensive and integrated set of programs similar to those proposed by advocates of the Green New Deal in the United States and Europe and of the California Climate Jobs Plan. The original proponents of the Green New Deal consciously intended that name to allude to the sweeping economic and social policies of US President Franklin Delano Roosevelt's New Deal in response to the Great Depression of the 1930s and 1940s to reflect the broad scope of the proposal, as opposed to policy alternatives that focus more narrowly on economic stimulus measures or technical issues without including social provisions intended to also make this transition a just and politically possible one. Although various versions of such a policy have been proposed by individuals, think tanks, and political parties in Europe and the USA (such

Figure 9.9

How do we work to stop global heating?

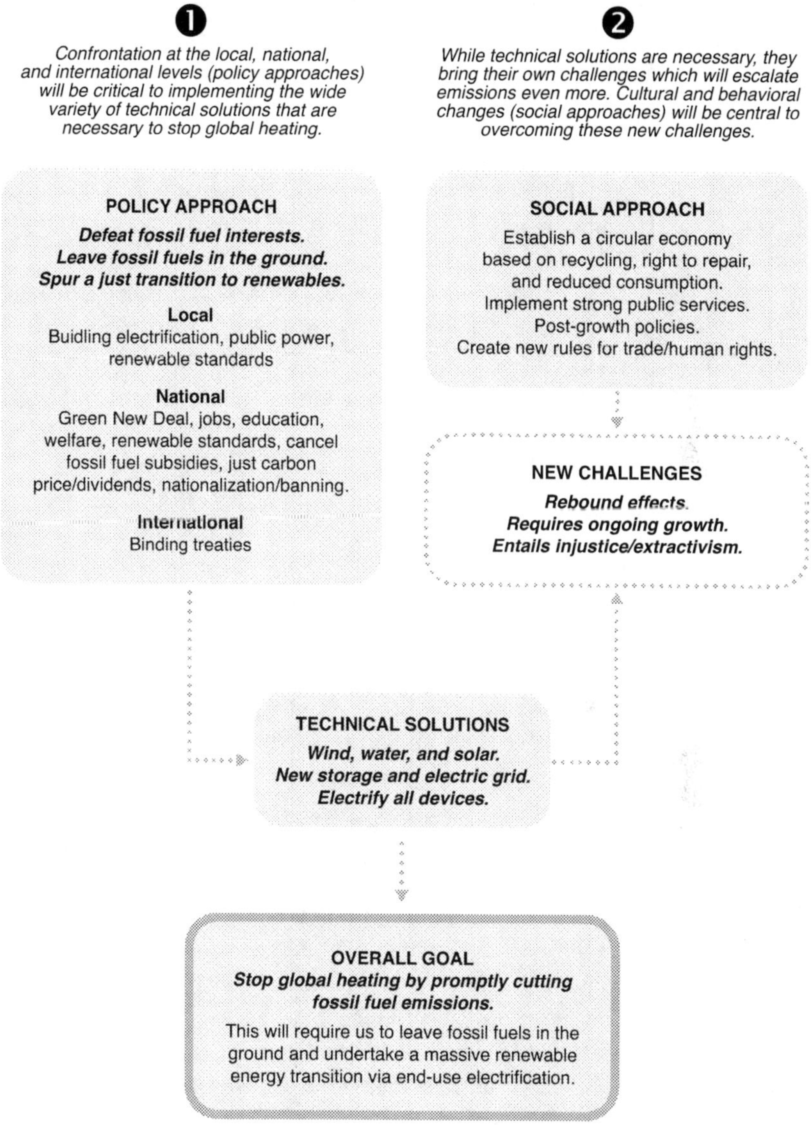

Confrontation at the local, national, and international levels (policy approaches) will be critical to implementing the wide variety of technical solutions that are necessary to stop global heating.

While technical solutions are necessary, they bring their own challenges which will escalate emissions even more. Cultural and behavioral changes (social approaches) will be central to overcoming these new challenges.

POLICY APPROACH

Defeat fossil fuel interests.
Leave fossil fuels in the ground.
Spur a just transition to renewables.

Local
Buidling electrification, public power, renewable standards

National
Green New Deal, jobs, education, welfare, renewable standards, cancel fossil fuel subsidies, just carbon price/dividends, nationalization/banning.

International
Binding treaties

SOCIAL APPROACH

Establish a circular economy based on recycling, right to repair, and reduced consumption.
Implement strong public services.
Post-growth policies.
Create new rules for trade/human rights.

NEW CHALLENGES

Rebound effects.
Requires ongoing growth.
Entails injustice/extractivism.

TECHNICAL SOLUTIONS

Wind, water, and solar.
New storage and electric grid.
Electrify all devices.

OVERALL GOAL

Stop global heating by promptly cutting fossil fuel emissions.

This will require us to leave fossil fuels in the ground and undertake a massive renewable energy transition via end-use electrification.

as House Resolution 109 in the US Congress, Senator Sanders' plan in 2020, President Biden's Build Back Better, and the California Climate Jobs Plan), they all include three major elements that proponents view as essential to a socially just and technically effective transition from fossil fuels to renewable sources of energy: economic support, regulations and policies, and social programs.

Economic Support

All Green New Deal–like proposals include large government investments to help fund a rapid mass mobilization to replace the current dependence upon fossil fuels (see Box 9.3 for a brief history of Green New Deal proposals). Although various plans differ in their particulars, these investments usually include support for R&D into renewable technologies, construction of solar and wind installations and new transmission lines, strengthening the power grid to facilitate power sharing across states or regions, supporting land restoration and regenerative farming practices, and easing the expense of retrofitting homes and businesses for better insulation and efficiencies.

Governments, and especially the federal government, have a number of financial tools at their disposal that could be used to rapidly scale up the transition to renewable resources, including direct grants and investments, loans and loan guarantees, and tax incentives. While these steps would undoubtedly require new spending and sources of funding, many of them could also be funded by shifting current subsidies and tax benefits from fossil fuels to renewables, granting loans that would eventually be repaid, and imposing higher taxes or penalties on polluters. None of these are new tools or practices, as the history of government support for various forms of energy goes back more than a century, including tax incentives to encourage oil exploration and financing the construction of hydroelectric dams and nuclear power plants. Proposals for accelerating the transition to renewables would simply speed up and expand a shift in such support and tax preferences from fossil fuels to renewable energy that has been underway since the 1970s.[46] As Griffith points in his book *Electrify*, for instance, much of what needs to happen in the USA to electrify everything must happen at the household level.[47] Because most low-income families, unlike more affluent households, are unlikely to be able make upfront capital investments to decarbonize their automobiles and appliances despite their eventual efficiency and cost savings, Griffith proposes public-private financing based on the New Deal model to provide low-interest loans to families to pay for what he estimates will cost about $40,000 per household.

Box 9.3 Brief history of the US Green New Deal

The start of the Green New Deal can be traced to proposals by presidential candidate Barack Obama in 2008 and the Europeans in 2009, which can be considered reformist in the sense that they included big government spending and technocratic planning with only a minor concern for promoting employment and reducing poverty. But more radical versions followed, first from the US Green Party in 2012, then in US House Resolution HR-109 in 2018, sponsored by congressional representative Alexandria Ocasio-Cortez and Senator Ed Markey. Although HR-109 did not directly address the need to disengage from fossil fuels, it did link the need for a massive mobilization to combat climate change with the need to also provide a just transition for frontline communities, including economic measures to redistribute wealth and provide for health care and education, as sociologist John Bellamy Foster has noted.[48]

In 2020, presidential candidate Bernie Sanders proposed a Green New Deal plan that provided specifics. His plan called for a shift to 100 percent renewable energy for electricity and transportation by 2030; allocated $16.3 trillion of public investment over ten years to displace fossil fuels; banned offshore drilling, fracking, and mountaintop removal coal mining; included $41 billion to help industrial agriculture make a shift to regenerative practices and $200 billion to support transformations in poorer countries; and proposed a just transition for workers, including wage guarantees, job placement assistance, and skill retraining. To help pay for these initiatives, he called for imposing a large tax increase on polluters and investors, eliminating subsidies to the fossil fuel industry, cutting back on military spending, and making corporations and the wealthy pay their fair share.

Following his election, Joe Biden proposed legislation that became known as Build Back Better that contained a number of Green New Deal elements, although its centerpiece, the Clean Electricity Performance Program, was soon stripped out during political haggling in Congress and the bill was blocked, very much reflecting the influence of fossil fuel interests.

A key concept at the core of the Green New Deal and other proposals such as the 2021 California Climate Jobs Plan is that of a **just transition**. This has been traced to the master union organizer Tony Mazzocchi of the Oil, Chemical and Atomic Workers Union in the 1980s, who argued that the winding down of industries that harmed the environment and workers must also safeguard livelihoods. For example, he proposed a kind of GI Bill for atomic workers who would otherwise be left unemployed by nuclear disarmament.

Two phases of the New Deal and the radical government intervention of World War II

Phase 1 of the New Deal was merely reformist, but the labor uprising of 1934 triggered a more radical, second phase. The World War II period saw radical government intervention in many areas of the economy.

Massive labor uprising (1934)
Laborers unite with the unemployed, students, African Americans, and farmers. The violent response by police and the National Guard sets the stage for the passage of the National Labor Relations Act, which guarantees the right to form trades unions and benefits the working class for decades.

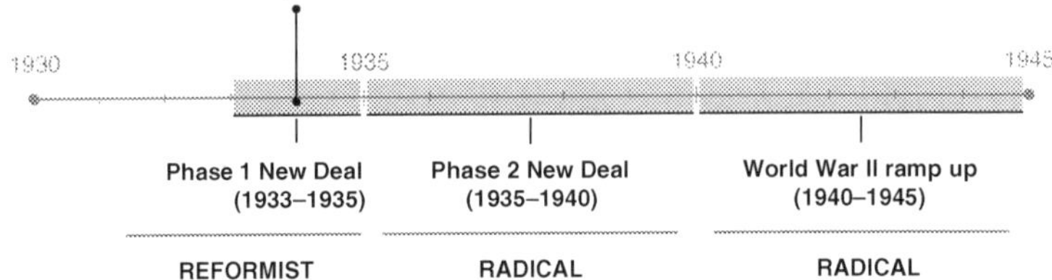

REFORMIST
Accommodates corporate interests through the failed National Recovery Act

RADICAL
Includes huge stimulus and overrides corporate interests via Fair Labor Standards Act

RADICAL
Government intercedes in many aspects of the economy via the War Production Board and fair-share rationing

A timeline of the Green New Deal (GND), from Mazzocchi's concept of a just transition to the President Biden era

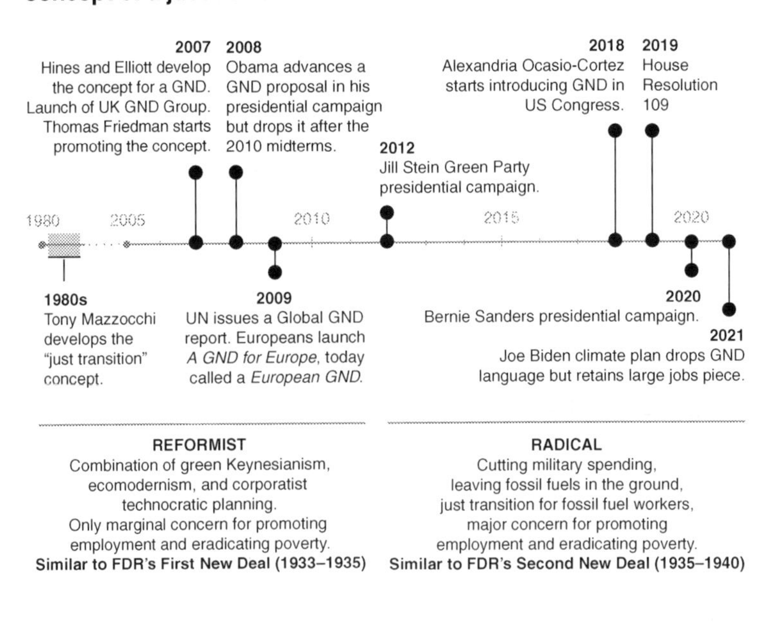

Regulations and Policies

In addition to providing funding, governments at various levels can hasten the transition to renewable sources of energy by passing new laws, policies, and regulations. As others have pointed out, such government attempts to reconcile their energy production policies with their climate goals can be split into demand-side and supply-side approaches.[49] Demand-side approaches, or policies intended to limit the demand for fossil fuels, could include the imposition of a carbon tax, which, as discussed in Chapter 8, has shown mixed success, but could also be designed to be more equitable and therefore popular. Other approaches would include such regulatory requirements as requiring utilities to shift to specific percentages of renewable sources by certain deadlines or car manufacturers to increase the fuel efficiency of their vehicles by certain dates. While such previous moves have been generally effective, they have not been severe or rapid enough to accomplish the dramatic reductions in emissions that will be necessary to meet most governments' stated climate goals.

As a result, governments should consider supply-side approaches that directly restrict the supply of fossil fuels. One such restriction would be to cancel fossil fuel subsidies, which in the USA are estimated, conservatively, to amount to at least $20 billion per year for production subsidies, about $15 billion per year in consumption subsidies, and another $12 billion for tax subsides. (Globally, it is estimated that as much as half of new oil and gas development would be unprofitable without such subsidies, with one estimate of the total sum of direct subsidies coming in at about $500 billion per year, as we saw.)[50] Another possible regulatory tool is to ban new extraction of fossil fuels, as has already been passed in a few countries such as Spain and New Zealand and the cities of Los Angeles and Quebec.[51] A third, as suggested by Saul Griffith, would be for governments to buy out fossil fuel companies, run them in the public interest until the transition to renewables is complete, and then dismantle them and leave the remaining fossil fuels in the ground, steps that such companies are very unlikely to take on their own given that the proven wealth to be extracted is as much as $100 trillion.[52]

Recognizing that much more radical supply-side actions may be required to leave fossil fuels in the ground, the ecologists Stan Cox and Larry Edwards proposed a "Cap-and-Adapt" plan.[53] Based on the planned economy put into place in the USA in response to World War II, this plan would decisively phase out fossil fuels in two steps. During the Cap stage, Congress would put a cap on the nation's total supply of fossil fuels, ratcheting down the amount of oil, gas, and coal that would be permitted to be extracted by fixed amounts each year until the supply reached zero.[54] During the Adapt stage, all private gas and electric utilities would be converted into public utilities that would now be under national oversight, much like the War Production Board of the 1940s (discussed in Chapter 4), which would funnel the diminishing amounts of fossil fuels into essential sectors while the renewable energy transition takes place.[55] Because nationwide energy shortages would occur during the transition, price controls and rationing (just like the World War II era) would be needed to make sure there is equitable access to electricity.

Social Programs

What most distinguishes comprehensive transition programs, such as the Green New Deal and the California Climate Jobs Plan, from earlier energy programs and policies is their recognition that a rapid transition to renewable forms of energy will also require social programs and investments to ease the inevitable economic and human dislocations it will involve. The purpose of such programs would be to ensure that the costs, or externalities, of the transition from fossil fuels will be widely borne by the society at large and not unfairly impact workers and regions that are currently economically dependent upon fossil fuel extraction and related industries.

A central aspect of any such plan, therefore, would include programs to retrain current workers in fossil fuel industries, such as construction workers, miners, pipefitters, transport drivers, and technicians, for jobs in renewable technologies or other industries and to support them and their families financially as they make this transition to other jobs. It might also include subsidizing the development of new industries and business opportunities outside the energy sector in some regions so that displaced workers can remain in their communities. Such policies would also include regulations ensuring the right to organize and advocate for higher wages within these replacement industries. As a specific example, the California Climate Jobs Plan estimates that 112,000 workers in the fossil fuel sector (0.6 percent of the state's workforce) would lose their jobs and that even those who found replacement jobs in the clean energy sector would at least initially take a considerable loss in wages, and accordingly it proposes a generous transition package to help workers make and weather that shift.[56]

Yet, as many advocates of such proposals assert, social programs aimed only at workers and regions involved in the energy sector will not be sufficient to make the transition to renewable sources of energy truly just, equitable, or sustainable. The authors of HR-109 (the congressional Green New Deal resolution), for instance, called for a plan that would not only achieve net zero emissions, create millions of new jobs, invest in sustainable energy infrastructure, and ensure a safe and healthy environment, but also "promote justice and equity by stopping current, preventing future, and repairing historic oppression of Indigenous peoples, communities of color, migrant communities, deindustrialized communities, depopulated rural communities, the poor, low-income workers, women, the elderly, the unhoused, people with disabilities, and youth."[57] As the authors of that bill recognized, many vulnerable communities and minority populations were left out of the New Deal programs that led to the vast expansion of the US middle class, a mistake they wanted to make sure would not happen again.[58] Incorporating this principle into all policies related to the energy transition also ensures that members of such communities will not be forced to bear an inordinate part of the burden, whether in the form of low wages, involuntary relocation, or the loss of their land or communities to new infrastructure, mines, or waste disposal. Steps taken to create a more just and equitable society are also likely to build more widespread support for these and similar policies that promote the public good over the economic enrichment of the wealthiest portion of the population. Despite the scope and costs of these social programs, they will be essential to gaining support from workers, unions, business interests, and eventually perhaps, if enough people value or will advocate for them, politicians who are currently intertwined with fossil fuel interests, such as Democrat Senator Joe Manchin of West Virginia and Republican Senator and Minority Leader Mitch McConnell of Kentucky (both from coal states).

Yet, according to social scientist Matt Huber, just transition renewable energy proposals, so far, have been based on wishful thinking – they insisted on the scientific urgency to promote a large-scale social transformation "without first getting that organization in place with working class support." [59] Turning to the history of the New Deal, he noted that only after the workers got organized in the late 1930s and allied with other mass social movements for radical action, was Roosevelt forced to make the truly dramatic legislative changes of the latter part of the New Deal (see again Box 9.3). For Huber, organizing for the just transition must be centered in the working class, and especially among unions. Of note, in California the leaders of three fossil fuel–related unions, recognizing the current demise of oil, chose to join seventeen other unions in supporting passage of the California Climate Jobs Plan in 2021,[60] yet wider union support from some of the key stakeholders in the building trades and wider fossil fuel industry is still elusive.

Other social policies could also be employed to advance the transition to renewable energy by helping to reduce emission levels. Projects to build or retrofit affordable housing, for instance, could be located close to public transit, include all-electric appliances and plugins, be insulated for efficiency, and use district heating. It has been calculated that upgrades to residential energy efficiency in the USA could avoid emitting 0.5 gigatons of CO_2 per year by 2050, about one tenth of the US total.[61] Around the globe, ongoing rapid urbanization[62] could be accompanied by public transportation, on-street electric charging infrastructure, rooftop solar, public power utilities, district heating or advanced temperature control, and locally sourced food.

Strategies for Combating the Compulsion to Consume

Even if all of the technical, political, and economic changes outlined in this chapter were to be rapidly enacted, there is reason to believe that they would not be sufficient to reduce emissions adequately without some reduction in consumption, especially by the wealthiest consumers. As Stan Cox writes in *The Path to a Livable Future*, "[e]xcessive energy consumption is locked in by our society's dependence on cargo hauling, farming equipment, and manufacturing steel and concrete, all of which are powered via direct burning of fossil fuels and required by excessively large, electricity-hungry, houses; by over-size, tightly sealed office blocks, big box stores, and other commercial spaces that require around-the-clock energy-intensive climate control; by suburban and exurban sprawl fed by the slow crawl of commuter traffic . . . ; and above all, by an economic system that cannot function without overproduction."[63] As Cox suggests, if we want to leave a habitable world behind for following generations, we will need not only to adopt cleaner and more efficient sources of energy but to also find ways to consume less and differently. This follows on from the discussion in Chapter 4 which pointed out that decoupling – growing the economy while reducing emissions at the same time – is unlikely to be generally successful in most countries, and especially not at levels sufficient to achieve emissions reductions at the rate that is needed, and such concerns apply even to much narrower proposals such as the California Climate Jobs Plan, which is likely to suffer from the rebound effects we saw earlier, where efficiency savings from electrification are channeled into new carbon-intensive services and products.[64] Chapter 4 also questioned the ideological commitment to unrestrained wealth accumulation, arguing that we must put aside the pursuit of growth as a thing in itself so that we can end ecological destruction while also providing sufficiency for everyone. As this section discusses, some of the needed shifts can be made by us individually, while others will require more systemic changes within our communities and society beyond the large-scale addition of public housing, public transit, and district heating efficiencies discussed earlier.

One of the fundamental ways in which capitalism creates the compulsion to consume and grow is to create artificial scarcity. As Jason Hinkel explains in

Less Is More, one form of artificial scarcity is planned obsolescence, such as the way computer or smartphone manufacturers design and update their products so that buyers need to buy a new one every few years.[65] The material consequences are large – an iPhone, for instance, is packed with hard-to-access metals which are extracted from mines mostly in the Global South, and producing each one generates as much as 176 pounds of CO_2, even before shipping.[66] Programmed obsolescence could be reduced through mandatory warranties, building products with a right to repair, and building products to last. Indeed, right-to-repair regulations are currently working their way through US state legislatures. These would require equipment manufacturers to provide consumers and independent repairers access to repair documentation, tools, diagnostics, firmware, and service parts.[67] Consumers can also support companies such as Patagonia that have adopted a different business model in which their products are more expensive but they offer to repair all products, free of charge. Another example is the Fairphone smartphone company, which provides a device that is meant to last for many years, uses more recycled components, and ships with a screwdriver so you can replace components yourself.[68] Although some consumers are already willing to pay more for such products, the general success of such companies in a competitive environment would be enhanced by regulations that encourage such business practices and secure greater consumer rights.

Because the compulsion to consume upon which so much of the current capitalist system depends is deliberately generated by advertising, laws could be passed to restrict the advertising of products that have been shown to have a negative effect on the environment, as has already been done for tobacco, alcohol, salty foods, and other products known to negatively affect health.[69] Some cities have gone so far as to ban advertising altogether in public spaces, and other steps to reduce government support for unnecessary or damaging forms of consumption are possible.[70] For instance, the European meat industry received nearly a third of the European Commission's five-year promotion budget to help it counteract the decline in meat sales following an increase in the number of vegetarians and vegans across Europe, an expenditure of public funds that would seem to present a clear target for policy change.[71]

Wasteful and ecologically damaging foodways offer another realm in which changes in production and consumption practices could help reduce emissions. Currently about 50 percent of all the food grown in the world ends up wasted each year, incurring an enormous ecological cost in emissions that serves no human good. Changes at every stage of the supply chain – from sourcing food locally to serving smaller portions in restaurants to finding ways to reuse leftovers and food scraps – could in theory reduce the agriculture industry by half and reduce global emissions of greenhouse gases by as much as 7 percent.[72] Cattle grown for meat and milk are estimated to account for 4 percent of all the greenhouse emissions in the USA per year,[73] and increases in the global appetite for beef (which has been actively promoted through advertising and trade deals) have led to massive deforestation in South America and elsewhere to graze cattle and grow crops for animal feed (overall, animal agriculture amounts to a colossal 14 percent of greenhouse gas emissions).[74] Beyond consumers shifting their eating habits to obtain more of their calories from plant sources, ending subsidies to beef farmers and taxing red meat in high-income countries could help reduce demand for and production of more resource-intensive and ecologically damaging food sources.

As Chapter 4 showed, the compulsion to consume is also related to job market pressures upon workers to stay competitive with each other, requiring them to purchase cars, smartphones, and fast food to "save time." Some of this competitive pressure could be reduced by providing a universal basic income, an idea raised during the US presidential election of 2020 and which has been tried in a few US states and has been most developed in Norway.[75] Others have pointed out that merely reducing the US work week to European levels would reduce the country's energy needs by 20 percent.[76]

Another way to combat the compulsion to consume would be to encourage communal and cooperative approaches to ownership. Around the globe, numerous people and communities have created new ways to function for mutual benefit outside the capitalist mindset and often out of respect for the Earth, from community Wi-Fi systems in Spain to community forests in India, from neighborhood nursing teams in the Netherlands to the Open Science framework way of doing science transparently and sharing the data publicly.[77] One aspect of this movement that could have a dramatic effect on reducing consumption is encouraging a more general shift from ownership to usership, such as sharing garden shed tools, which sit idly 99 percent of the time, or sharing a car with neighbors for shopping runs. Car-sharing and bike-sharing programs have been adopted in hundreds of locations in the USA and around the world, as have online sites for reselling or giving away personal possessions to extend their useful life. All of these measures can enable people to live well without requiring perpetual growth in the economy or in individual consumption, and some of these ideas have even gained traction at the city or municipal level. The city of Amsterdam, for instance, has adopted a circular economy concept intended to reuse raw and other materials multiple times throughout their life cycle. Expanding upon the familiar three R's of the global recycling movement – *reduce*, *reuse*, and *recycle* – the Amsterdam program calls upon its services and residents to *refuse*, *rethink*, and *reduce* in the design

Table 9.1. *A framework for demand-side reductions: Avoid-Shift-Improve*

	Avoid	Shift	Improve
Transport	New urban public planning Teleworking	Shift from cars to cycling, walking, or public transit	EVs, especially buses Smaller, lighter vehicles
Buildings	Change thermostat set-points	Heat pumps and district heating	Insulation Energy-efficient end-use devices
Food	Reduce food waste	Shift from meat to plant-based protein	Reuse food waste Shift to fresh rather than processed food

Source: Adapted from Creutzig et al., "Towards Demand-Side Solutions."[78]

and use of goods; to *reuse*, *repair*, *refurbish*, and *remanufacture* to prolong the life of goods; and to *repurpose*, *recycle*, and *recover* goods at the end of their useful life.[79] A similar program has been adopted in Barcelona, and in the USA, citizens of various cities have organized cooperative movements to encourage the sharing of resources and reducing waste and fossil fuel use, including Seattle, Washington, and Jackson, Mississippi (in the USA some of these have been part of the social ecology movement).[80] Another framework for thinking about such demand-side ways to reduce consumption, called Avoid-Shift-Improve, can be seen in Table 9.1. Strikingly, in 2022, the IPCC report included some of these ideas for the first time in a chapter dedicated to "demand, services and social aspects of mitigation." It noted that behavioral and cultural changes represent an "overlooked strategy" and that there was high agreement among scientists that such changes "hold up to a gigaton-scale CO_2 emissions reduction potential."[81]

More radical possibilities exist and may prove to be necessary if we stay on the current high-emissions pathway and global heating reaches 1.5°C as soon as 2030, as predicted by the CMIP6 modeling we saw in Chapter 3. During World War II, the US capitalist system demonstrated that it was able to regulate production and consumption in all sorts of ways and with public support in the face of a widely recognized national priority. A similar response to the climate and ecological crisis could include a comprehensive industrial policy to direct the use of energy and other resources away from wasteful production and toward the essential goods and services needed for basic sufficiency for everyone. In a recent example of a strong but narrow policy related to emissions, the French government made financial aid to Air France during the pandemic conditional on the company's cutting the number of domestic flights to force customers to favor train travel.[82] And, as mentioned earlier, myriad and very dramatic state interventions were temporarily achieved in the USA and other countries in response to the Covid-19 pandemic, including the injection of trillions of dollars of economic stimulus, the provision of housing for the homeless, and compensation for millions of unemployed people, demonstrating that such interventions and policies are possible and can gain public support in the face of a perceived emergency. In 2022, President Biden invoked the Defense Production Act to address a baby formula shortage,[83] showing yet again that the government is quite capable of evoking war powers to intervene in domestic production in the public interest if it only has the will.

Conclusion

The only effective pathway to stopping global heating requires leaving fossil fuels in the ground and undertaking a transition to electrification with a near-total reliance on renewable sources of energy. This chapter has argued that the technical vision behind such a transition by 2050 is feasible in a country such as the USA, and surely many other countries too, but that achieving it will require a political framework that includes government and private economic investments, substantial government regulations to accelerate growth in the renewable energy sector, and a set of social programs to transition fossil fuel workers and build a new public urban environment around affordable housing and transit. But as this chapter has also argued, taking these necessary steps toward electrification will not ultimately be sufficient without contending with the tendency of capitalist systems to grow, which will require social, individual, and cultural changes to reduce consumption. Given that many of these changes will be met with considerable resistance from the fossil fuel industry and other current interests, making these necessary changes will require action by all of us and the advocacy of a vigorous and substantial grassroots social movement, which is the topic of Chapter 10.

10
Building and Taking Collective Action

> Walk the street with us into history. Get off the sidewalk.
>
> Dolores Huerta

As Chapter 9 argued, the kinds of changes that will be necessary to complete the transition away from fossil fuels in time to avoid the worst consequences of global heating are unlikely to take place without concerted government oversight and action, which in turn is unlikely to take place unless national decision makers are compelled to act by pressure from below. Although that pressure can take different forms, including advocacy by individual people and large nongovernmental organizations, the greatest impetus for significant change is likely to be grassroots collective action, which operates relatively free of elite or institutional control and derives its politics from the willingness to disrupt established institutional functions. As earlier chapters showed, action to cut emissions must reject false solutions and technofixes and focus instead on leaving the remaining fossil fuels in the ground and achieving a technical transition to renewable energy. While grassroots climate organizations have already offered several far-reaching proposals for achieving those goals and have had some notable successes in changing both public perceptions and institutional and governmental policies, the urgent need to accelerate confrontation and meaningful action across many levels of government and realms of society will require building a much broader and more powerful movement. This chapter explores strategies for growing and empowering such a movement, looking first at insights offered by the sociology of social movements and the psychology of collective action, then at the histories and tactics of three prominent grassroots climate groups in the Global North. Finally, the chapter turns to the various kinds of advocacy that can be undertaken by many more members of society more widely, recognizing that only a small proportion will join the grassroots.

Activating Grassroots Collective Action

As described earlier, the climate and environmental justice movement has been engaged in myriad struggles all over the world, some of which it has won. Just in late 2021, for instance, the Los Angeles County board of supervisors finally bowed to grassroots pressure and pledged to phase out all existing oil wells in the county; and the province of Quebec responded to the demands of a vigorous movement by declaring it would ban new oil and gas exploration.[1] But despite such victories, the larger movement still seems far too small,[2] considering the rapidly narrowing time frame within which serious emissions cuts must be achieved to prevent major disruptions in organized human existence. This section examines what lessons can be gained from research into social movements and the psychology of collective action about how to grow active participation in this movement.

Social Movement Theory

Social movement theory is an interdisciplinary field that explores the factors that influence the emergence, development, and effectiveness of social movements. Insights offered by practicing organizers and academics from the fields of history, sociology, and political science can help us better understand the main forms of organizing that activists may engage in, the different types of struggle they can wage, the varied ways in which they may frame the wider meaning of their struggle, and the locus at which social change is focused, each of which is considered in this section.

Forms of Organizing

Successful social movements vary their strategies for recruiting members and organizing their activities. Among researchers who have studied such movements, Mark and Paul Engler have identified three main approaches to such social organizing – community-based, mass mobilization, and momentum-based organizing – each of which offers different strategies for growing grassroots climate movements.[3]

Movements that follow the approach of **community-based organizing** (also called structure-based organizing), which was set out most famously in Saul Alinsky's 1971 *Rules for Radicals*, tend to be pragmatic, nonpartisan, and ideologically diverse.[4] As Alinsky noted, this approach is slow: "To build a powerful organization takes time. It is tedious, but that's the way it is played – if you want to play and not just yell."[5] Such organizations are focused on goals defined from the bottom up so as to meet the immediate needs of the local community rather than high-profile national goals. Alinksy framed the goals of such organizing in terms of people's self-interest (getting more local public transit) instead of lofty moral ideals, and he was suspicious of volunteer activists who were motivated by ideology. Alinksy's framing is an important insight for today's climate movements, which, as we will see, tend to be disproportionally made up of college students or middle-class professionals who are sometimes untethered from local communities. Despite being slow and requiring a long and constant local presence, community-based organizing also offers the advantages of being useful in building coalitions with other organizations and being more sustainable over time.

Another major approach to organizing, **mass mobilization**, utilizes the power of disruption by quickly drawing together a lot of people and leaving the establishment scrambling. As scholars Frances Fox Piven and Nelson Cloward have described, historical examples of this approach include actions by unemployed workers during the Great Depression, the industrial strikes that gave rise to unions in the 1930s, and the civil rights movement in the South in the 1950s and 1960s.[6] An advantage of this approach is that it can come together quickly and makes it possible for people who have few resources and little regular political influence to attract attention and force change. While this method of quickly organizing on a large scale could be key in specific political battles over reducing greenhouse gas emissions, a disadvantage of this method for movements that want to have a lasting impact is that its energy can quickly dissipate unless it is accompanied by a sustained organizational structure.

A third type of approach is **momentum-based organizing**, a hybrid form that attempts to combine both the short-term explosive potential of disruptive action (to bring in more people and gain attention) with the benefits of a strong leadership and administrative structure to keep those participants engaged. While organizations may experience some tension between adherents of slow, structure-based organizing and advocates of quick mobilization, research and the histories of the climate change groups discussed later show that groups that practice both of these approaches together are likely to be most effective in achieving their intended goals.

Types of Struggle

Scholars such as the Englers have also identified two different types of struggle that social movements engage in. One type is a **transactional struggle**, which typically involves working toward concrete legislative and legal victories, such as pressuring the governor of California to stop issuing permits for new oil and fossil gas extraction or pressuring a city to adopt a policy to require all new buildings to run on electric power rather than fossil gas. In contrast, the goal of

transformational struggle is to shift public opinion, often as a prelude to a later transactional win. The power of engaging in transformational struggle can be seen in numerous historical examples in which people with very few material resources managed to create change that many high officials considered absurd right up to the moment that those changes became seen as common sense. In the case of same-sex marriage, for instance, victories in state legislatures and courts and the eventual ruling in its favor by the US Supreme Court in 2015 reflected the end point of a person-by-person, family-by-family change of opinion that occurred over a decades-long struggle for LGBT rights.

As Figure 10.1 shows, achieving the transformational aims of a grassroots social movement requires shifting, or even pulling down, the specific pillars of support for the status quo that the movement hopes to change. These might include the media, business leaders, churches, labor, the civil service, the education establishment, and the courts. In one such example, the 2020 campaign to get the ten-campus University of California system to stop burning fossil fuels for heating and electric power generation focused on the pillars of students, staff, and faculty opinion, as well as the attitudes of campus and system administrators, including sustainability officers, chancellors, the president, and the board of regents.[7] This campaign began with an energy petition that was signed by 3,500 staff, faculty, and students and supported by university unions representing 50,000 workers. Further organizing led to meetings with individual campus chancellors and sustainability officers, on-campus protests, op-eds in the media, and a social media presence. This emphasis on shifting the beliefs of key pillars of the university system, undertaken by a small group, produced some fairly rapid success: within a year, several of the ten campuses had allocated money to make plans to shift away from using fossil fuels, and the combined faculty were voting on a resolution to reduce on-campus fossil fuel combustion by 60 percent by 2030 and 95 percent by 2035.[8]

As such examples show, shifting the pillars of support requires a core of energized, active supporters engaged in collective action who are willing to show up at rallies, protests, and meetings, to persuade others around them, and to act independently wherever they are positioned in society so as to push against the pillars they are closest to. As impossible as it might seem to wage a struggle against formidable fossil fuel interests and their political and institutional allies, it is helpful to remember that the eventual victories against British colonialism in India, apartheid in South Africa, and lynching in the USA were in fact the culmination of a long line of events and pressures, many of which at the time appeared small and even unsuccessful.

Based on these prior social movements, it is important for participants in the climate movement to regard progress in the transformational struggle, as

Figure 10.1

Types of social movement and types of movement struggle

(A) Two classic approaches to social movements are community-based organizing and mass mobilization. A more recent approach called hybrid or momentum-based organizing tries to combine their best features.

(B) Two ways for a movement to struggle. Transactional struggle aims for concrete wins, while transformational works to shift hearts and minds. The latter has also been described as kicking away the pillars of support. For climate action, this would be to remove support for the fossil fuel industry and electric utilities from the judiciary, academia, churches, media, and business.

(A) Types of social movements

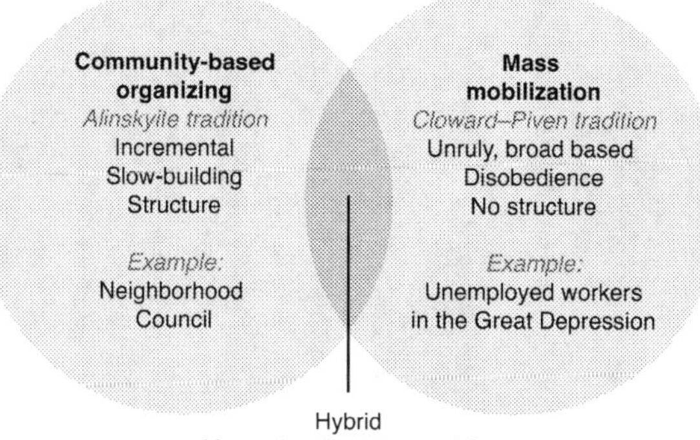

Hybrid
Momentum-based organizing
Has structure, also accommodates upwellings of protest
Example: Sunrise Movement

(B) Two ways for a movement to struggle

Transactional struggle
Win concrete goals such as legislation

Transformational struggle
Incremental struggle for public opinion
Kick away the pillars of support
(judiciary, academia, churches, military, business, and media)

partial as it sometimes seems, as a win, which should be celebrated as such. Take, for example, the city of San Diego's decision in early 2021 to approve a ten-year deal with the subsidiary of a fossil fuel company to continue providing electricity utility services, which on a transactional level was a resounding loss for the city's climate movements. At the transformational level, however, the campaign was at least a partial win: the relentless campaigning by different climate groups led to a lot of media coverage and wider recognition among the city's citizens of problems with the fossil fuel subsidiary and the need for publicly owned power providers, and the campaign efforts helped the local organizations build a coalition for future struggles. It can be highly motivating to climate activists to recognize that the process of attempting to achieve specific goals, even when not immediately successful at the transactional level, can help shift the pillars supporting the destructive status quo.

Frames of Meaning

As Paul and Mark Engler point out, transformational struggles often require a narrative about their moral significance. In the US South, for example, the Montgomery bus boycott in the 1950s began with a narrow demand for racial desegregation on city buses but soon became widely seen as part of a larger struggle for human dignity that energized its participants, attracted national attention and support, and provided moral legitimacy for its disruptive strategies. This additional element in the success of social movements is what social movement theorists such as Doug McAdam refer to as the **frames** they utilize: the shared meanings and cultural understandings that bind people together in a movement and create resonances among larger parts of the public.[9]

Specific frames that often motivate participation in social movements include shared feelings of grievance, threat, and anger, such as outrage at the poor treatment of animals, or, in the environmental case, the impact of fossil fuel extraction on the health of local communities. Larger **master frames** may also be shared across social movements, such as the notions of human rights or opposition to colonialism or globalization (such as the protests against the WTO mentioned in Chapter 4).[10] The existing literature has shown that movements that create persuasive and coherent frames are more likely to be successful and to persist over time and that master frames can be especially useful for coalition building across movements.

As McAdam has noted, one problem facing the climate change movement is that for most people in the USA, the issue is not linked to a salient collective identity in the same way that, for example, the Me Too and Black Lives Matter movements have been to many people. With the exception of a few serious climate activists, McAdam argues, climate identity is not the most important issue in the lives of most people, for whom the salience of the climate issue varies over time in relation to many issues, including the economy, weather patterns, media coverage, and their material needs to provide for households. As a result, no substantial group "owns" the issue. To change this, he advises climate change organizers to find ways "to establish a clear, compelling connection between the issue and one or more highly salient identities, thereby conferring ownership of the issue to those groups."[11]

One successful example of framing the climate issue to forge a stronger link to other salient identities has been the fossil fuel divestment movement on college campuses (which is discussed in more detail later in this chapter). Inspired in part by the earlier anti-apartheid boycott movement on US campuses, this movement has managed to link university community members' identification with their institutions to concerns about how that institution's money is invested in ways that impact climate change, thereby elevating the issue to an ethical and personal one.

Another attempt to connect the climate issue with identity is the environmental justice frame which tries to forge broader alliances between middle-class white environmentalists and people of color and Indigenous groups who are victims of environmental injustice, which arises from their exposure to toxic air pollution from fossil fuel infrastructure or from having oil pipelines built across their lands. Another current frame is the notion of a just transition, which, as Chapter 9 showed, tries to link climate and environmental struggles with workers' rights and interests, especially among union workers within the fossil fuel industry, some of whom face the demise of coal and oil.

At a higher level, a master frame that has been employed by the climate movement is the notion of climate justice. As Chapter 8 showed, climate justice has rich resonances, encompassing the historical responsibility of different countries, the disproportionate emissions of a tiny global elite, the exposure of false technocratic solutions, and a shift away from the technicalities of cutting emissions toward a focus on the poor and marginalized people who bear the greatest climate impacts while having the least responsibility for emissions.[12] While advocates see this master frame (of the ethical and political implications of climate change) as having enormous potential for motivating action, similar to a broad appeal to human rights, there are also potential problems. One problem is that the linkage of climate action with social justice sometimes encourages institutions to skirt the issue: for example, some institutions now go about addressing the climate concern with strategies to increase diversity and inclusion instead of dealing substantively with their ongoing burning of fossil fuels. Another problem is that the climate justice framing may not be inclusive enough. The broad political and social framework of effective action that was considered in Chapter 9 requires a politics of huge public investment, green urbanization, rewilding and afforestation, and a focus on human flourishing aside from growth per se. As the writer James Butler put it, "any such programme would need to garner the support not only of metropolitan liberals and the young, but to penetrate and revive the atrophied organisations of the old working class, to appeal ruthlessly to the desire of parents to hand on a better world to their children, and to recruit one pillar of the community for every activist or street prophet. It would need many more leaders and allies, interpreters and defenders at every level of culture."[13] This potentially broad alliance of which Butler speaks – which in the USA would include not only Indigenous peoples but also current and former fossil fuel industry, electrification, and other union workers, and presumably part of the wider white working class and college and high school students – certainly needs a master frame, but to date it is not yet clear what that is. A different argument for the climate justice frame to be more inclusive was made by Matt Huber in his book *Climate Change as Class War*. He faults the typical climate justice frame for making the problem too much about the source and impact of emissions (for instance, that the Global North caused the emissions, rich people exacerbate them, and the Global South and marginalized people experience the impacts) instead of focusing the struggle against the class that controls, owns, and profits from fossil fuel capital.[14] He argues it is a mistake to assume that the real environmental struggle will only emerge from those with a direct material relationship to land and pollution (the Indigenous and marginalized), instead of appealing to working-class people who make up the vast majority of the population and who live from the market and *not* the land and who face little directly apparent environmental threat to their livelihoods. He suggests the electricity sector as a place where workers could organize to confront capital and to improve their material interests in a way that could also win climate policy.

Locus at which Social Change Is Focused

Although confronting fossil fuel interests and making the energy transition will require significant action at the national level in the USA, as discussed in Chapter 9, several aspects of the current political landscape present huge obstacles for a movement hoping to achieve goals through legislation, including partisanship, gridlock, and the influence of enormous amounts of money in US politics, particularly from fossil fuel and electric utility interests. A 2014 study of ninety-one climate change countermovements funded by conservative causes in the USA found that they had a combined annual budget of more than $900 million, dwarfing the funding of even the best-funded environmental action organizations,[15] and contributions from fossil fuel interests have undoubtedly increased significantly since the Citizens United ruling allowed untold amounts of "dark money" to flow to politicians. The influence of fossil fuel interests has similarly held back the renewable energy transition in many other countries, including Canada, Australia, and Germany, and at the level of international negotiations. The UN Conference of the Parties 26 in Glasgow in

Figure 10.2

How support and opposition for climate reforms relates to the left–right political spectrum and the interests of capital and labor

Economic winners for climate policy are businesses and workers in the low-carbon sectors, while economic losers are businesses and workers in the high-carbon sectors. There is larger support for climate policy reforms from the left, but not entirely; there are supporters and opponents on both the left AND the right.

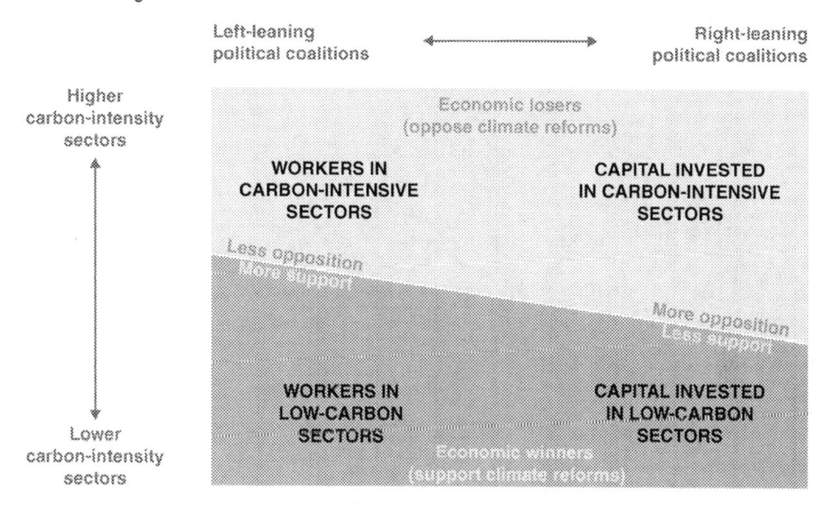

Adapted from Matto Mildenberger, Carbon Captured: How Business and Labor Control Climate Politics. MIT Press, 2020.

2021, for instance, was attended by more than 500 fossil fuel industry executives among the nations' delegates, which exceeded the combined representation of all the Indigenous representatives from around the globe.[16]

Further, as Matto Mildenberger points out in his book-length analysis of energy policy in the USA, Norway, and Australia, the political problem facing the climate change movement is more complex than simply whether the governments in those countries are currently under the control of liberal or conservative parties, as carbon polluters have considerable influence within all major parties, including Republicans and Democrats in the USA, albeit somewhat less on the latter (Figure 10.2).[17]

Although none of this means it is pointless for grassroots social movements to try to influence national policy directly, the difficulties involved and the history of earlier social change efforts suggest that grassroots movements may need to focus most of their energy on the tried-and-trusted strategy of making local change first. Indeed, many successful attempts to create social change, such as the movement to legalize same-sex marriage or raise the minimum wage to $15, did begin locally and then spread to many states before becoming national policy.[18] And the basic environmental laws that many take for granted in the USA didn't come out of nowhere, they were issued by President Richard Nixon in the early 1970s. Yet Nixon, who was certainly no environmentalist but instead an incredibly cynical and conniving politician from the beginning to end of his career, ended up signing the Clean Air Act, the Clean Water Act, the Endangered Species Act, and other environmental laws, only because he was compelled to by the political climate that arose from many years of mobilization at the local level, the passage of local anti-pollution measures throughout the country, and huge lawsuits against polluting corporations.[19]

Social Psychology Theory

Whereas social movement theory carries lessons for how to grow the grassroots movement at a group or social scale, the field of psychology carries lessons for how to motivate individuals. Specifically, it might help us understand why some people are more likely to enter into these movements and what strategies might make the movement more successful at attracting and retaining them. As discussed in Chapter 7, psychology research has shown that people with stronger perceptions of the risk of climate change have higher biospheric values, and these in turn are shaped in part by personality (especially openness at the aesthetic level) and early life experiences, such as role models and exposure to nature. Yet, beyond this, research in the specific area of social psychology, which deals with social effects on the individual, could reveal psychological variables that specifically affect the decision to join collective action.[20] As some of these influences might be more modifiable than personality or early life experience, a better understanding of which variables are key could be useful for growing grassroots social movements

While many studies in social psychology have looked at collective action over several decades, they mostly focused on the anti-discrimination, anti-nuclear, and broader environmental movement rather than climate action, and the research was mostly based on survey reports about what people said they would do, or what they had done, rather than objective verification of their activist behavior.[21] A 2008 meta-analysis of 182 such studies showed that two psychological variables – identity and efficacy – relate to self-reported engagement in collective action.[22] The first of these, identity, comes in different forms: there is self-identity, which is the sum of all one's attributes (e.g., father, musician, New Yorker), and **collective identity**, which is related to one's sense of membership of a social group.[23] When a group to which one belongs locates an external body, authority, or enemy that represents power and against which the group feels a grievance, that collective identity may become politicized. For example, although any group of women may develop a collective identity, once their identity as women becomes politicized they may become feminists.[24] Having developed a politicized identity, the group may now choose to engage in collective action. The second variable identified by the meta-analysis is efficacy. This also comes in different forms: **self-efficacy**, which refers to an individual's belief that they can accomplish whatever they want to and that they can use their skills to perform a behavior that will lead to a desired outcome,[25] and **collective efficacy**, which refers to an individual group member's belief about what the group can do, which could be summed up by the saying, "we can do this, people!" [26]

Moving past the 2008 meta-analysis of 182 studies on collective action, more recent research, which included a focus on climate action, suggested the importance of additional psychological variables such as social norms (i.e., what the people around one are doing or one thinks they expect one to do) and **participatory efficacy**, which leads a group member to turn up for a climate rally or meeting with city officials because they believe that their participation is essential to the group's effectiveness.[27]

Although there has been scant further psychological research on what drives people to join collective action specifically on the climate crisis,[28] earlier chapters of this book strongly suggest additional variables beyond the biospheric values and open personality mentioned above, such as level of knowledge about the human cause and high threat perception.[29] Other variables that are also likely to be important are one's beliefs about how change is made. For example, someone who thinks climate action is relevant only for China to undertake (as the biggest current emitter) is less likely to want to take action in the USA, while someone who thinks climate action is the province of national policy will be less likely to act locally. Indeed, having a **theory of change** about the centrality of local action is probably a necessary criterion for joining grassroots action of any kind. So too, perhaps, is having the correct level of **faith in institutions**. As illustrated in Figure 10.3, too much faith may lead individuals to overly trust government and decision makers to make necessary change and therefore to not want to do anything themselves, whereas too little faith may make them too cynical or skeptical of the prospects for change to act. Instead, the "sweet spot" is probably an intermediate level of belief. This would be consistent, for instance, in believing that, based on its own history, the US Democratic Party is not by itself likely to produce serious climate policy, but is amenable to taking

Figure 10.3

To increase grassroots action, target people who have low/medium faith in current political institutions: they doubt action can happen now, but they believe it can with a push

Strong skeptics of capitalism such as revolutionary Marxists, who do not believe that any elite government will be capable of making social change, will not waste their time with grassroots climate action – they are working for a wider workers' revolution. At the other extreme, people who have strong faith in political institutions such as the German Greens, British Labour Party and the US Democratic Party might be unwilling to engage in grassroots action, beyond perhaps getting their preferred party into power. The middle level of faith in institutions is a "sweet spot" for grassroots climate activism.

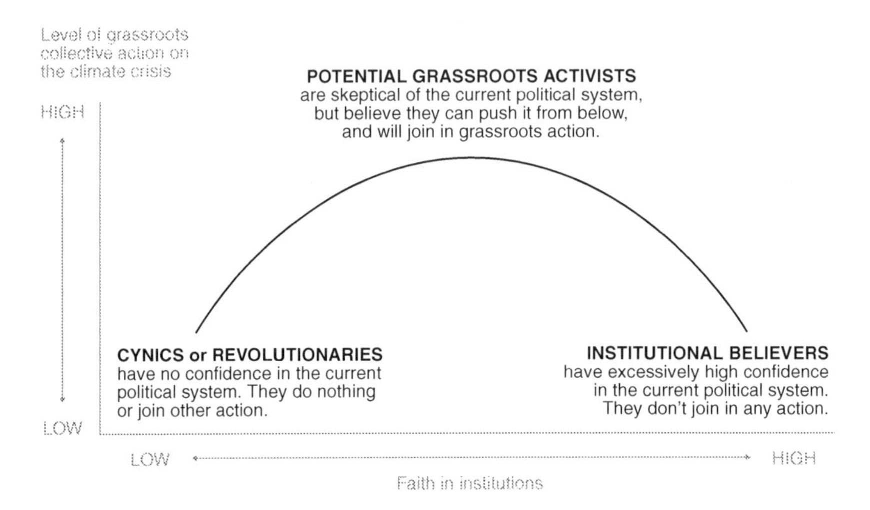

bold action if grassroots pressure grows strong enough (which was vindicated by passage of the Inflation Reduction Act of 2022).

Although, as noted earlier, biospheric values arising from personality and early life experience are unlikely to be modifiable, many of these other psychological variables probably *are* modifiable. Understanding which to focus on, and how to boost them with educational interventions, is an important practical question facing social science, given the stakes for life on Earth and given how few sustained activists are currently organizing in the climate movements. What is needed is a research program that tests which specific psychological variables are causal to joining grassroots climate action, and also objectively verifies whether study participants did so. Figure 10.4 shows an example of a possible experimental design to increase the psychological factors underlying grassroots activism. Figure 10.5 provides an overview of all the psychological factors discussed in this section of the book

Figure 10.4

How could we design a study to test what interventions increase grassroots activism and what psycho-social factors predispose to it?

The figure shows a hypothetical study based on one that was recently run. Two groups of human participants have their attitudes and activist behavior measured at baseline. Group 1 then gets a real intervention, such as videos or other kinds of training about the climate crisis and activism, and Group 2 does not. Then attitudes and activist behavior are measured again. The prediction is that the real intervention will boost activist behavior; if it does, it will be possible to discover which psychological factors change and how they were affected by the intervention.

Using interventions to increase grassroots activism

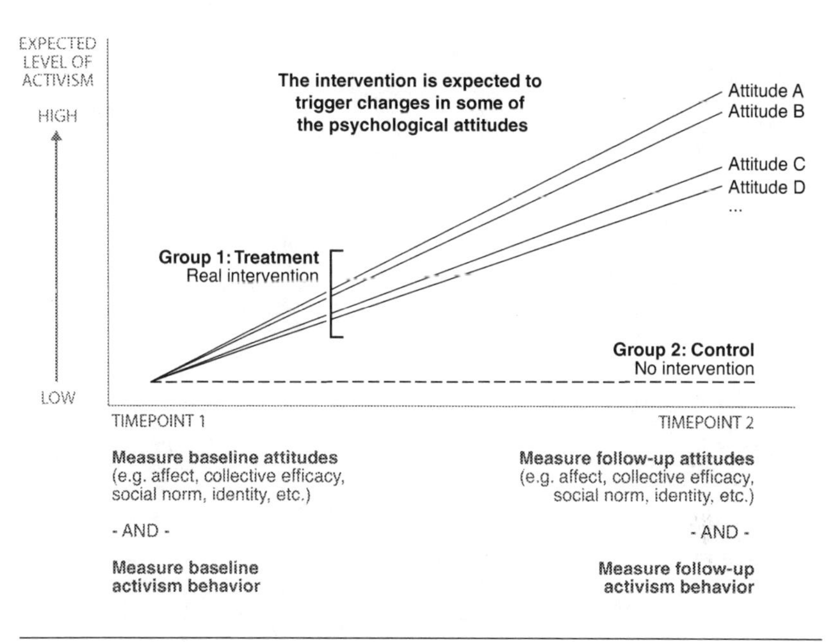

Adapted from Castiglione A et al. (2022). Royal Society - Open Science

Examples of Climate Activist Movements

A better understanding of how to grow grassroots action can surely benefit from examining some of the main existing climate change groups to see what appears to have worked well and what does not. For that purpose, this section examines three of the most prominent such groups currently operating in the Global North: 350.org, Extinction Rebellion, and the Sunrise Movement. As it will show, these groups have taken quite different approaches, reflect different central features of social movements, and have received different sorts of criticism, as summarized in Table 10.1.

Figure 10.5

What does psychology tell us about how we can encourage collective action?

Examples of variables which can lead to collective action for grassroots climate groups.

Established in literature | Hypothesized by author

Self identity
"I am a person who believes that climate change is an issue that needs to be addressed."

Collective identity
"I feel connected to my group, which is struggling for action on climate change."

Politicized collective identity
"My group confronts those in power to remove obstacles to climate action."

Self-efficacy
"I believe that my actions will increase the chances of the government changing its approach to climate change."

Collective efficacy
"I believe that my group's actions will increase the chances of the government changing its approach to climate change."

Participatory efficacy
"I feel as though the climate change movement will not do as well without me."

Sense of injustice
"I am angry because climate change impacts are felt unequally."

Social norms
"Others around me are concerned about climate change."

Overall effect
"I feel that climate change is very unpleasant/ unfavorable/negative."

Threat perceptions
"I believe that climate change is a threat."

Faith in institutions
"I believe that our institutions cannot address climate change in their current state."

Theory of change
"I believe that local change is the critical foundation for climate change action."

Biospheric values
"I am connected to the earth, and things that affect it will affect me."

Open personality
"I am good at abstract thinking, appreciate variety, and am open to unusual experiences."

COLLECTIVE ACTION FOR GRASSROOTS CLIMATE GROUPS
You struggle as a group to push for fossil free energy and fossil free finance

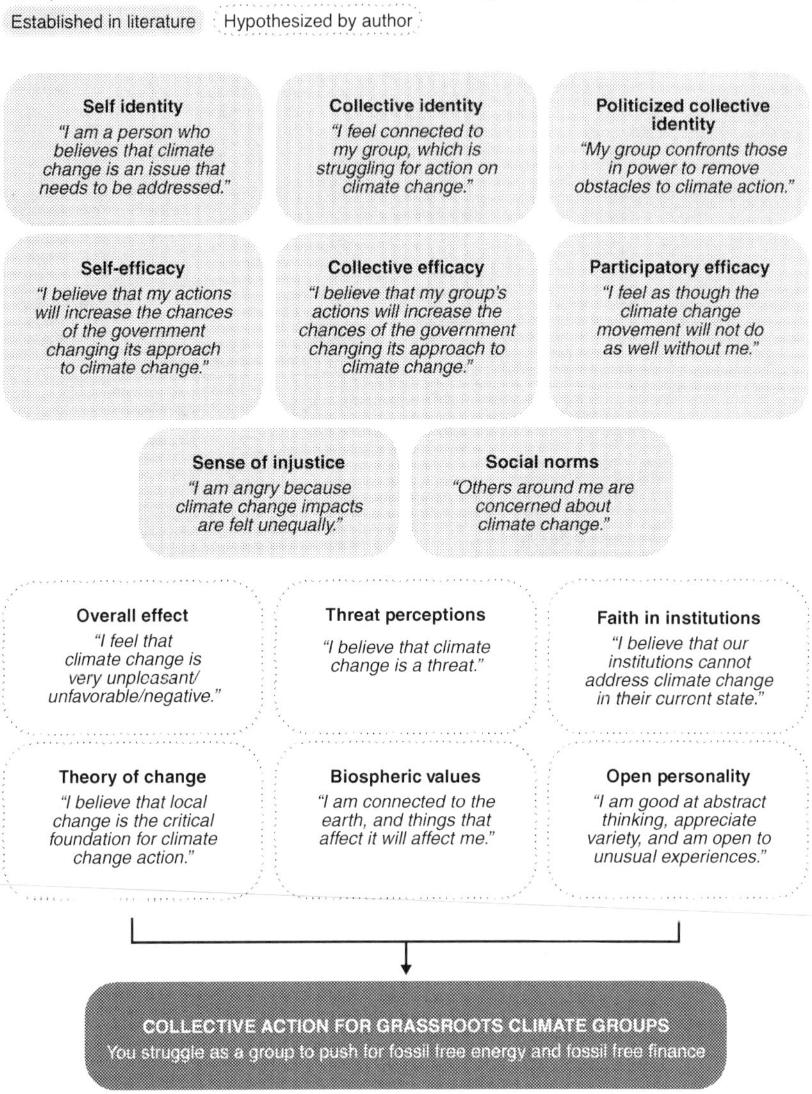

Table 10.1. *Comparison of three grassroots social movements*

	Type of movement	Type of struggle	Frames	Theory of change	Critiques
350.org	Mostly structure-based organizing; occasional protest	Mostly transformational struggle; some transactional campaigns	International climate justice; fossil fuel divestment (ethical issue)	Mostly local action through chapters	In the USA, needs to diversify beyond middle-class, white, older people
Extinction Rebellion	Mostly rebellions/ protest; some structure-based organizing	Mostly a transformational struggle to tell the truth and develop a moral perspective	Extinction; Tell the Truth	Initially focused on national policy shift, big actions in London	Needs to move beyond the mass arrest strategy, focus on local finance, energy infrastructure; in the UK, needs to include more racially diverse and working-class members; needs to focus more on "solutions" than criticism
Sunrise Movement	Tries to have both structure and fast mobilizations	More emphasis on transactional struggle; tries to win legislation in Congress and state legislatures	Just transition; Green New Deal; intergenerational justice	Initially focused on national policy shift (Green New Deal legislation); later chapter work focused at state and city level	Needs to be more rooted in local communities, which is challenging for a mostly middle-class, student group that is mobile; needs to connect with union organizing

350.org

The climate action organization 350.org was founded in 2007 by a group of students and the American environmentalist and author Bill McKibben.[30] The name 350.org stands for 350 parts per million of CO_2, which some scientists have identified as the safe upper limit to avoid disruptive climate impacts.[31] Since then, 350.org has grown into an international organization whose goal is to help end the use of fossil fuels and hasten the transition to renewable energy by building a global grassroots movement. It has since mobilized thousands of people through the activities of multiple chapters in nearly 200 countries.

In what CNN called at the time "the most widespread day of political action in the planet's history,"[32] 350.org mobilized people in 181 countries to come together ahead of the COP 15 meeting in Copenhagen in 2009 with a single message for world leaders: they must produce a fair, ambitious, and binding climate treaty to stay below 350 parts per million of CO_2 in the atmosphere.

The group has initiated many campaigns but perhaps the best known, which started in 2012, was the fossil fuel divestment movement. The organizers argued that if the existing fossil fuel industry stockpiles of coal, gas, and oil were actually burned, we would blow through our remaining carbon budget five times over.[33] The resulting campaign soon galvanized local groups at campuses across North America, and later more widely, to demand that their universities take their financial holdings out of fossil fuel stocks and bonds.[34] To date, dozens of universities have claimed they have, or will, remove their investments from fossil fuel extraction, including the University of California, Oxford, Harvard, and Columbia, albeit the last two with such a late date, 2050, as to be practically meaningless.[35] While these institutions have usually not framed their action in terms of the moral imperative of divestment, preferring to call it **derisking**,[36] (which means they could in principle buy the stocks and bonds back when the financial risk changes), and while it is usually not possible to transparently verify their actions or to establish a bottom-line impact,[37] the divestment movement has clearly had important effects. For example, it has focused the attention of hundreds of thousands, if not millions, of faculty, staff, and students on the need to leave fossil fuels in the ground, which is part of a transformational struggle that has also spread to other campaigns, such as those focused on state pensions, whose investments in fossil fuel extraction in California alone total over 80 billion.[38] The fossil fuel divestment campaign has also brought a lot of people into the wider climate movement, including the youth organizers Sara Blazevic and Varshini Prakash, who went on to found the Sunrise Movement, and it led 350.org to develop a later campaign known as Stop the Money Pipeline, focused on banks

Figure 10.6

Since the Paris agreement of 2015, the top banks have poured money into financing the fossil industry

Individuals and institutions need to shift away from using the large banks and insurers that finance climate chaos to instead use credit unions/alternative banks and alternative insurers. Raising consciousness about this is a valuable target for activism.

Fossil fuel financing (2016–2020)

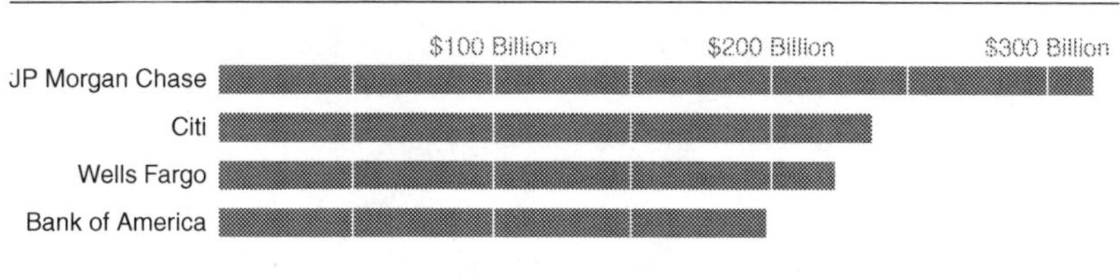

Adapted from Rainforest Action Network. Banking on climate chaos – Fossil fuel finance report 2021. Available online at https://www.ran.org/bankingonclimatechaos2021/

and insurers. This new campaign noted that, since the Paris Accord in 2015, JP Morgan Chase had financed the fossil fuel industry by over $400 billion, with Bank of America, Citibank, Wells Fargo, and other banks not far behind (Figure 10.6).[39] While these banks financed fracking, Arctic oil and gas exploration, coal mining, and tar sands oil, insurance companies, such as Liberty Mutual, were behind the Transmountain pipeline and other projects. So far, Chase responded to the campaign by taking the relatively minor step of announcing in February 2020 that it would stop financing new oil and gas drilling in the Arctic, but a more notable response, also in early 2020, was from the money manager BlackRock, which announced changes to its investment approach, including disclosing climate-related risks.

Thus, 350.org achieved international prominence and reach, which reflected its strong structure-based organization strategy. It developed dozens of chapters in the USA, some quite large, where it engaged in a mix of both transformational struggle, such as to shift attitudes and beliefs around the Stop the Money Pipeline campaign, and transactional struggle, such as partnering with other organizations to win local legislative changes by pressuring state officials to cancel new drilling. The frames it used were climate justice, environmental justice, and fossil fuel divestment as an ethical approach. Its theory of change appeared to be focused mostly on creating local shifts to undergird national policy later.

Members of 350.org have engaged in diverse actions, including civil disobedience outside Chase Bank locations, organizing marches, providing training, pressuring local congresspeople, engaging in outreach to students in high schools, and working with local cities to develop clean energy plans.[40] This diversity of approaches is a strength in providing opportunities for different kinds of people to be involved. One possible weakness that limited 350.org's growth potential was the perception that its volunteer base was mostly older, white, middle-class retirees, although, to be sure, those members were also hugely valued for providing a dedicated and sustained community-based substrate for further organizing, and meanwhile many chapters were making progress in diversity and inclusion.

Extinction Rebellion

An altogether quite different grassroots movement is Extinction Rebellion, which first burst into public awareness in London in October 2018, when it received major attention from the news media for blocking access to Parliament, for using music, dance, and elaborate costumes, and for adopting a mass arrest strategy.

The group was founded by a small number of activists with previous experience in climate and human rights organizing and the Occupy movement.[41] The group adopted a highly confrontational direct action approach that followed a long tradition of transgressive environmental action, such as anti-nuclear protests, but was also novel and particularly effective in its emphasis on grief about ecological and human loss and its alarmism about the severity and speed of the crisis. As a result, the group was able to quickly mobilize tens of thousands of people, some for the first time, to take illegal disruptive protest actions, such as blocking bridges and roads.

The group's approach was inspired by earlier transformative civil rights movements, such as those associated with Gandhi and Martin Luther King, and particularly influenced by co-founder and Kings College London researcher Roger Hallam's studies of effective civil resistance against authoritarian regimes.[42] Specifically, the mass arrests at the center of the group's strategy were designed to overwhelm police resources, escalate costs, and fill up jail cells to create a level of disruption sufficient to force the government to take the climate crisis seriously.

In April 2019, Extinction Rebellion staged what it referred to as a second rebellion with actions at multiple sites in London and the wider UK, which led to more than a thousand "rebels" being arrested. As a result of these actions, discussion of the climate crisis penetrated many levels of civic society, media, and politics. The following month, Parliament declared a "climate emergency," making it the first country to do so. In October, Extinction Rebellion launched what it termed an International Rebellion, staging actions in sixty countries, which in London included protests against institutions for their financing of the fossil fuel sector and a faux funeral march down a major London street that attracted about 20,000 participants.[43]

Thus, at least in its first two years, and centered mostly in London, Extinction Rebellion had a small core of structure-based organizers but also relied strongly on mass mobilization. Its goal for struggle was mostly transformational, raising awareness by using the frames of extinction and climate and ecological emergency, and its theory of change appeared to be to jolt political change through raising moral awareness.

Although Extinction Rebellion has clearly achieved its first stated goal, which was to force more people in the UK, and especially within the government, to tell the truth about the climate emergency, it has made much less progress toward its second goal of decarbonization or third goal of making changes to the democratic process via citizens' assemblies. Although many institutions, cities, and countries across the world have now declared climate emergencies, such declarations do not themselves lead to genuine emissions reductions. Those disappointments experienced by Extinction Rebellion and some internal conflicts about strategy, control, and racial politics led to the splitting away of some of the founding members and internal reflection about strategies and tactics.

Specifically, some critics of Extinction Rebellion noted that the organization failed to attract large numbers of nonwhite participants and that it had not dealt with its class privilege; nor had it developed close contacts with the communities whose support it would need to become a mass movement,[44] echoing a more general critique of most environmental organizations in the Global North, including the other two discussed here. Resolving this problem may perhaps require applying a more Alinsky-like approach to slow, structure-based organizing that centers a movement in the surrounding community of stakeholders. Another critique, made by journalist Nafeez Ahmed, was that its reliance on mass arrests directed at overwhelming the police represents a misreading of social science research.[45] Ahmed argued that while the mass arrest strategy may have worked for the American Civil Rights Movement and Gandhi's movement, it may be fundamentally out of place in the climate movement, as this model simply cannot be transplanted to the modern Western context, where the structures of power are much more complex, the repression much more invisible, and the institutions being targeted are only indirectly related to the problem being addressed.

As it emerged from the Covid-19 pandemic, Extinction Rebellion appeared to shift tactics by broadening its effects beyond the state to a focus on companies, banks, IT, agribusinesses, and the news media, such as disrupting the distribution of multiple newspapers owned by Rupert Murdoch's News Corp, the same organization that controls Fox News in the USA, in response to its failure to adequately report on the climate emergency.[46] Indications were that Extinction Rebellion in the UK was attempting to increase its diversity and to link its struggles to class inequality: for example, as noted earlier, in 2021, it conducted an action that blockaded the private Farnborough Airport near London with signs declaring that "1% of people cause 50% of aviation emissions." [47] Further, responding to the criticism that it was all about castigation and not about "solutions," Extinction Rebellion launched the Insulate Britain campaign.[48] Activists blocked motorways and planned repeated actions until arrest to draw attention to their demand for the government to make the investments to insulate housing, a very substantive way to reduce emissions through increased thermal efficiency, and also an equity issue that might draw in more people from the center and center right who were concerned about the cost of electricity and gas bills. In 2022, members of Extinction Rebellion also launched the campaign Just Stop Oil, which aimed to draw attention specifically to fossil fuel infrastructure, for example by trying to block trucks from oil terminals.[49] In perhaps a sign that these tactics of blockade and civil disobedience were having an effect, the government proposed a repressive protest crackdown bill which would obligate the courts, which had often been sympathetic, to start jailing activists.[50]

Meanwhile, hundreds of Extinction Rebellion chapters have flourished around the world, from Paris to Lagos, from Delhi to Cape Town – indeed, Extinction Rebellion in Cape Town managed to temporarily, at least, stop Shell Oil from prospecting for fossil fuels off the South African coast.

Sunrise Movement

Unlike 350.org and Extinction Rebellion, which had an international dimension, and includes a wide range of ages, the Sunrise Movement has been tightly focused on electoral politics in the USA and is deliberately a youth movement.

Sunrise was launched in the USA in 2017 by students who got their start in climate change activism in the fossil fuel divestment movement on college campuses and in the pipeline protest movement.[51] Its initial goal was to elect proponents of renewable energy in the 2018 midterm elections, when, indeed, half of the group's first twenty endorsements won their elections.[52] After that election, the group shifted its focus to gaining a consensus in the Democratic Party in support of the Green New Deal, a plan that, as discussed in Chapter 9, included the core principles of decarbonization, jobs, and justice. By late 2019, Sunrise leaders estimated that around 15,000 young people had shown up at in-person actions across the USA and 80,000 had participated in mailings and less direct actions.[53]

The tactics that have been used by Sunrise at the US capitol, state houses, and Democratic National Committee meetings are reminiscent of those of the civil rights movement of the 1960s: singing loudly, then standing quietly as officers tied their hands for arrest. Sunrise also mastered digital organizing, using the social media program Slack and Google documents to host digital field offices that have no walls and no set hours, which is easier for students and gig workers. Sunrise uses carefully crafted imagery and short and punchy videos, a striking example of which was a confrontation between middle-school students and Senator Diane Feinstein of California.[54] Sunrise clearly understands well the impact of youth speaking to power, especially on the topic of intergenerational justice, emphasized by youth telling personal stories of how they feel and what they expect to experience.

In perhaps the most famous Sunrise action to date, in November 2018 members occupied the office of Nancy Pelosi, the top Democrat in Congress, and demanded that all members of the Democratic leadership refuse donations from the fossil fuel industry and that Pelosi herself work to build consensus in the Congress over Green New Deal legislation. About 250 members of Sunrise used their loud tactics to disrupt the office of the speaker of the house and persisted even after fifty-one were arrested. Within weeks, their ambitious demand for a Green New Deal was on the lips of every congressional staffer and progressive candidate for president in the country, and the HR-109 Green New Deal resolution was presented to Congress.[55] By August 2020, President Biden had proposed a $2 trillion climate program that, although not called the Green New Deal, was the largest climate plan yet proposed by an administration.

Sunrise initially appeared to be stunningly successful in what it set out to do: a few thousand young supporters, distributed around the USA, were able to help make the climate crisis a central topic of discussion in the Democratic Party and to make it politically relevant. Part of this success may be attributable to Sunrise's observing one of the key principles of good movement organizing discussed above: framing. As one of the founders, Varshini Prakash, observed, few people get excited about the topic of decarbonization in its own right, so she and her colleagues at Sunrise developed a vision for solving climate change through the creation of green jobs, which "are things people intuitively understand because they relate to their everyday lives." Although "climate activists are always frustrated people don't care about their issue," she believed that "the real problem is we're not listening to what people care about."[56]

Thus, Sunrise soon achieved national prominence by reflecting a mix of structure-based organizing and explosive mobilization/protest: the momentum-based approach. Its original impetus was an openly transactional struggle to produce specific national policy outcomes, and it used the frames of inter-generational justice, the just transition, and the Green New Deal.

But by mid-2021, some critics had begun to point out that the group's activities were having little impact on the few Democrat senators who were holding out on climate legislation, that media attention to Sunrise was generally waning, and that the kind of multiracial populism it was advocating was out of touch with the realities of the American body politic.[57] These criticisms echo the broader critique leveled at the other groups discussed in this section: that if a grassroots climate change movement is to become successful, it will need to penetrate a much broader cross section of the population than its mostly middle-class, well-educated members from metropolitan areas. For all of these groups, becoming more effective is likely to require the development of a mass base of support around the country and across current class and racial divides.[58] A possible remedy for the Sunrise Movement might be to shift its focus from influencing politicians directly to building mass support among working people, including fossil fuel workers and union workers, by helping them see the renewable energy transition as in their own best material interests.

All Together Now

Overall, 350.org, Extinction Rebellion, and the Sunrise Movement have achieved strong success in different ways but are still struggling to increase their numbers. Perhaps the common denominator for the future of all three is to find ways to make their climate and ecological struggle of interest beyond the relatively tiny number of environmentalists who have the class, race, and leisure-time privileges to engage in it. (See, for example, the comments of an expert organizer in Box 10.1.) This may require different kinds of framing and perhaps a renewed focus on community-based organizing among communities of color and union workers. Another very promising direction is coalition building with other organizations involved in climate, environmental, and social justice.[59] One example was the formation of the San Diego Green New Deal Alliance during the Covid-19 pandemic in California in 2020. A big-tent alliance of more than sixty local organizations soon sprung up

Box 10.1 Interview with Masada Disenhouse

Masada Disenhouse is a nonprofit administrator and program manager who for more than a decade has empowered people to organize, advocate, and campaign to build grassroots political power, most recently as executive director of SanDiego350.

AA: Can you trace what it was in your early life experiences that set you up to be a grassroots organizer, and specifically one so focused on the climate crisis?

MD: I feel like a lot of it just comes out of my personality. I have a strong sense of right and wrong and moral outrage. And I think I got that from my dad, who is also an organizing type personality. And then I think I got a real love of nature from my family, which used to camp and hike. I do *not* get my politics from my family, which is very conservative – I've always been very progressive.

I started being an activist in high school. I distinctly remember reading the summary of the second IPCC report which had just come out, and I just couldn't believe that there was this huge problem that nobody was talking about and seemingly nobody was doing anything about, and I was kind of freaked out by the whole thing.

Working on Ralph Nader's Green Party presidential campaign in 2000, I really saw the value of long-term movement building versus short-term political campaigns. If you're not doing long-term movement building between campaigns, you don't have a base to draw from. It showed me that if you want to get people civically engaged, you need to create the structures and the resources that let people plug in when they're available and for the small amounts of time that they're available. It's not just about climate – I developed a passion for helping people understand that they *do* have political power and how to use it on any issue that they might be interested in.

AA: When some students in my university get their eyes on the climate and ecological crisis, they want to do something about it. But they often say they don't know what to do. What are your recommendations for people just starting to get involved in the climate movement? As an organizer, how do you match individual people with what they can do?

MD: Start by talking to everybody you know about why you care about this issue and what you're personally doing about it. Movement building is related to how much people get the word out and how much you convince regular people that this is an important issue, that it's a problem that affects them, and that something needs to be done. Everybody needs to realize that they're an influencer.

Whether it's on social media or talking to people in person, everybody has their own networks. My Step 2 is always going to be joining a group. In this country, unless you are a rock star or have a ton of money, pretty much the only avenue that's left to you for exercising serious political power is through organizing, and that means working with other people in some sort of structure. So I really recommend joining a group and learning how you can contribute. For it to be sustainable, for you to continue to contribute to that cause, you must do work that's meaningful and rewarding for you. There are a lot of different ways that people can contribute. You don't have to do everything.

About how to match people, the role of the organizer is to talk to people and find out where their passion is and what they like to do. I like to listen, then make some suggestions of things to try out, so they see if it's a good fit. I also do like to push people to try things out of their comfort zone. I think that current political power in this country relies a lot on people accepting the status quo, and there are a lot of social norms around what's acceptable and what's not and how to engage as an activist and advocate, and change doesn't happen without disrupting those norms.

AA: Your organizing work is mostly at the local, regional, and state levels. Yet we need a huge energy shift, ideally driven by national policy. How does your local work connect with the national and international levels? How do you motivate someone new to join the grassroots when the changes that are needed are so much bigger than what we can accomplish locally?

MD: First of all, I think that the local matters a lot because it's where you have the most influence, right? It's where you understand how politics works. Getting things done in cities and regional agencies like SANDAG [San Diego County's transportation agency] is where we can build power and really affect things. Also, accomplishments in a city can serve as a model. For example, San Diego passed one of the most aggressive and accountable climate action plans back in 2015. It was one of the few big city models that were available at that time, so it had a far-reaching impact on other cities. I also think belonging to a network is important. My local organization is affiliated with the national and international organization 350.org, and we also build relationships with lots of other local organizations. Having those networks is really important, so that occasionally you can bring everybody together and bring all that power to bear on something specific. For example, we've participated in big days of action, where people show power in the streets. You can get lots of people to sign petitions or make phone calls, you can get a significant number of people from each area to participate in key moments, like the Keystone XL protest at the White House or the Dakota access pipeline protests. I also think that sharing information and resources across networks is very important, so everybody doesn't have to reinvent the wheel. Just look at the Black Lives Matter movement this last year. People turned out in their own city, and some were able to get changes to the policy in their own city, but you can see that the whole nation shifted on this issue, and similarly on other issues. I like to give the example of same-sex marriage. It went from being something that was unthinkable when I was in high school, a hot-button issue that nobody would talk about, to something that successfully passed, and is now completely supported by a lot more than 50 percent of the people in this country. Within a generation, not a long time. And the reason it became so acceptable was mostly because people decided that they were going to go and talk to their families, right? It's hard to tell someone you don't know about the right to same-sex marriage, but it's a lot easier when you're a sibling or a child. I think that can be really powerful. And I think it goes back to the relevance of working locally. If we all push in the same direction, we can achieve that national-level shift.

AA: What role does money play in your grassroots work? Can you define grassroots? Do you need more funding, and what would you do with it? At what point might your approach be compromised by (large) sources of money?

MD: Grassroots organizing is all about having accountability to the community that you work in and being driven by the people in that community as opposed to top down. Our organization is very volunteer-led. We have about twenty different volunteer-led teams and decisions are mostly made by the people in the teams, as long as what they're doing is consistent with our mission. And we bring team representatives together to make organizational decisions. In terms of money, I would say that, yes, it's very helpful for getting things done. It's really important that most of our money comes from individuals, though some of it comes from individuals who have a lot of money and give us a good chunk of money, and we definitely try to stay in their good graces. In the non-profit universe there are grants, but there are very few foundations who will give money toward disruptive political work. So organizations that rely on grants often do end up being influenced by that, and that's one of the reasons that we really value donations from individuals, because it gives us the freedom to be disruptive. The funds primarily allow us to provide the structure that enables a lot more people to get involved. Our staff make sure that there's onboarding and follow-up for new people, that we have tech, and we provide training and mentorship to volunteers as they're developing their skills. So I think you can ideally balance your independence and commitment to grassroots organizing with being able to raise the money to support that work. I think a lot of people were inspired by Senator Sanders' campaign, which was able to raise a lot of money from small donations and that's the model that we use.

AA: To go up against the fossil fuel industry, we need massive political power and the largest grassroots big-tent movement possible. What are some of the factions or divisions in the climate movement? How do you join forces with progressive allies whose focus isn't climate change? How do you determine who to ally with? What can we do to ensure we're all pushing in the same direction?

MD: Well, first of all, I would say, if I knew all the answers to this, I would probably win the Nobel Prize! There are legitimate differences of opinion about what is politically viable, and I think in every movement, one of the classic divisions is between people who want to do incremental work versus people who are more radical and push for the full changes that are needed, and there's a lot of friction that's caused by that. But I don't think there's an easy solution. As an example, within the climate movement, there is the Citizens' Climate Lobby and they've put forward a bill for a carbon tax that gets paid back as a dividend. This is an incredibly incremental, and I would say, conservative-leaning approach. And then, on the other side, I have a friend who was one of the valve turners. They identified the five pipelines that transport oil between Canada and the US, and they went out one day and cut the bolts on the fences and turned off those valves. So there are a lot of narratives out there. I'm going to sound like a broken record, but it really comes back to how do you shift the narrative? You need the wider public to be outraged by it and think of it like a moral justice issue that they need to pay attention to. We haven't gotten there on climate, not yet. You know, I think of our local work at the city council. On the one hand, it will give us a climate action plan and at the same time, it will shift the way people think about the issue. You're trying to accomplish both at the same time. And I think if people in the wider movement agree on shifting the narrative then they are allies, rather than people to be fought.

I also believe that we need to escalate, to disrupt things and bring pressure to get people to see that moral outrage, so I wouldn't make any apologies about that. And something else is trigger points that suddenly raise massive public awareness and concern, like the murder of George Floyd. That was one of those moments that made people focus and come out in a way that they had never come out before. You can't predict when those triggers will occur, but you can get ready for them.

You also asked me about who to work with outside of climate groups. I would just say that, first of all, the most important thing to know is that environmentalists are too small a percentage right now to win anything on our own. So we have no choice but to go out and partner with people whose main issue is something else. And I think we have some natural allies, since climate is clearly a justice issue. Lower-income people and communities of color are way more impacted by climate and way less able to deal with the pollution that's caused by drilling for and burning oil and gas. So I think social justice groups are kind of a natural ally for us on this. And we've worked with labor, housing groups, faith groups, and others. I think it's really about broadening that coalition. The other benefit is familiarizing other people with your issues. I would partner with anybody pretty much as long as I feel like we can get to common ground. Partnering means having a genuinely reciprocal and respectful relationship.

AA: I want to follow up on the labor issue. Some have said that the unions are not with the shift to renewable energy.

MD: I think that the reality is that there are a lot of legitimate concerns by people in labor. For one thing, I'll say that labor has gotten the short end of the stick on politics a lot in the last several decades. There's been a concerted attack on labor since before I was born, and union memberships have gone down tremendously. And I think that one thing to recognize is that labor has on the whole been playing a very defensive game for a very long time now, and they are very vested into hanging on to what they have. And that's legitimate, they've been attacked. Another issue is that there is little overlap between the people in the environmental movement and unions. Environmentalists tend to be more professional and often misunderstand what the unions are for and how they work. The main job of the union, really, is to look after their members, to make sure that they're getting treated fairly and paid reasonable wages and benefits. To underestimate the value of unions is a mistake. There are a lot of people in this country who don't get paid enough to live on, who have to work multiple jobs, who don't

have child-care, who don't have health care, who don't have sick pay. Workers in this country, unionized or not, have been attacked for a long time. And if we're serious about justice and equity, we have to support unions and collective bargaining.
The unions are looking out for people now, so they may say no to things that are a long-term opportunity. For example, you might say to them, okay over the next ten years, things are going to really shift and a lot more jobs will be coming out of the renewable energy industry. But when they look at it, they see the choice between working now for SDG&E [local electric utility], which has been unionized for a long time, which they have a relationship with, and some political leverage over, or they can work for fifty or a hundred different solar companies, 80 percent of which are not unionized, probably will never be unionized, where they will completely lose all that power to get decent compensation for their members. So you need to try to understand where people are coming from and try to find solutions that work for them.

which was able to leverage substantial people power to shape climate policy at the city council and the regional transportation authority.

While grassroots social movements such as 350.org, Extinction Rebellion, and Sunrise were certainly effective in elevating truth-telling, in creating policies to end extraction, in pushing some universities and institutions to divest, or promise to divest, their holdings of fossil fuel stocks and bonds, and in creating local policies for renewable energy, the world was still on course to emit 42 gigatons of CO_2 in 2021 alone, and the COP 26 in Glasgow that year was widely recognized as a failure. The big stuff had not yet changed: the enormous fossil fuel subsidies provided by governments, the interference of fossil fuel interests in international talks and national policy decisions, the escalating extraction in many countries, the private jets, the enormous sales volumes of CO_2-spewing SUVs, which hit record sales in 2021,[60] and the official greenwashing strategies of carbon neutral and net zero, which amount to ongoing extraction now with a vague promise of action later.

Whether ongoing grassroots activity, including a scaling up along the principles discussed in this chapter, will be sufficient to stop this trajectory is obviously still an open question. The existential threat posed by global heating and the urgency of taking dramatic action to address it has led the human ecology writer Andreas Malm to counsel the wider climate movement to reevaluate its commitment to nonviolence against property.[61] Malm has acknowledged that such strategies as damaging SUVs, disabling extractive machinery, and shutting off oil pipelines could backfire on the movement more generally and that such actions would touch only a tiny fragment of the vast CO_2-emitting property of the global energy infrastructure. Nonetheless, based on his understanding of the role that a radical flank that engaged in property damage played in the suffragette movement, the civil rights movement in the US, and the anti-apartheid struggle in South Africa, he argued that similar actions by a radical flank within the environmental movement could discourage further investment in the fossil fuel industry and hasten the day when fossil gas, oil, and coal become stranded assets, leading to a huge plunge in the industry's stock prices.[62] Although Malm's position is not currently shared by many in the climate movement, even a commentator who has argued strenuously against it nevertheless concluded by saying, "[p]oliticians should take it as a warning: If governments cannot protect their citizens from fossil fuel oligarchs, then those citizens will turn to other means of self-protection – regardless of their strategic merit."[63]

Being Active without Being an Activist

Although the transition away from burning fossil fuels is unlikely to happen without a strong and growing grassroots movement to hold institutions and politicians to account, engaging in political or social activism is not always an option for a given person at a given point in their lives or the only way one can have an impact on the problem. This section thus explores some other ways in which individuals can support that transition (Figure 10.7).

One such way is through membership of other kinds of group than the grassroots organizations described above. For example, the larger nonprofit or nongovernmental organizations that are sometimes referred to as Big Green, such as the Sierra Club, have long been engaged in efforts that can be synergistic with those of grassroots climate change movements. While these larger groups often depend for their legitimacy and financial survival on their embeddedness in the established organizational structures of wider society, which can often limit or compromise some of the actions they can take, they also have much more financial clout, access to mainstream media, networks of millions of paying supporters, in-house lawyers, and access to politicians and powerbrokers.

Figure 10.7

Finding a way to engage: being an activist or an advocate

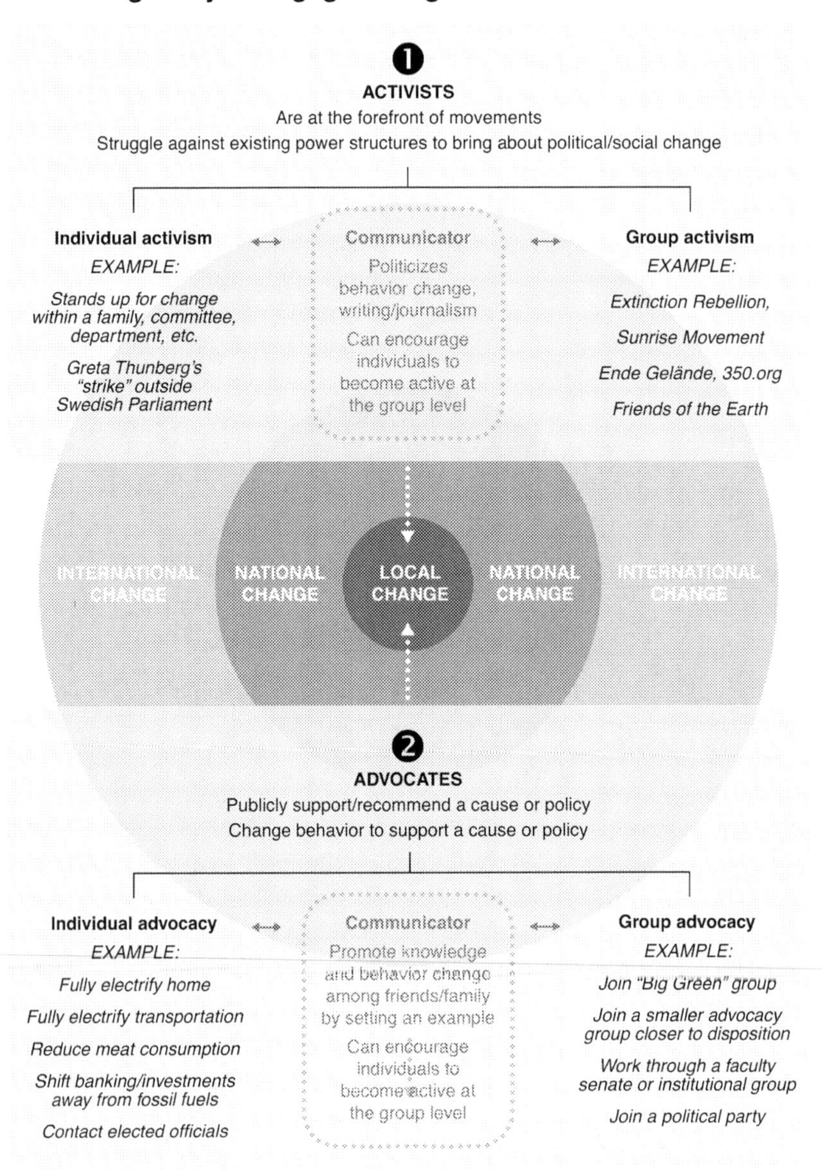

Individuals can also use their membership or role within an institution, such as serving on the faculty senate of a university, the budget committee of a church, or the investment committee of a condominium association, to influence that organization's energy use and financial decisions, such as shifting their banking services from banks that are financing the fossil fuel industry or moving their campus from burning fossil gas for power generation to renewable energy sources.

Another option is to take actions as an individual that will support and hasten the transition to renewable energy sources. It is now possible for even mid-income households to get off fossil fuels by using electricity to cook and heat water and to charge an electric car. (The price of an entry-level EVs is expected to hit parity with internal combustion engine vehicles by 2025, and they are cheaper to maintain – although recent supply chain issues cast this into some doubt.)[64] One can also choose to shift one's mortgage and personal banking away from Chase, Citibank, Bank of America, and Wells Fargo to a local credit union that does minimal financing of fossil fuel extraction.[65] There are also numerous ways in which individuals can contact city, university, health system, or state and national leaders about their climate policy positions.

Of course, these kinds of individual actions can become much more consequential when they are coordinated with other people, such as shifting one's personal banking in concert with the Stop the Money Pipeline campaign, advocating for similar actions among friends and family members, and timing one's contacts with officials to coincide with those of hundreds or thousands of others. Likewise, making changes in one's personal lifestyle is most effective when it helps serve as a lever for broader society-wide changes. As environmental writer Sami Grover observes, "[w]hen we ride our bikes, our power lies not in cutting our personal travel footprint – an impact that seems trivial when surrounded by gigantic, diesel-chugging trucks. Instead, it is in creating a space where politicians and planners feel confident investing in bike-friendly infrastructure and policies."[66] Better yet is to also team up with or donate to a local bicycle-promoting group to help make their advocacy within the city more powerful.

As this chapter has shown, a big part of this struggle is transforming people's attitudes and shifting social norms, which almost anyone can help do by sharing knowledge about the science and impacts of global heating and the technical feasibility of the energy transition with friends and family and modeling such behavior changes as flying less, driving less, and reducing meat-eating (as Chapter 9 pointed out, animal agriculture amounts to about

Table 10.2. *Examples of ways to make a difference within different jobs and roles*

Job/role	Type of climate action
Consumer	Replace combustion vehicles and gas-burning appliances with electric, install rooftop solar
Farmer	Practice regenerative agriculture that does not rely on fossil fuel fertilizers and industrial machinery
Lawyer	Work on litigation against fossil fuel interests
Investor	Invest in companies that are carbon free
Architect	Design more fuel-efficient buildings, and new ways of powering and sharing space
Teacher	Share the facts of global heating and the kind of transition that is needed to leave fossil fuels in the ground
Union representative	Anticipate a huge number of new jobs in green energy and advocate for a just transition
Engineer or tech worker	Help develop the infrastructure, technology, and efficiencies of an electrified future, from energy to transmission and to control systems and delivery
Health professional	Publicize the enormous human costs of pollution from fossil fuels, such as cancers and lung disease, and the health impact of sedentary car-based cultures
Artists of all types	Music and visual arts can be used to move people and help them imagine new futures

Source: Adapted from Griffith, *Electrify*.

13 percent of global emissions). Elementary, high school, and college teachers can introduce elements of the climate crisis into their classes, or even create classes directly focused on the topic. Academic researchers can choose to reorient their research to topics that can advance our ability to understand and meet the technological, political, social, and psychological challenges of the shift to renewable energy.[67] Artists can volunteer their services to help produce posters, websites, and murals promoting climate awareness and advocacy, and see Table 10.2 for examples of ways that other careers provide opportunities for climate action.

System change is fundamentally important, but we will get there partly by personal changes made along the way. It is a false dichotomy that personal changes and system changes are separate from each other (see Table 10.3 for this point and other "discourses of climate delay"). As activist and climate scientist Peter Kalmus has succinctly written, "[t]here is no collective action without individuals choosing to contribute to it. There is no cultural shift without individuals leading the way." In short, he argues, "[t]here is no bright line between systems change and individuals acting according to their principles."[68]

Table 10.3. *Common discourses of climate delay (forms of response skepticism)*

Response skeptic	Intended meaning	Type of discourse	Answer
What about China?	It is pointless to cut emissions elsewhere because China emits more now	Redirecting responsibility – someone else should act first	China's emissions are currently greater than those of the USA annually, but not per capita or historically, and much of China's emissions arise from products consumed by the Global North. China has already installed over 1,000 GW of renewable energy, more than ten times the total energy generation of California. As the major military, economic, and political power, the USA has a responsibility to lead in emissions reductions.
Individuals and consumers are responsible for action *or* evil corporations are at fault and must change	The focus on individuals lets the system and fossil fuel industry off the hook; the exculpatory focus on corporations lets individuals off the hook	Redirecting responsibility – someone else should act first	The focus solely on individuals or corporations is a false dichotomy. There is no clear dividing line between the need for systems change and for individual action – e.g., if many individuals shift their personal banking from banks that finance fossil fuel extraction, those banks will change their actions.
What about lithium?	Shorthand for a legitimate concern about extractivism related to the transition to renewable energy and EVs	Emphasizing the downside – change will bring new disruptions	This legitimate concern could be accommodated with a commitment to consumption reduction, more public transit, increased recycling, substitute materials, and new ethical standards for minerals sourcing.
Renewable energy is greenwashing	Greenwashing refers to creating a false impression that one is taking environmental action when one is not; the concern here is	Emphasizing the downside – change will bring new disruptions	A strong version of this claim seems like nihilism, as there is no way to stop global heating without abandoning fossil fuels and transitioning to renewable energy. A weaker version intended to criticize boosterism for renewable energy without

Table 10.3. (*cont.*)

Response skeptic	Intended meaning	Type of discourse	Answer
	that renewable energy is yet another form of exploitation		cognizance of its problems is legitimate but can be accommodated with a commitment to consumption reduction, public transit, and minimization of extractivism.
It's too late	Catastrophic climate change is already baked in no matter what we do	Fatalism, doomism, or defeatism – denying possibility of mitigating climate change	If we reach tipping points, it may become too late, but until then, they provide only more reason to act. Although some amount of heating is baked in no matter what we do, there is a huge difference between that and projected rises if we do not act immediately. Every fraction of a degree has enormous implications for billions of people.
It's hopeless	Because of capitalism or "selfish" human beings, nothing we can do will matter	Fatalism, doomism, or defeatism – denying possibility of mitigating climate change	This position is too easy for comfortable people to take while hundreds of millions in other parts of the world face mounting difficulties because of the ongoing emissions created by rich countries and people. History has shown that social change is possible despite entrenched interests and injustices and that industrial economic policy is not incompatible with capitalism.
The renewable energy transition generates lots of emissions itself	It's pointless advocating for renewable energy	Fatalism, doomism, or defeatism – denying possibility of mitigating climate change	The facts rely on life-cycle assessments of how many emissions are created by building wind and solar plants, relative to the fossil fuel energy status quo. A large wind turbine does require mining, steel, and concrete, but about six to nine months of operation is enough to compensate for the emissions from construction, leaving it virtually carbon free for the next 20 years.[69]

Our climate plan is carbon neutrality	Carbon neutrality typically means to keep on emitting and to rely on purchases of carbon offsets	Pushing nontransformative approaches – disruptive change is not needed	By continuing to burn fossil fuels one is maintaining the economic and political might of fossil fuel interests and continuing to do damage to the biosphere. Meanwhile, carbon offsets are usually impossibly cheap, often neocolonial, often uncertain whether they will deliver the claimed benefit, and hard to validate regarding additionality.
We need to rely on carbon capture and hydrogen produced from methane	The fossil fuel industry is "part of the solution," "low-carbon" fossil fuels are bridge fuels	Pushing nontransformative approaches – disruptive change is not needed	Some continued use of fossil fuels may be necessary to manufacture petrochemicals for specific kinds of plastics, but other uses of fossil fuels must be ended quickly. The promise of technical carbon capture and of "low-carbon" fuels are simply ways for the industry to attempt to maintain its enormous economic and political power.
We must develop new technologies	New nuclear, fusion, geoengineering, BECCS, and direct air capture will provide solutions in the future	Pushing nontransformative approaches – disruptive change is not needed	Most of the technology needed for the renewable transition already exists, although further developments in substitute materials and better batteries requiring less extractivism are likely and welcome. But promises of grand new technology constitutes a form of technological solutionism that simply delays the cultural, behavioral, and political changes that are really needed.

Source: Inspired by Lamb et al., "Discourses of Climate Delay."

Conclusion

The chapters in Part III have outlined what kind of action will be necessary to achieve a transition to renewable forms of energy and to permanently lower CO_2 concentrations to manageable levels. As the book as a whole has made abundantly clear, however, accomplishing this urgent and existential goal will require that decision makers be compelled by pressure from below to change course. In this, every one of us has a role if we want one. We all have the choice and the power to take actions that can make a difference, including making individual lifestyle changes that reduce emissions, voting for candidates committed to eco-friendly policies, changing one's bank, becoming a climate change advocate within our institutional settings, joining or donating to a Big Green organization, or engaging in grassroots collective action. And all of us can choose to have conversations with ourselves and the people around us that can create a wider awareness of the facts about the human causes of global heating, its current and predicted impacts, the speed and scale of emissions reduction that is needed, and the essential actions that we must take right now.

The long story of skepticism, inaction, and outright opposition to climate action and justice detailed in this book is admittedly discouraging and often maddening. But as this chapter has also shown, history and research both tell us that transformational changes in attitudes, policies, and practices are indeed possible when such individual actions are accompanied by coalition building, community activism, and a grassroots movement willing to disrupt established institutional functions by protesting, engaging in civil disobedience, and questioning the social license to pollute. Although concerted pressure must continue to be exerted upon government officials at the national and international levels, occasional wins against fossil fuel interests and for renewable energy in cities, regions, and states remind us that social change can be made by relatively small groups of people working together.

Conclusion

In Australia, a team of researchers, architects, and designers are building a steel vault, the size of a school bus, which they are calling Earth's Black Box (Figure C.1). Like the black box flight recorder in an airplane that records crucial data that allow investigators to reconstruct events in case of a crash, this vault will receive daily measurements of land and ocean temperatures, atmospheric CO_2 levels, and biodiversity loss, and collect data regarding leaders' actions and inactions by scouring the internet for keywords related to climate change.[1] The intention behind its construction is that if catastrophically bad climate outcomes ensue, the box will provide the objective data that will be needed to piece together the story of how that happened and who should be held responsible. As chilling a possibility as such a possible outcome may be, this book argues that there is still time for us to take action to prevent the worst effects of global heating and stave off such a collapse.

The previous chapters suggest three main takeaways about how to accomplish that. The first of these is that, based on the available evidence, taking effective action now appears to require reversing course and shifting most of our efforts from the international to the national level, before expecting internationally binding agreements later. The most recent COP 26 in Glasgow in November of 2021 made clear that fossil fuel interests had again sabotaged any possibility of concerted international action. As with so many earlier such meetings, the end result was little more than another set of weaselly words and promises, leading at least one naturalist to wonder if fossil fuel industry executives, who have shown themselves willing to sacrifice the health of the planet for short-term profits, might someday be charged with committing ecocide, even genocide, by future climate courts.[2] But even without such corporate interference, the conference made clear that most nation states, barring Denmark and Costa Rica, had little interest in committing themselves to leaving fossil fuels in the ground, thereby continuing a pattern

Figure C.1

Earth's Black Box, to be installed in Australia

From Earth's Black Box, https://www.earthsblackbox.com

that, as we have seen, was set decades ago. Indeed, the Biden administration recently issued fossil fuel extraction licenses on federal lands at levels not seen since the George W. Bush administration; Canada has chosen to exploit tar sands oil, the dirtiest kind of extraction, and pipe it to the USA over Indigenous lands; Germany, its renewable transition having stalled, continues to mine lignite coal, the dirtiest kind, and until the outbreak of war in Ukraine planned to rely on the Nord Stream 2 pipeline to carry fossil fuels from Russia. With so many nations still deeply committed to fossil fuel use and extraction, we cannot expect them to agree to internationally binding agreements to significantly reduce emissions until the struggle against fossil fuel interests has first been substantially won at the national level. Winning such a battle in the USA will be especially critical, given the enormous influence its military, financial, and organizational might plays in international affairs.

The second main takeaway of this book is that averting the most catastrophic impacts of global heating will require forging a new and powerful agreement within high-emitting nations themselves to leave fossil fuels in the ground while achieving a near-total transition to renewables along with the electrification of most end uses. As this book has argued, such a near-total renewable transition, while undoubtedly expensive, is technically feasible, requiring perhaps 0.5 to 1 percent of land and a vast integrated grid with storage to deal with intermittency. The technology to do this already exists, even though we can expect that, over time, batteries and wind turbines will be improved, costs reduced, and problematic materials substituted. The 2022 IPCC report on mitigation was also notably upbeat on the in-principle prospect of achieving decarbonization of energy systems, transportation, and industry with current technology if there were the will. Yet, as this book has argued, this technical transition, embodied in the spectacle of vast wind farms and solar arrays and millions of EVs humming down freeways, will require colossal amounts of new mining, which could easily reproduce human exploitation and ecological destruction. And while electrification brings large efficiencies, those may not immediately translate into greenhouse gas emissions reductions because of rebound effects, especially if nations remain committed to keep growing their economies. For all these reasons, as discussed in previous chapters, the transition from fossil fuels to electrification must also be accompanied by investment in public goods such as transit and housing and policies that reduce consumption, such as circumventing planned obsolescence through right of repair, encouraging sharing instead of owning, and providing a universal basic income. And even though the renewable transition has accelerated based primarily on private investment, so far renewable fuels and technologies have been largely treated as merely additions to rather than replacements for fossil fuels. This lack of vision and will reflects the ongoing policy impacts of fossil fuel interests and is unlikely to change without the eventual strong intervention of governments, including the revoking of huge taxpayer-funded fossil fuel subsidies. Thus, this book argues, the renewable transition must take place within a political framework of economic support, regulations, and social policies that simultaneously achieves the energy transition, reduces energy use and consumption, confronts fossil fuel interests, and creates a popular base of support. While such frameworks have been proposed at both the national and regional levels, in most cases they have not been enacted because politicians have not been compelled to take such action by strong public support.

The third and final major takeaway of this volume is that gaining widespread public support for such a framework will require strong individual and collective advocacy and active grassroots mobilization, especially at the local level. Whereas ongoing attempts to achieve a political consensus on the need for such a framework in the USA have been thwarted by political polarization and the influence of the fossil fuel industry and other economic interests on both major political parties, this book has also shown that meaningful and sometimes even massive changes at the national level have indeed been catalyzed by local actions and by shared perceptions of existential risk. As previous chapters have shown, achieving this much higher level of social mobilization will require finding ways to overcome both epistemic and response skepticism and move people to action. Epistemic skepticism, which remains especially strong in fossil fuel extraction countries such as the USA, Canada, and Australia, can be countered by remedying the knowledge deficit (such as communicating Global Heating 1-2-3) and elevating risk perceptions through fiction, nonfiction, and data visualization. Other strategies, such as in-group messaging and presenting climate policy in terms of its co-benefits, can be deployed to counter worldviews, values, and motivated cognition that are mostly immune to climate-specific knowledge. The co-benefits approach especially holds strong promise, as it can be used to build support for leaving fossil fuels in the ground without requiring people to profoundly change their beliefs about climate change or social identity. Evidence has shown, for instance, that fossil fuel industry workers, increasing numbers of whom are coming to see the writing on the wall regarding the future of their industries, can be encouraged to support just transition policies, a shift that can lead to changed voting patterns in unions, parties, and legislatures at the local level that could undergird national shifts later. In addition to the prospect of new jobs, other co-benefits to cutting fossil fuel use are reducing air pollution (itself a massive problem) and achieving energy security in geopolitical terms, an issue that suddenly became highly salient to Western European countries with Russia's invasion of Ukraine in 2022. Time will tell whether the European Union doubles down on mining more coal and importing more fossil gas from the USA and the Middle East or makes good on its goals to escalate renewables,[3] and whether the UK keeps expanding its offshore wind capacity or takes up offshore gas drilling anew.

While these strategies for shifting the epistemic skeptic could translate into changes in voting behavior, becoming stronger advocates and even activists requires also overcoming response skepticism about whether anything can actually be done about global heating and the value of one's own efforts. As this book has argued, such response skepticism partly reflects the obstacles produced by the neoliberal form of capitalism, which in numerous countries has countered solidarity by undermining unions, privatizing public goods, and fetishizing private choice and hyperconsumption. To counter these effects, advocates and activists can encourage people to experience the satisfaction and even joy of engaging together in such collective action as planning campaigns together, emailing their friends and officials, and adding their voice in local committees and meetings in their workplaces. One antidote to hopelessness can be to focus on one's own institution, such as pressuring one's university to switch from burning fossil fuels to using electricity sourced from renewables, encouraging one's vast public pension fund to divest its holdings from fossil fuel extraction companies, shifting one's own finances out of fossil fuel–financing banks, and encouraging everyone one knows to do the same thing, and advocating within one's own city for a regional decarbonization plan and expanded public transit.

While everyone can be an advocate, probably only a small subset of people is likely to become activists in grassroots social movements. As this book has shown, while grassroots movements can be effective even when their numbers are low, growing those movements substantially is now critical. One way to do that is to find ways to forge wider alliances, including linking the efforts of unionized workers advocating for a just transition, college students insisting on change, middle-class retirees in environmental organizations, and environmental justice advocates for the Indigenous and other communities on the frontlines of fossil fuel pollution.

In 2022, the IPCC released its Working Group II report on climate impacts, which UN Secretary-General António Guterres referred to as "an atlas of human suffering." Calling the report "a damning indictment of failed climate leadership," he concluded that "the world's biggest polluters are guilty of arson on our only home."[4] As this book has shown, those arsonists include not just countries but local decision-making bodies such as cities and large institutions that have so far mostly chosen to maintain the status quo. Yet, as the following IPCC Working Group III report showed, we already have most of the technology necessary to dramatically cut emissions now by leaving fossil fuels in the ground, and doing so would likely cost less than the global economic benefit of limiting heating to 2°C.[5]

As this book went to press, the first major climate legislation in the USA, the Inflation Reduction Act, squeaked through Congress. This included nearly $400 billion over 10 years for investments in renewable energy, tax credits, and other climate-related provisions. While the investments were not nearly large enough and the legislation included measures to expand the fossil fuel industry there was reason to hope that it may serve as a first step in engaging more constituents to shift the politics in favor of more ambitious policies.

As the hour grows late in our collective efforts to confront the global heating crisis, such shifts in the political landscape along with the efforts of grassroots climate movements might together have the capacity to disrupt the status quo in ways that will become increasingly necessary to impel decision makers to directly confront fossil fuel energy and fossil-related finance.

Notes

Chapter 1

1 https://ieep.eu/news/more-than-half-of-all-co2-emissions-since-1751-emitted-in-the-last-30-years; William Ripple et al., "World Scientists' Warning of a Climate Emergency," *BioScience* 71, no. 9 (2021): 894–8.
2 www.skepticalscience.com/4-Hiroshima-bombs-worth-of-heat-per-second.html.
3 www.ipcc.ch/report/ar6/wg1/ and Ripple et al., "World Scientists' Warning."
4 www.ipcc.ch/report/ar6/wg1/.
5 Ibid.
6 www.ipcc.ch/sr15/.
7 Dan Tong et al., "Committed Emissions from Existing Energy Infrastructure Jeopardize 1.5 C Climate Target," *Nature* 572, no. 7769 (2019).
8 www.ipcc.ch/report/ar6/wg1/.
9 www.nrc.gov/docs/ML1209/ML120960701.pdf.
10 Timothy Mitchell, *Carbon Democracy* (Verso, 2013), chapter 1.
11 https://theconversation.com/john-tyndall-the-forgotten-co-founder-of-climate-science-143499; Roland Jackson, "Eunice Foote, John Tyndall and a Question of Priority," *Notes and Records* 74, no. 1 (2020).
12 www.scientificamerican.com/article/discovery-of-global-warming/, www.easterbrook.ca/steve/2015/08/who-first-coined-the-term-greenhouse-effect/.
13 Nikolay L Mihaylov and Douglas D Perkins, "Local Environmental Grassroots Activism: Contributions from Environmental Psychology, Sociology and Politics," *Behavioral Sciences* 5, no. 1 (2015).
14 Nathaniel Rich, "Losing Earth: The Decade We Almost Stopped Climate Change," August 18, 2018, www.nytimes.com/interactive/2018/08/01/magazine/climate-change-losing-earth.html.
15 Naomi Oreskes and Erik M Conway, Merchants of Doubt: How a Handful of Scientists Obscured the Truth on Issues from Tobacco Smoke to Global Warming (Bloomsbury Publishing USA, 2011).
16 Philip Shabecoff, "Global Warming Has Begun, Expert Tells Senate," *New York Times*, June 24, 1988, p. 1.
17 Some of this history of IPCC efforts is based on
B Tokar, Toward Climate Justice: Perspectives on the Climate Crisis and Social Change (New Compass Press, 2014).

18 www.newscientist.com/article/2093579-was-kyoto-climate-deal-a-success-figures-reveal-mixed-results/ and *The Emissions Gap Report 2012*, a UNEP Synthesis Report: www.unep.org/publications/ebooks/emissionsgap2012/.
19 Tokar, Toward Climate Justice.
20 www.nytimes.com/2008/12/11/business/worldbusiness/11carbon.html.
21 E.g., Tokar, *Toward Climate Justice* and www.propublica.org/article/cap-and-trade-is-supposed-to-solve-climate-change-but-oil-and-gas-company-emissions-are-up.
22 https://ec.europa.eu/clima/sites/clima/files/ets/docs/clean_dev_mechanism_en.pdf.
23 https://therevolvingdoorproject.org/wp-content/uploads/2022/03/Industry-Agenda_-Carbon-Offsets.pdf.
24 http://adamwelz.wordpress.com/2009/12/08/emotional-scenes-at-copenhagen-lumumba-di-aping-africa-civil-society-meeting-8-dec-2009/ and www.theguardian.com/environment/2009/dec/08/copenhagen-climate-summit-disarray-danish-text.
25 www.ipcc.ch/report/ar6/wg1/downloads/report/IPCC_AR6_WGI_SPM.pdf; https://interactive-atlas.ipcc.ch. Indeed, in 2019, an analyst at a climate research group in South Africa showed that the amount of heating experienced by South Africa was double that of the global average in recent years, although there was substantial year-by-year variation which makes it uncertain whether it is a true doubling. http://www.csag.uct.ac.za/2019/09/25/twice-the-global-rate/.
26 Senator Sheldon Whitehouse, "Dark Money," August 11, 2021, US Congress. https://youtu.be/Nny7SnBkbTM.
27 www.theguardian.com/environment/2015/dec/12/james-hansen-climate-change-paris-talks-fraud.
28 www.newscientist.com/article/2269432-we-are-nowhere-near-keeping-warming-below-1-5c-despite-climate-plans/ and https://unfccc.int/news/full-ndc-synthesis-report-some-progress-but-still-a-big-concern.
29 www.ipcc.ch/sr15/.
30 https://nca2018.globalchange.gov.
31 www.jacobinmag.com/2021/06/joe-biden-climate-policy-drilling-trump and www.bloomberg.com/news/articles/2021-01-20/why-the-world-frets-over-russian-nord-stream-pipeline-quicktake.
32 www.ipcc.ch/report/ar6/wg1/downloads/report/IPCC_AR6_WGI_SPM.pdf.
33 https://ourworldindata.org/grapher/co2-mitigation-2c www.carbonbrief.org/unep-1-5c-climate-target-slipping-out-of-reach.
34 www.iea.org/news/global-carbon-dioxide-emissions-are-set-for-their-second-biggest-increase-in-history; Jeff Tollefson, "Covid Curbed Carbon Emissions in 2020 – but Not by Much," *Nature* 589, no. 7842 (2021).

Chapter 2

1 J Bennett, *The Global Warming Primer (Big Kid Science, 2016).*
2 www.howglobalwarmingworks.org/35-words.html.
3 James Hansen et al., "Climate Sensitivity, Sea Level and Atmospheric Carbon Dioxide," *Philosophical Transactions of the Royal Society A: Mathematical, Physical and Engineering Sciences* 371, no. 2001 (2013).
4 CO_2 levels collected from the Mauna Loa Observatory in Hawaii can be tracked at this site: www.esrl.noaa.gov/gmd/ccgg/trends/.
5 M Rubino et al., "A Revised 1000 Year Atmospheric $\Delta 13C–CO_2$ Record from Law Dome and South Pole, Antarctica," *Journal of Geophysical Research: Atmospheres* 118, no. 15 (2013).

6 Earlier studies found a consensus of 97 percent among all earth scientists and closer to 100 percent among climate scientists. John Cook et al., "Consensus on Consensus: A Synthesis of Consensus Estimates on Human-Caused Global Warming," *Environmental Research Letters* 11, no. 4 (2016). With the IPCC 2021 report statement on "unequivocal" evidence for human-caused heating, we can now regard the consensus as 100 percent among climate scientists.
7 Now consider that while the Earth absorbs 240 W/m^2 from the Sun, the variation of the Sun, known as the sunspot cycle, only changes its brightness by about 0.1 percent: so the range of the Sun's forcing is only about 0.24 W/m^2. Meanwhile, it has been calculated that the increase of CO_2 from the pre-industrial level of 280 ppm to the 2018 amount of 407 ppm corresponds to a forcing of about 2 W/m^2. So any change owing to an increased forcing from the Sun (variation of 0.24 W/m^2) is about 10 times less than the forcing from CO_2 since the industrial revolution (2 W/m^2).
8 From Bennett "Global Warming Primer" and IPCC 2021 report: www.ipcc.ch/report/ar6/wg1/.
9 www.carbonbrief.org/qa-how-is-arctic-warming-linked-to-polar-vortext-other-extreme-weather.
10 www.carbonbrief.org/state-of-the-climate-2020-on-course-to-be-warmest-year-on-record; http://berkeleyearth.org/global-temperature-report-for-2020/.
11 Almut Arneth et al., "Historical Carbon Dioxide Emissions Caused by Land-Use Changes Are Possibly Larger than Assumed," *Nature Geoscience* 10, no. 2 (2017).
12 Marielle Saunois et al., "The Global Methane Budget 2000–2012," *Earth System Science Data* 8, no. 2 (2016). D'Aspremont et al. show that deliberate releasing of methane by fossil extraction operators amounts to 12 percent of global methane emissions. Alexandre d'Aspremont et al., "Global Assessment of Oil and Gas Methane Ultra-Emitters," *Science* 375, no. 6580 (2021).
13 Cecelia Macfarling Meure et al., "Law Dome CO_2, CH_4 and N_2O Ice Core Records Extended to 2000 Years BP," *Geophysical Research Letters* 33, no. 14 (2006).
14 www.vox.com/energy-and-environment/2019/5/30/18643819/climate-change-natural-gas-middle-ground.
15 SC Sherwood et al., "An Assessment of Earth's Climate Sensitivity Using Multiple Lines of Evidence," *Reviews of Geophysics* 58, no. 4 (2020). Also IPCC 2021 report: www.ipcc.ch/report/ar6/wg1/.
16 Bennett, Global Warming Primer.
17 www.climate.gov/news-features/understanding-climate/climate-change-atmospheric-carbon-dioxide.
18 Bärbel Hönisch et al., "Atmospheric Carbon Dioxide Concentration across the Mid-Pleistocene Transition," *Science* 324, no. 5934 (2009).
19 Oana A Dumitru et al., "Constraints on Global Mean Sea Level during Pliocene Warmth," *Nature* 574, no. 7777 (2019).
20 For more on plants and animals not being able to adapt, see https://skepticalscience.com//Can-animals-and-plants-adapt-to-global-warming.htm.
21 Kevin D Burke et al., "Pliocene and Eocene Provide Best Analogs for near-Future Climates," *Proceedings of the National Academy of Sciences* 115, no. 52 (2018).
22 Ripple et al., "World Scientist's Warning."
23 Lomborg has written two books downplaying the risks of global heating and often comments in the media; for more information on his claims, see www.desmog.com/bjorn-lomborg/. The quoted exchange comes from www.foxnews.com/us/ipcc-climate-change-reports-claims-challenged-by-skeptics and his reference to "global greening" is taken from Shilong Piao et al., "Characteristics, Drivers and

Feedbacks of Global Greening," *Nature Reviews Earth & Environment* 1, no. 1 (2020).

24 Meetpal S Kukal and Suat Irmak, "Us Agro-Climate in 20th Century: Growing Degree Days, First and Last Frost, Growing Season Length, and Impacts on Crop Yields," *Scientific Reports* 8, no. 1 (2018).

25 Noah S Diffenbaugh, Daniel L Swain, and Danielle Touma, "Anthropogenic Warming Has Increased Drought Risk in California," *Proceedings of the National Academy of Sciences* 112, no. 13 (2015).

26 https://agfax.com/2018/07/02/climate-change-longer-growing-season-more-heat-115-years-of-data/.

27 www.mckinsey.com/business-functions/sustainability/our-insights/will-the-worlds-breadbaskets-become-less-reliable; Mueller, Brigitte, Mathias Hauser, Carley Iles, Ruksana Haque Rimi, Francis W Zwiers, and Hui Wan, "Lengthening of the Growing Season in Wheat and Maize Producing Regions," *Weather and Climate Extremes* 9 (2015).

28 www.theguardian.com/world/2020/jan/05/russia-announces-plan-to-use-the-advantages-of-climate-change.

29 www.dni.gov/files/documents/climate2030_russia.pdf.

30 http://riskybusiness.org.

31 https://eciu.net/analysis/briefings/climate-impacts/climate-economics-costs-and-benefits.

32 www.cnbc.com/2019/02/14/climate-disasters-cost-650-billion-over-3-years-morgan-stanley.html.

33 Daoping Wang et al., "Economic Footprint of California Wildfires in 2018," *Nature Sustainability* 4, no. 3 (2021).

34 www.desmogblog.com/tony-abbott.

35 www.climatechangenews.com/2017/10/09/climate-change-probably-good-says-former-australian-pm-abbott/.

36 https://nymag.com/news/features/67285/index4.html.

37 www.desmog.com/2015/02/05/exclusive-bjorn-lomborg-think-tank-funder-revealed-billionaire-republican-vulture-capitalist-paul-singer/.

38 www.forbes.com/sites/bjornlomborg/2014/10/24/its-time-to-give-up-the-two-degree-target/?sh=3e8be4701e7c.

39 www.nytimes.com/2021/04/22/climate/climate-change-economy.html.

40 Steve Keen, "The Appallingly Bad Neoclassical Economics of Climate Change," *Globalizations* (2020).

41 Andrew Dessler, *Introduction to Modern Climate Change* (Cambridge University Press, 2015).

42 www.theguardian.com/environment/2017/dec/14/eu-must-not-burn-the-worlds-forests-for-renewable-energy; James Hansen et al., "Young People's Burden: Requirement of Negative CO_2 Emissions," *Earth System Dynamics* 8, no. 3 (2017).

43 Hansen, "Young People's Burden."

44 Jeff Tollefson, "How Hot Will Earth Get by 2100?" *Nature* 580, no. 7804 (2020).

45 www.ipcc.ch/report/ar6/wg1/.

46 Fourth National Climate Assessment, 2018, US Government: https://nca2018.globalchange.gov.

47 Christopher R Schwalm, Spencer Glendon, and Philip B Duffy, "RCP8: 5 Tracks Cumulative Co2 Emissions," *Proceedings of the National Academy of Sciences* 117, no. 33 (2020).

48 www.carbonbrief.org/global-carbon-project-coronavirus-causes-record-fall-in-fossil-fuel-emissions-in-2020.

49 www.iea.org/news/global-carbon-dioxide-emissions-are-set-for-their-second-biggest-increase-in-history.
50 Norman G Loeb et al., "Satellite and Ocean Data Reveal Marked Increase in Earth's Heating Rate," *Geophysical Research Letters* 48, no. 13 (2021).
51 Veerabhadran Ramanathan et al., *Bending the Curve: Climate Change Solutions* (2019). Regents of the University of California.
52 Mike Hulme, *Why We Disagree about Climate Change: Understanding Controversy, Inaction and Opportunity* (Cambridge University Press, 2009).
53 Salvador Herrando-Pérez et al., "Statistical Language Backs Conservatism in Climate-Change Assessments," *BioScience* 69, no. 3 (2019).
54 Keynyn Brysse et al., "Climate Change Prediction: Erring on the Side of Least Drama?," *Global Environmental Change* 23, no. 1 (2013).
55 https://insideclimatenews.org/news/08092017/climate-change-ipcc-scientists-reports-overhaul-inefficient-hayhoe-stocker/; Jem Bendell and Rupert Read, *Deep Adaptation: Navigating the Realities of Climate Chaos* (John Wiley & Sons, 2021).
56 Tong et al., Committed Emissions; M Lynas, Our Final Warning: Six Degrees of Climate Emergency (4th Estate 2020).
57 www.columbia.edu/~jeh1/mailings/2018/20181206_Nutshell.pdf.
58 Andrew H MacDougall et al., "Is there Warming in the Pipeline? A Multi-model Analysis of the Zero Emissions Commitment from CO_2," *Biogeosciences* 17, no. 11 (2020). And see www.carbonbrief.org/explainer-will-global-warming-stop-as-soon-as-net-zero-emissions-are-reached.
59 Thorsten Mauritsen and Robert Pincus, "Committed Warming Inferred from Observations," *Nature Climate Change* 7, no. 9 (2017); Chen Zhou et al., "Greater Committed Warming after Accounting for the Pattern Effect," *Nature Climate Change* 11, no. 2 (2021).
60 www.globalcarbonproject.org.

Chapter 3

1 Huihui Feng and Mingyang Zhang, "Global Land Moisture Trends: Drier in Dry and Wetter in Wet over Land," *Scientific Reports* 5, no. 1 (2015).
2 www.carbonbrief.org/jet-stream-is-climate-change-causing-more-blocking-weather-events.
3 https://abcnews.go.com/Technology/global-warming-climate-scientists-effect-weather-steroids/story?id=15534047.
4 www.carbonbrief.org/mapped-how-climate-change-affects-extreme-weather-around-the-world.
5 Swain et al., "Attribute Extreme Events"; Daniel L Swain et al., "Attributing Extreme Events to Climate Change: A New Frontier in a Warming World," *One Earth* 2, no. 6 (2020).
6 www.cnbc.com/2019/02/14/climate-disasters-cost-650-billion-over-3-years-morgan-stanley.html.
7 Daoping Wang et al., "Economic Footprint of California Wildfires in 2018," *Nature Sustainability* 4, no. 3 (2021).
8 www.scientistswarning.org/2020/09/07/new-book-stresses-urgency-of-action-on-climate-crisis/.

9 Diego G Miralles et al., "Mega-heatwave Temperatures Due to Combined Soil Desiccation and Atmospheric Heat Accumulation," *Nature Geoscience* 7, no. 5 (2014).
10 www.carbonbrief.org/mapped-how-climate-change-affects-extreme-weather-around-the-world.
11 Jean-Marie Robine et al., "Death Toll Exceeded 70,000 in Europe during the Summer of 2003," *Comptes Rendus Biologies* 331, no. 2 (2008).
12 www.indiaenvironmentportal.org.in/files/file/2003%20European%20heatwave.pdf.
13 https://storage.googleapis.com/lancet-countdown/2019/10/2018-lancet-countdown-policy-brief-usa.pdf.
14 www.bbc.com/news/science-environment-49753680; www.carbonbrief.org/climate-change-made-europes-2019-record-heatwave-up-to-hundred-times-more-likely.
15 www.abc.net.au/news/2019-12-20/finger-pointed-at-climate-change-as-heatwave-smashes-records/11817884.
16 www.scientificamerican.com/article/pacific-northwest-heat-wave-killed-more-than-1-billion-sea-creatures/.
17 www.worldweatherattribution.org/wp-content/uploads/NW-US-extreme-heat-2021-scientific-report-WWA.pdf.
18 www.nytimes.com/2022/05/17/opinion/india-heat-wave-pakistan-climate-change.html; https://insideclimatenews.org/news/07052022/heatwave-india-pakistan-deaths-health-risks/.
19 www.metoffice.gov.uk/binaries/content/assets/metofficegovuk/pdf/research/climate-science/attribution/indian_heatwave_2022.pdf.
20 Diffenbaugh et al., "Anthropogenic Warming"; www.scientificamerican.com/article/climate-change-has-helped-fuel-a-megadrought-in-the-southwest/.
21 www.weforum.org/agenda/2019/08/cape-town-was-90-days-away-from-running-out-of-water-heres-how-it-averted-the-crisis/.
22 https://edition.cnn.com/2019/06/20/world/chennai-satellite-images-reservoirs-water-crisis-trnd/index.html.
23 Evidence for the Syria connection can be found in Gleick et al., Kelly et al., and Cook et al., but see Burrows and Kinney for a different view; Peter H Gleick, "Water, Drought, Climate Change, and Conflict in Syria," *Weather, Climate, and Society* 6, no. 3 (2014); Colin P Kelley et al., "Climate Change in the Fertile Crescent and Implications of the Recent Syrian Drought," *Proceedings of the National Academy of Sciences* 112, no. 11 (2015); Benjamin I Cook et al., "Spatiotemporal Drought Variability in the Mediterranean over the Last 900 Years," *Journal of Geophysical Research: Atmospheres* 121, no. 5 (2016); Kate Burrows and Patrick L Kinney, "Exploring the Climate Change, Migration and Conflict Nexus," *International Journal of Environmental Research and Public Health* 13, no. 4 (2016).
24 Nick Watts et al., "The 2019 Report of the Lancet Countdown on Health and Climate Change: Ensuring that the Health of a Child Born Today Is Not Defined by a Changing Climate," *The Lancet* 394, no. 10211 (2019).
25 www.carbonbrief.org/factcheck-how-global-warming-has-increased-us-wildfires.
26 https://english.pravda.ru/news/hotspots/142605-fire_siberia/.
27 Frank J Wentz et al., "How Much More Rain Will Global Warming Bring?" *Science* 317, no. 5835 (2007).
28 Erich M Fischer and Reto Knutti, "Anthropogenic Contribution to Global Occurrence of Heavy-Precipitation and High-Temperature Extremes," *Nature Climate Change* 5, no. 6 (2015).
29 www.climatecentral.org/gallery/maps/extreme-precipitation-events-are-on-the-rise.

30 www.scientificamerican.com/article/no-end-in-sight-for-record-midwest-flood-crisis/; www.nationalgeographic.com/environment/2019/06/midwest-rain-climate-change-wrecking-corn-soy-crops/.
31 www.nytimes.com/2020/01/02/world/asia/indonesia-jakarta-rain-floods.html.
32 https://en.wikipedia.org/wiki/2020_Kyushu_floods.
33 www.aljazeera.com/news/2020/7/14/grim-china-battles-record-flooding-after-torrential-downpours.
34 www.ucsusa.org/resources/toxic-relationship; Timothy B Erickson et al., "Environmental Health Effects Attributed to Toxic and Infectious Agents following Hurricanes, Cyclones, Flash Floods and Major Hydrometeorological Events," *Journal of Toxicology and Environmental Health, Part B* 22, no. 5–6 (2019).
35 www.theguardian.com/environment/2019/mar/21/climate-change-could-make-insurance-too-expensive-for-ordinary-people-report.
36 www.gfdl.noaa.gov/global-warming-and-hurricanes/; James P Kossin et al., "Global Increase in Major Tropical Cyclone Exceedance Probability over the Past Four Decades," *Proceedings of the National Academy of Sciences* 117, no. 22 (2020).
37 Kieran T Bhatia et al., "Recent Increases in Tropical Cyclone Intensification Rates," *Nature Communications* 10, no. 1 (2019).
38 www.ipcc.ch/report/sixth-assessment-report-working-group-i/.
39 www.realclimate.org/index.php/archives/2018/05/does-global-warming-make-tropical-cyclones-stronger/.
40 Iam-Fei Pun, I-I Lin, and Min-Hui Lo, "Recent Increase in High Tropical Cyclone Heat Potential Area in the Western North Pacific Ocean," *Geophysical Research Letters* 40, no. 17 (2013).
41 https://coast.noaa.gov/states/fast-facts/hurricane-costs.html.
42 https://en.wikipedia.org/wiki/Cyclone_Idai.
43 Lijing Cheng et al., "Record-Setting Ocean Warmth Continued in 2019" (Springer, 2020).
44 Penn, Justin L, and Curtis Deutsch, "Avoiding Ocean Mass Extinction from Climate Warming," *Science* 376, no. 6592 (2022).
45 Andrew Shepherd et al., "Mass Balance of the Greenland Ice Sheet from 1992 to 2018," *Nature* 579, no. 7798 (2020).
46 www.climate.gov/news-features/featured-images/2019-arctic-sea-ice-extent-ties-second-lowest-summer-minimum-record.
47 Eric Rignot et al., "Four Decades of Antarctic Ice Sheet Mass Balance from 1979–2017," *Proceedings of the National Academy of Sciences* 116, no. 4 (2019).
48 Grossman, "Emergency on Planet Earth," www.scientistswarning.org/2020/09/07/new-book-stresses-urgency-of-action-on-climate-crisis/.
49 Frank Pattyn and Mathieu Morlighem, "The Uncertain Future of the Antarctic Ice Sheet," *Science* 367, no. 6484 (2020).
50 Ripple et al., "World Scientist's Warning."
51 https://theconversation.com/climate-change-is-triggering-a-migrant-crisis-in-vietnam-88791.
52 Simon Albert et al., "Interactions between Sea-Level Rise and Wave Exposure on Reef Island Dynamics in the Solomon Islands," *Environmental Research Letters* 11, no. 5 (2016).
53 https://edition.cnn.com/2019/11/13/europe/venice-flooding-state-of-emergency-intl-hnk/index.html.

54 https://coast.noaa.gov/states/fast-facts/hurricane-costs.html; www.climatecentral.org/news/coastal-flooding-us-cities-18148.
55 IPCC 6th Assessment Report 2021: www.ipcc.ch/report/sixth-assessment-report-working-group-i/.
56 Watts et al., "The 2019 Report of the Lancet Countdown on Health and Climate Change: Ensuring that the Health of a Child Born Today Is Not Defined by a Changing Climate," *The Lancet* 394, no. 10211 (2019).
57 www.unwater.org/publications/world-water-development-report-2018/.
58 Giovanni Benelli and Heinz Mehlhorn, *Mosquito-Borne Diseases* (Springer, 2018).
59 Watts, "2019 Lancet Countdown."
60 www.cdc.gov/zika/reporting/2017-case-counts.html; www.ecdc.europa.eu/en/publications-data/rapid-risk-assessment-local-transmission-dengue-fever-france-and-spain; Giovanni Benelli, Marco Pombi, and Domenico Otranto, "Malaria in Italy: Migrants Are Not the Cause," *Trends in Parasitology* 34, no. 5 (2018).
61 https://insideclimatenews.org/news/22042022/valley-fever-climate-change/.
62 www.bbc.com/earth/story/20170504-there-are-diseases-hidden-in-ice-and-they-are-waking-up.
63 www.hsph.harvard.edu/c-change/subtopics/coronavirus-and-climate-change/; www.acmicrob.com/microbiology/the-impact-of-climate-change-and-other-factors-on-zoonotic-diseases.php?aid=220.
64 Colin J Carlson, Gregory F Albery, Cory Merow, Christopher H Trisos, Casey M Zipfel, Evan A Eskew, Kevin J Olival, Noam Ross, and Shweta Bansal, "Climate Change Increases Cross-Species Viral Transmission Risk," *Nature* (2022).
65 www.eia.gov/energyexplained/gasoline/gasoline-and-the-environment.php.
66 https://electrek.co/2020/01/09/california-looking-to-ban-gas-powered-lawnmowers-leaf-blowers-sf-chronicle/.
67 Philip J Landrigan et al., "The Lancet Commission on Pollution and Health," *The Lancet* 391, no. 10119 (2018); Karn Vohra et al., "Global Mortality from Outdoor Fine Particle Pollution Generated by Fossil Fuel Combustion: Results from Geos-Chem," *Environmental Research* 195 (2021); GBD Risk Factors Collaborators, "Global, Regional, and National Comparative Risk Assessment of 79 Behavioural, Environmental and Occupational, and Metabolic Risks or Clusters of Risks in 188 Countries, 1990–2013: A Systematic Analysis for the Global Burden of Disease Study 2013," *Lancet* 386, no. 10010 (2015).
68 www.reuters.com/article/us-global-climatechange-health-idUSKCN1NX2ZX.
69 http://saveoursoils.com/userfiles/downloads/1368007451-Soil%20Erosion-David%20Pimentel.pdf.
70 Watts, "Lancet Countdown."
71 www-cdn.oxfam.org/s3fs-public/file_attachments/rr-impact-russias-grain-export-ban-280611-en_3.pdf.
72 www.agriculture.gov.au/sites/default/files/documents/EffectsOfDroughtAndClimateVariabilityOnAustralianFarms_v1.0.0.pdf.
73 Walter W Immerzeel et al., "Importance and Vulnerability of the World's Water Towers," *Nature* 577, no. 7790 (2020).
74 Josh M Maurer et al., "Acceleration of Ice Loss across the Himalayas over the Past 40 Years," *Science Advances* 5, no. 6 (2019).
75 https://oceanservice.noaa.gov/facts/coralreef-climate.html#:%7E:text=Climate%20change%20leads%20to%3A,to%20the%20smothering%20of%20coral.
76 www.fs.usda.gov/ccrc/topics/bark-beetles-and-climate-change-united-states.
77 Grossman, "Emergency on Planet Earth."

Sandra Díaz et al., "Assessing Nature's Contributions to People," *Science* 359, no. 6373 (2018).
78 Stuart L Pimm et al., "The Biodiversity of Species and Their Rates of Extinction, Distribution, and Protection," *Science* 344, no. 6187 (2014).
79 Gerardo Ceballos, Paul R Ehrlich, and Rodolfo Dirzo, "Biological Annihilation via the Ongoing Sixth Mass Extinction Signaled by Vertebrate Population Losses and Declines," *Proceedings of the National Academy of Sciences* 114, no. 30 (2017).
80 www.ucsusa.org/resources/whats-driving-deforestation.
81 Thomas E Lovejoy and Carlos Nobre, "Amazon Tipping Point: Last Chance for Action" (American Association for the Advancement of Science, 2019).
82 http://pdf.wri.org/world_greenhouse_gas_emissions_flowchart.pdf.
83 www.metoffice.gov.uk/about-us/press-office/news/weather-and-climate/2021/2c-rise-to-put-one-in-eight-of-global-population-at-heat-stress-risk.
84 www.mckinsey.com/~/media/mckinsey/business%20functions/sustainability/our%20insights/will%20india%20get%20too%20hot%20to%20work/will-india-get%20too-hot-to-work-vf.pdf.
85 The wet-bulb temperature is the temperature read by a thermometer wrapped in wet cloth over which air is passed. The cloth acts as a proxy for human skin: if the water evaporates, the thermometer is cooled and the wet-bulb temperature will be lower than the air temperature. But in high humidity, with less evaporation, this doesn't happen. Overall, wet-bulb temperature can be thought of as a measurement of not only how hot it is but how well humans could cope. The theoretical human limit for wet-bulb temperature is 35°C, which is close to human skin temperature. Any hotter than that, and the body will overheat, even in the shade, after four to five hours.
86 www.mckinsey.com/business-functions/sustainability/our-insights/will-mortgages-and-markets-stay-afloat-in-florida.
87 www.nytimes.com/2021/09/24/climate/federal-flood-insurance-cost.html.
88 Chi Xu et al., "Future of the Human Climate Niche," *Proceedings of the National Academy of Sciences* 117, no. 21 (2020).
89 www.carbonbrief.org/cmip6-the-next-generation-of-climate-models-explained.
90 Lynas, Our Final Warning.
91 An even more detailed imagining of this scenario firmly rooted in science can be found in David Wallace-Wells, *The Uninhabitable Earth.* Columbia University Press, 2019.
https://nymag.com/intelligencer/2017/07/climate-change-earth-too-hot-for-humans.html.
92 Grossman, "Emergency on Planet Earth."
93 Hansen, "Climate Change in a Nutshell." www.columbia.edu/~jeh1/mailings/2018/20181206_Nutshell.pdf.
94 Ibid.
95 https://epic.awi.de/id/eprint/36677/1/bg-11-6573-2014.pdf.
96 Anton Vaks et al., "Speleothems Reveal 500,000-Year History of Siberian Permafrost," *Science* 340, no. 6129 (2013).
97 Monique Brouillette, "How Microbes in Permafrost Could Trigger a Massive Carbon Bomb," *Nature* 591, no. 7850 (2021).
98 Robert B Jackson et al., "Increasing Anthropogenic Methane Emissions Arise Equally from Agricultural and Fossil Fuel Sources," *Environmental Research Letters* 15, no. 7 (2020); Merritt R Turetsky et al., *Permafrost Collapse Is Accelerating Carbon Release* (Nature Publishing Group, 2019).
99 SE Chadburn et al., "An Observation-Based Constraint on Permafrost Loss as a Function of Global Warming," *Nature Climate Change* 7, no. 5 (2017).

100 www.carbonbrief.org/explainer-nine-tipping-points-that-could-be-triggered-by-climate-change; Grossman, "Emergency on Planet Earth."
101 TM Lenton et al., "Climate Tipping Points: Too Risky to Bet against," *Nature* 575, no. 7784 (2019).
102 www.nytimes.com/interactive/2021/03/02/climate/atlantic-ocean-climate-change.html.
103 Levke Caesar et al., "Current Atlantic Meridional Overturning Circulation Weakest in Last Millennium," *Nature Geoscience* 14, no. 3 (2021).
104 The IPCC 2021 report had this to say about tipping points: "Low-likelihood outcomes, such as ice sheet collapse, abrupt ocean circulation changes, some compound extreme events and warming substantially larger than the assessed very likely range of future warming cannot be ruled out."
105 www.nature.com/articles/d41586-020-00508-4.
106 Ripple, "World Scientists' Warning."
107 https://voxeu.org/article/why-are-economists-letting-down-world-climate-change.
108 Richard SJ Tol, "The Economic Impacts of Climate Change," *Review of Environmental Economics and Policy* (2020).
109 Lynas, Our Final Warning.
110 Keen, "Appallingly Bad."
111 Ibid., 9.
112 Benjamin Franta, "Weaponizing Economics: Big Oil, Economic Consultants, and Climate Policy Delay," *Environmental Politics* (2021).
113 www.swissre.com/media/news-releases/nr-20210422-economics-of-climate-change-risks.html.
114 Ibid.
115 www.ipcc.ch/report/ar6/wg3/resources/spm-headline-statements/.

Chapter 4

1 This increase in population was particularly impactful in the richer countries such as the Organization of Economic Cooperation and Development (OECD) and Brazil, Russia, India, China and South Africa (the BRICS countries). Yet we should recognize that increasing population in itself does not necessarily do much to increase emissions; rather, the level of emissions depends on how that population lives. As shown later in the book, by far the majority of the emissions are from developed countries, and especially from elites in both developed and developing countries. A rich person in India, for example, produces several orders of magnitude more emissions than a poor Indian.
2 Jason Hickel, *Less Is More: How Degrowth Will Save the World* (Random House, 2020).
3 N Klein, This Changes Everything: Capitalism vs. the Climate (Simon and Schuster, 2014).
4 As pointed out by Weintrobe, although globalization is important for trade, problem solving (such as the Montreal Protocol concerning ozone depletion), and growing interconnectivity, it did not have to happen in the neoliberal way. Weintrobe argues that those who built this system were transfixed by an ideology that encouraged them to dissociate themselves from planetary limits and boundaries, leading them to discount true costs and to ignore ecological and social costs; Sally Weintrobe,

Psychological Roots of the Climate Crisis: Neoliberal Exceptionalism and the Culture of Uncare (Bloomsbury Publishing USA, 2021).

5 Economic liberalism is an economic philosophy based on strong support for a market economy and private property and a system in which the greatest number of decisions are made by individuals or households rather than collective institutions or the state. Often contrasted to socialism and planned economies, economic liberalism is the economic expression of political liberalism, which arose out the rejection of kings and authoritarianism in Europe, leading liberals to argue that people have rights to life, liberty, and property and that governments must not violate these rights. This classic sense of liberalism – economic and political – is a bit confusing to some in the USA, where "liberal" is typically used to connote social liberalism, or really social democracy, which entails more state regulation of the market for welfare and the common good.

6 In Sweden, for example, a plan was put forward to buy out the owners' share in their own businesses and to turn the country into a worker/share-owner democracy. www.jacobinmag.com/2017/08/sweden-social-democracy-meidner-plan-capital.

7 Maher, *Apollo in the Age of Aquarius*, reviewed at https://harvardpress.typepad.com/hup_publicity/2017/03/the-space-race-and-the-grassroots-neil-maher.html.

8 D Harvey, *A Brief History of Neoliberalism* (Oxford University Press, 2005); www.pewresearch.org/fact-tank/2018/08/07/for-most-us-workers-real-wages-have-barely-budged-for-decades/.

9 https://monthlyreview.org/2020/04/01/how-long-can-neoliberalism-withstand-climate-crisis/.

10 www.statista.com/chart/16675/us-carbon-emissions/; Michael R Raupach et al., "Global and Regional Drivers of Accelerating CO_2 Emissions," *Proceedings of the National Academy of Sciences* 104, no. 24 (2007).

11 https://ieep.eu/news/more-than-half-of-all-co2-emissions-since-1751-emitted-in-the-last-30-years.

12 HE Daly, "The Perils of Free Trade in Scientific American," (November, 1993).

13 https://ourworldindata.org/is-globalization-an-engine-of-economic-development. Advocates of neoliberal globalization argue that falling wages in high-wage countries will only be temporary (which is clearly not true in the USA, where real wages have barely budged in decades), and that once the middle class grows in developing countries, that group will start to bring in positive environmental changes. www.pewresearch.org/fact-tank/2018/08/07/for-most-us-workers-real-wages-have-barely-budged-for-decades/.

14 For example, China, the main developing country, emitted over 14 gigatons of CO_2 in 2019. www.reuters.com/article/climate-change-china-emissions/china-carbon-emissions-exceeded-oecd-total-in-2019-research-idUSL1N2MT031; https://wiiw.ac.at/how-economic-globalisation-affects-income-inequality-n-431.html.

15 www.chinawaterrisk.org/resources/analysis-reviews/2019-state-of-ecology-environment-report-review/; www.cfr.org/backgrounder/china-climate-change-policies-environmental-degradation; www.iisd.org/articles/toxic-soil-china.

16 Klein, This Changes Everything, 77.

17 www.iatp.org/documents/trans-pacific-partnership-undermines-global-efforts-address-climate-change; www.sierraclub.org/sites/www.sierraclub.org/files/uploads-wysiwig/dirty-deal.pdf.

18 Trade rulings of this kind are significant blocks to climate policy, since more localized systems bring large benefits. One study showed that community-owned wind energy confers as much as eight times the financial benefit to a local community compared to outside ownership, and this translates into more jobs, which is part of a just transition

to a low carbon economy. See Katharina Wolf and Craig Morris, "Ownership Matters: Local Added Value from a Community Wind Farm," *Renewables International*, June 27, 2016, www.renewablesinternational.net/local-added-value-from-a-community-wind-farm/150/537/96249/; www.iatp.org/sites/default/files/2016_09_06_ClimateCostFreeTrade.pdf.

19 Ken Conca, "The WTO and the Undermining of Global Environmental Governance," *Review of International Political Economy* 7, no. 3 (2000).

20 www.iatp.org/sites/default/files/2016_09_06_ClimateCostFreeTrade.pdf.

21 www.epi.org/blog/naftas-impact-workers/.

22 Reported in Klein, This Changes Everything.

23 Ibid, 84.

24 Ibid. As Clinton said at the time, "[w]e are now on the verge of a global economic expansion ... Already the confidence we've displayed by ratifying NAFTA has begun to bear fruit. We are now making real progress toward a worldwide trade agreement so significant that it could make the material gains of NAFTA for our country look small by comparison." Years later it was argued that NAFTA had caused the loss of some 700,000 US jobs as production moved to Mexico, with most of those workers suffering a permanent loss of income, and also that it had a deeply destructive effect on Mexican agricultural and local business, with several million displaced, along with their families, generating a dramatic increase of undocumented workers to the US. www.epi.org/blog/naftas-impact-workers/. By 2017, US voters were split 46 to 48 percent whether NAFTA was bad or good for them, https://news.gallup.com/poll/204269/americans-split-whether-nafta-good-bad.aspx.

25 https://indypendent.org/2011/12/seattle-wto-shutdown-99-to-occupy-organizing-to-win-12-years-later/.

26 Karl Gerth, *Unending Capitalism: How Consumerism Negated China's Communist Revolution* (Cambridge University Press, 2020).

27 This reduction in poverty went from 60 percent in 2006 to 35 percent in 2017 (and extreme poverty from 38 percent to 15 percent). www.theguardian.com/world/2019/mar/07/how-a-populist-president-helped-bolivias-poor-but-built-himself-a-palace; https://nacla.org/blog/2015/06/15/morales-greenlights-tipnis-road-oil-and-gas-extraction-bolivia's-national-parks; www.theguardian.com/environment/andes-to-the-amazon/2015/jun/05/bolivia-national-parks-oil-gas.

28 https://jacobinmag.com/2014/01/managing-bolivian-capitalism.

29 Harvey, Brief History; Weintrobe, Psychological Roots.

30 Weintrobe, *Psychological Roots*, 51.

31 Ibid.

32 Quotes from Harvey, A Brief History of Neoliberalism.

33 Klein, This Changes Everything, 62.

34 Kate Raworth, Doughnut Economics: Seven Ways to Think like a 21st-Century Economist (Chelsea Green Publishing, 2017).

35 Murray Bookchin, *The Ecology of Freedom* (New Dimensions Foundation, 1982).

36 Thorstein Veblen, *The Theory of the Leisure Class* (Houghton Mifflin Boston, 1973).

37 Luigi Esposito and Fernando Pérez, "The Global Addiction and Human Rights: Insatiable Consumerism, Neoliberalism, and Harm Reduction," *Perspectives on Global Development and Technology* 9, no. 1–2 (2010).

38 Thomas Wiedmann et al., "Scientists' Warning on Affluence," *Nature Communications* 11, no. 1 (2020).

39 First quote is from page 56, second is from page 53 of

Noam Chomsky, Robert Pollin, and CJ Polychroniou, *Climate Crisis and the Global Green New Deal: The Political Economy of Saving the Planet* (Verso, 2020).

40 www.irena.org/-/media/Files/IRENA/Agency/Publication/2021/Apr/IRENA_RE_Capacity_Statistics_2021.pdf; www.theguardian.com/power-of-green/2020/dec/17/uk-offshore-wind-global-renewable-future.

41 www.tni.org/en/publication/reclaiming-public-services.

42 Stan Cox, The Green New Deal and beyond: Ending the Climate Emergency While We Still Can (City Lights Books, 2020).

43 Stan Cox, The Path to a Livable Future: A New Politics to Fight Climate Change, Racism, and the next Pandemic (City Lights Books, 2021).

44 www.nytimes.com/2021/12/28/climate/chile-constitution-climate-change.html.

45 Ripple et al., "World Scientist's Warning."

46 Wiedmann et al., "Scientists' Warning."

47 R Pollin, "De-growth vs. a Green New Deal," *New Left Review* (2018).

48 Hickel, *Less Is More*; Jason Hickel, "What Does Degrowth Mean? A Few Points of Clarification," *Globalizations* (2020); Jason Hickel et al., "Urgent Need for Post-growth Climate Mitigation Scenarios," *Nature Energy* 6, no. 8 (2021).

49 In "Drivers of Declinining CO_2," Le Quéré et al. found that some of the eighteen developed economies they studied had managed to reduce emissions while still growing their economies through a shift from coal to fossil gas, increased efficiencies in buildings, more support for renewables, and similar steps, even while apparently taking into account the production of emissions in other countries that are then consumed in the developed country. Yet these reductions of emissions also occurred partly in a period when there was actually less economic growth, and the reduction was nowhere near enough to meet Paris or IPCC targets. Corinne Le Quéré et al., "Drivers of Declining CO_2 Emissions in 18 Developed Economies," *Nature Climate Change* 9, no. 3 (2019).

50 Helmut Haberl et al., "A Systematic Review of the Evidence on Decoupling of GDP, Resource Use and GHG Emissions, Part II: Synthesizing the Insights," *Environmental Research Letters* 15, no. 6 (2020).

51 Hickel et al., "Urgent Need for Post-Growth."

52 Paul E Brockway et al., "Energy Efficiency and Economy-Wide Rebound Effects: A Review of the Evidence and Its Implications," *Renewable and Sustainable Energy Reviews* (2021).

53 Wiedmann et al., "Scientists' Warning"; www.oxfam.org/en/press-releases/carbon-emissions-richest-1-percent-more-double-emissions-poorest-half-humanity.

54 Matthew T Huber, "The Case for Socialist Modernism," *Political Geography* (2021).

55 Hickel, *Less Is More*.

56 Hickel et al., "Urgent Need." Note that, for the USA, GDP, adjusted for inflation, has increased 800 percent in the last 70 years while 14 percent of people live beneath the poverty line (similar to the 1970s) and an estimated 35 million experienced food insecurity even before the Covid-19 pandemic. "Enough for Everyone," August 10, 2021: www.yesmagazine.org/issue/how-much-is-enough/2021/08/10/what-is-enough.

57 Raworth, Doughnut Economics.

58 Millward-Hopkins et al., show that the annual energy use of hunter-gatherers was about 5 GJ per person annually, which by 1850 rose to about 20 GJ, and today to about 80 GJ (80×10^9). Therefore, today, multiplying by nearly eight billion people, total annual energy use is about 600 EJ (exajoules are 10^{18}). The authors

then computed the Decent Living Energy for the entire world for nine billion people in 2050. Contrary to the oft-repeated cliché that environmentalists want people to return to caves, they showed that each person could have highly efficient facilities for cooking, storing food, and washing clothes; low-energy lighting; 50 liters of clean water per day; 15 liters for a warm bath per day; an air temperature maintained at about 20°C through the year; a networked computer; access to an extensive public transportation network providing 5,000–15,000 km of transport per year; and universal healthcare and education for everyone between five and nineteen years old. Overall, the total annual energy use was estimated at about 150 EJ, compared to more than 600 EJ on the business-as-usual trajectory. Joel Millward-Hopkins et al., "Providing Decent Living with Minimum Energy: A Global Scenario," *Global Environmental Change* 65 (2020).

59 www.ipcc.ch/report/ar6/wg3/resources/spm-headline-statements/.

60 www.iea.org/news/global-carbon-dioxide-emissions-are-set-for-their-second-biggest-increase-in-history; www.reuters.com/business/environment/europe-miss-2030-climate-goal-by-21-years-current-pace-study-2021-09-04/.

Chapter 5

1 https://climatecommunication.yale.edu/publications/climate-change-in-the-american-mind-march-2021/2/.

2 Stuart Bryce Capstick and Nicholas Frank Pidgeon, "What Is Climate Change Scepticism? Examination of the Concept Using a Mixed Methods Study of the UK Public," *Global Environmental Change* 24 (2014).

3 https://climatecommunication.yale.edu/visualizations-data/ycom-us/.

4 www.abc.net.au/news/science/2019-09-10/climate-of-nation-australia-attitudes/11484690; www.smh.com.au/environment/climate-change/a-record-share-of-australians-say-humans-cause-climate-change-poll-20190328-p518go.html.

5 www.europeansocialsurvey.org/docs/findings/ESS8_toplines_issue_9_climatechange.pdf; Hao Yu et al., "Public Perception of Climate Change in China: Results from the Questionnaire Survey," *Natural Hazards* 69, no. 1 (2013).

6 Matthew J Hornsey et al., "Meta-analyses of the Determinants and Outcomes of Belief in Climate Change," *Nature Climate Change* 6, no. 6 (2016).

7 Objective knowledge was verified by testing knowledge of facts, and subjective knowledge measured by participants' ranking of their level of knowledge.

8 Dunlap et al., "Measuring Endorsement." The New Ecological Paradigm is one of the most widely used surveys in the environmental psychology literature. This scale includes a set of questions that emphasizes the fragility of the environment and the importance of minimizing human impact on it, and although it does not directly mention climate change, it has been shown to relate more strongly to belief in climate change than any other factor. Some of the items it includes are the following: (1) We are approaching the limit of the number of people the Earth can support; (2) Humans have the right to modify the natural environment to suit their needs; (3) When humans interfere with nature it often produces disastrous consequences; (4) Human ingenuity will insure that we do not make the Earth unlivable; (5) Humans are seriously abusing the environment. R Dunlap et al., "Measuring Endorsement of the New Ecological Paradigm: A Revised NEP Scale," *Journal of Social Issues* 56, no. 3 (2000).

9 Quotations from Matthew J Hornsey and Kelly S Fielding, "Understanding (and Reducing) Inaction on Climate Change," *Social Issues and Policy Review* 14, no. 1 (2020).
10 Matthew J Hornsey, Emily A Harris, and Kelly S Fielding, "Relationships among Conspiratorial Beliefs, Conservatism and Climate Scepticism across Nations," *Nature Climate Change* 8, no. 7 (2018).
11 www.gov.uk/government/news/end-of-coal-power-to-be-brought-forward-in-drive-towards-net-zero.
12 www.theguardian.com/environment/2022/apr/06/uk-more-oil-gas-drilling-north-sea-energy-security-strategy-kwasi-kwarteng-net-zero-targets.
13 Hornsey et al., "Relationships among Conspiratorial Beliefs."
14 www.theatlantic.com/daily-dish/archive/2010/05/denialism-vs-skepticism/186824/.
15 Exxon's own CO_2 research program in the 1970s noted that "a doubling of carbon dioxide is estimated to be capable of increasing the average global temperature by from 1 to 3°C, with a 10°C rise predicted at the poles." At a 1980 meeting of the American Petroleum Institute, representatives from Exxon, Texaco, and Shell concluded that this likely impact would be 1°C by 2005, 2.5°C by 2038 (creating "major economic consequences"), and 5°C by 2067 (leading to "globally catastrophic effects"). https://climaterealityproject.org/blog/climate-denial-machine-how-fossil-fuel-industry-blocks-climate-action.
16 A trove of fossil fuel industry documents is held at UCSF: www.industrydocuments.ucsf.edu/fossilfuel/blog/. Video interviews can be seen at: https://theconversation.com/a-brief-history-of-fossil-fuelled-climate-denial-61273. Also see: www.climatechangecommunication.org/wp-content/uploads/2019/10/America_Misled.pdf.
17 Leah Cardamore Stokes, Short Circuiting Policy: Interest Groups and the Battle over Clean Energy and Climate Policy in the American States (Oxford University Press, 2020).
18 https://climaterealityproject.org/blog/climate-denial-machine-how-fossil-fuel-industry-blocks-climate-action.
19 https://influencemap.org/reports/Reports?type=287770&year=382457.
20 Senator Whitehouse explains Dark Money after Citizens United: www.youtube.com/watch?v=MoY7I7PEAWw.
21 Jordi Xifra, "Climate Change Deniers and Advocacy: A Situational Theory of Publics Approach," *American Behavioral Scientist* 60, no. 3 (2016).
22 Oreskes and Conway relate that one such dissenter is Fred Seitz, who received $45 million in funding from the RJ Reynolds tobacco company to obscure smoking's risks and then, with funding from the fossil fuel companies, became the highest-profile climate denier in the USA, penning op-eds for the *Wall Street Journal* and other leadings news outlets. Naomi Oreskes and Erik M Conway, *Merchants of Doubt: How a Handful of Scientists Obscured the Truth on Issues from Tobacco Smoke to Global Warming* (Bloomsbury Publishing USA, 2011).
23 https://royalsociety.org/-/media/Royal_Society_Content/policy/publications/2006/8257.pdf.
24 www.ucsusa.org/resources/how-fossil-fuel-lobbyists-used-astroturf-front-groups-confuse-public.
25 https://insideclimatenews.org/news/22102015/Exxon-Sowed-Doubt-about-Climate-Science-for-Decades-by-Stressing-Uncertainty/. Of course, this kind of denial did not end with G.W. Bush's leaving office. In 2016, Rex Tillerson, was named Donald Trump's secretary of state; Tillerson, as CEO of ExxonMobil, had announced at its

annual meeting in 2015 that the company believed it would be best to wait for more solid science before acting on climate change.

26 Riley E Dunlap and Robert J Brulle, "Sources and Amplifiers of Climate Change Denial," in *Research Handbook on Communicating Climate Change* (Edward Elgar, 2020).

27 And a new campaign is underway, to persuade the public that the fossil industry is part of the "solution" to climate change using "low carbon" methods.

28 https://slate.com/human-interest/2021/11/oil-gas-companies-climate-education.html.

29 https://apps.publicintegrity.org/oil-education/.

30 www.statista.com/chart/15195/wind-turbines-are-not-killing-fields-for-birds/.

31 Emily M Eaton and Nick A Day, "Petro-Pedagogy: Fossil Fuel Interests and the Obstruction of Climate Justice in Public Education," *Environmental Education Research* 26, no. 4 (2020).

32 Anthony E Ladd, "Priming the Well: 'Frackademia' and the Corporate Pipeline of Oil and Gas Funding into Higher Education," *Humanity & Society* 44, no. 2 (2020).

33 www.theguardian.com/environment/climate-consensus-97-per-cent/2017/mar/13/the-fossil-fuel-industrys-invisible-colonization-of-academia.

34 https://ui.adsabs.harvard.edu/abs/2017AGUFMPA33C0375F%2F/abstract; https://insideclimatenews.org/wp-content/uploads/2015/12/Global-Climate-Science-Communications-Plan-1998.pdf.

35 Hertsgaard and Pope, Columbia Journalism Review, www.cjr.org/special_report/climate-change-media.php.

36 Maxwell T Boykoff and Jules M Boykoff, "Balance as Bias: Global Warming and the US Prestige Press," *Global Environmental Change* 14, no. 2 (2004).

37 David M Romps and Jean P Retzinger, "Climate News Articles Lack Basic Climate Science," *Environmental Research Communications* 1, no. 8 (2019).

38 A striking example of good climate coverage in 2021 was the *New York Times* multimedia special coverage on climate impacts in 193 countries: www.nytimes.com/interactive/2021/12/13/opinion/climate-change-effects-countries.html.
A recent report of climate coverage from 2005 to 2019 in the UK, Australia, Canada, New Zealand, and the USA suggested it had gotten more accurate. Lucy McAllister et al., "Balance as Bias, Resolute on the Retreat? Updates and Analyses of Newspaper Coverage in the United States, United Kingdom, New Zealand, Australia and Canada over the Past 15 Years," *Environmental Research Letters* 16, no. 9 (2021).

39 A powerful explanation of the reasons for the limitations of the news media to challenge the status quo and to critique power were provided in Edward S Herman and Noam Chomsky, *Manufacturing Consent: The Political Economy of the Mass Media* (Random House, 2010). A more recent short account is here: www.theguardian.com/commentisfree/2021/dec/09/whats-really-wrong-mainstream-media.

40 MJ Hornsey and KS Fielding, "Attitude Roots and Jiu Jitsu Persuasion: Understanding and Overcoming the Motivated Rejection of Science," *American Psychologist* 72, no. 5 (2017).

41 For example, one research study that surveyed a large sample of visitors to climate blogs found that their level of skepticism regarding climate change correlated with an endorsement of a cluster of conspiracy theories. S Lewandowsky, K Oberauer, and GE Gignac, "NASA Faked the Moon Landing – Therefore, (Climate) Science Is a Hoax: An Anatomy of the Motivated Rejection of Science," *Psychological Science* 24, no. 5 (2013).

42 Karen M Douglas and Robbie M Sutton, "Climate Change: Why the Conspiracy Theories Are Dangerous," *Bulletin of the Atomic Scientists* 71, no. 2 (2015).
43 https://en.wikipedia.org/wiki/Climatic_Research_Unit_email_controversy.
44 Charles G Lord, Lee Ross, and Mark R Lepper, "Biased Assimilation and Attitude Polarization: The Effects of Prior Theories on Subsequently Considered Evidence," *Journal of Personality and Social Psychology* 37, no. 11 (1979).
45 André Mata et al., "Strategic Numeracy: Self-serving Reasoning about Health Statistics," *Basic and Applied Social Psychology* 37, no. 3 (2015).
46 Peter Kalmus, *Fly Less to Convey Urgency* (American Association for the Advancement of Science, 2019).
47 Hornsey and Fielding, "Attitude Roots."
48 George Marshall, Don't Even Think about It: Why Our Brains Are Wired to Ignore Climate Change (Bloomsbury Publishing USA, 2015).
49 Marshall, Don't Even Think about It.
50 Holly Jean Buck, Ending Fossil Fuels: Why Net Zero Is Not Enough (Verso, 2021).
51 Marshall, Don't Even Think about It.
52 Ullrich KH Ecker et al., "The Psychological Drivers of Misinformation Belief and Its Resistance to Correction," *Nature Reviews Psychology* 1, no. 1 (2022).
53 www.huffpost.com/entry/sebastian-gorka-ocasio-cortez-green-new-deal-cpac_n_5c78657ee4b087c2f294a61f.
54 Hornsey and Fielding, "Attitude Roots."
55 Irina Feygina, John T Jost, and Rachel E Goldsmith, "System Justification, the Denial of Global Warming, and the Possibility of 'System-Sanctioned Change'," *Personality and Social Psychology Bulletin* 36, no. 3 (2010).
56 Kelly S Fielding et al., "Using Ingroup Messengers and Ingroup Values to Promote Climate Change Policy," *Climatic Change* 158, no. 2 (2020).
57 https://clcouncil.org.
58 Paul G Bain et al., "Co-benefits of Addressing Climate Change Can Motivate Action around the World," *Nature Climate Change* 6, no. 2 (2016).
59 Vohra et al., "Global Mortality."
60 Anatol Lieven, Climate Change and the Nation State: The Case for Nationalism in a Warming World (Oxford University Press, 2020); www.nbcnews.com/think/opinion/u-s-military-terrified-climate-change-it-s-done-more-ncna1240484.
61 P Wesley Schultz et al., "The Constructive, Destructive, and Reconstructive Power of Social Norms," *Psychological Science* 18, no. 5 (2007).
62 https://behavior.rare.org/wp-content/uploads/2020/07/Social-Influences.-Opower.7.8.pdf.
63 Kimberly S Wolske, Kenneth T Gillingham, and P Wesley Schultz, "Peer Influence on Household Energy Behaviours," *Nature Energy* 5, no. 3 (2020).

Chapter 6

1 Two detailed reviews are Ecker et al., "The Psychological Drivers of Misinformation Belief and Its Resistance to Correction"; Stephan Lewandowsky, "Climate Change Disinformation and How to Combat It," *Annual Review of Public Health* 42 (2021).
2 Based on the theory of motivated cognition discussed in Chapter 5, some researchers have urged science communicators to focus on ways to present information

regarding climate change that are likely to be persuasive to various cultural groups. Advocates of this approach (e.g., Kahan et al., "Cultural Cognition") assume that differences in skepticism about climate change stem not from a lack of information about climate science or the impacts of global warming but from how they are inclined to weigh that information, such as how they evaluate the relative risks of unmitigated climate change and of the impact of emissions cuts on economic activity. Other researchers, however, have pushed back against the claim that cultural cognition has such a strong influence and argue instead that knowledge itself has an important influence on beliefs. (See, e.g., Van Der Linden et al., "Culture versus Cognition"; Lewandowsky, "Climate Change Disinformation".) But as Ranney and Clark ("Climate Change") concluded – and as this book contends – it is probably as wrong to assume there is a simple dichotomy between culture and knowledge as there is to assume one between nature and nurture. Sander Van Der Linden et al., "Culture versus Cognition Is a False Dilemma," *Nature Climate Change* 7, no. 7 (2017); Lewandowsky, "Climate Change Disinformation and How to Combat It"; Dan M Kahan, Hank Jenkins-Smith, and Donald Braman, "Cultural Cognition of Scientific Consensus," *Journal of Risk Research* 14, no. 2 (2011); Michael Andrew Ranney and Dav Clark, "Climate Change Conceptual Change: Scientific Information Can Transform Attitudes," *Topics in Cognitive Science* 8, no. 1 (2016).

3 Ranney and Clark, "Climate Change."

4 Oliver Taube et al., "Increasing People's Acceptance of Anthropogenic Climate Change with Scientific Facts: Is Mechanistic Information More Effective for Environmentalists?" *Journal of Environmental Psychology* 73 (2021). Susan Joslyn and Raoni Demnitz, "Communicating Climate Change: Probabilistic Expressions and Concrete Events," *Weather, Climate, and Society* 11, no. 3 (2019).

5 John D Sterman, "Risk Communication on Climate: Mental Models and Mass Balance," *Science* 322, no. 5901 (2008); John D Sterman and Linda Booth Sweeney, "Understanding Public Complacency about Climate Change: Adults' Mental Models of Climate Change Violate Conservation of Matter," *Climatic Change* 80, no. 3 (2007).

6 This section drawn from Kara Weisman and Ellen M Markman, "Theory-Based Explanation as Intervention," *Psychonomic Bulletin & Review* 24, no. 5 (2017).

7 J Cook et al., "The Consensus Handbook" (2018), www.climatechangecommunication.org/all/consensus-handbook/.

8 Ibid.

9 Eric Plutzer et al., "Climate Confusion among US Teachers," *Science* 351, no. 6274 (2016).

10 https://climatecommunication.yale.edu/visualizations-data/ycom-us/.

11 Cook et al., "Consensus Handbook."

12 Sander L Van der Linden et al., "The Scientific Consensus on Climate Change as a Gateway Belief: Experimental Evidence," *PLoS ONE* 10, no. 2 (2015).

13 Sander van der Linden, Anthony Leiserowitz, and Edward Maibach, "The Gateway Belief Model: A Large-Scale Replication," *Journal of Environmental Psychology* 62 (2019).

14 Matthew H Goldberg et al., "The Experience of Consensus: Video as an Effective Medium to Communicate Scientific Agreement on Climate Change," *Science Communication* 41, no. 5 (2019); Sander van der Linden, Anthony Leiserowitz, and Edward Maibach, "Perceptions of Scientific Consensus Predict Later Beliefs about the Reality of Climate Change Using Cross-Lagged Panel Analysis: A Response to Kerr and Wilson (2018)" (2018).

15 Kahan and Carpenter flag limitations of these lab-based studies. First, experimental stimuli (such as the consensus statement) are stripped of information about the source of the information, although such information is often critical in the real world. Second, the study participants read or viewed the materials because they were conscripted to do so, usually with monetary incentives, but this doesn't happen in the real world where people are instead awash with information sources. Third, techniques that positively affect college students or online workers who make money from taking Qualtrics or Prolific surveys might have much weaker effects in real-world samples, particularly as those include climate skeptics. Finally, this kind of study only asks people about their beliefs; it doesn't measure their behavior. Dan M Kahan and Katherine Carpenter, "Out of the Lab and into the Field," *Nature Climate Change* 7, no. 5 (2017).
16 www.theguardian.com/environment/climate-consensus-97-per-cent/2014/may/23/john-oliver-best-climate-debate-ever.
17 Although awareness of that consensus increased about 33 percent in 2010 to 40 percent in 2015 to about 55 percent in 2020, about 45 percent of US adults still do not believe it exists. https://climatecommunication.yale.edu/visualizations-data/ycom-us/.
18 Sander Van der Linden et al., "Inoculating the Public against Misinformation about Climate Change," *Global Challenges* 1, no. 2 (2017).
19 www.petitionproject.org.
20 www.desmogblog.com/2016/11/29/revealed-most-popular-climate-story-social-media-told-half-million-people-science-was-hoax/.
21 In a fourth condition, called "enhanced inoculation," the effect of inoculation was even stronger when arguments were added to refute specific key assumptions of the petition (e.g., that some of the names suggested that many signers were fake, such as Charles Darwin and the Spice Girls), which is an example of debunking.
22 www.ibe.unesco.org/en/news/italy-first-country-require-climate-change-education-all-schools.
23 Ecker et al., "The Psychological Drivers of Misinformation Belief and Its Resistance to Correction."
24 www.climatechangecommunication.org/wp-content/uploads/2020/10/DebunkingHandbook2020.pdf.
25 Sander van der Linden et al., "How Can Psychological Science Help Counter the Spread of Fake News?" *The Spanish Journal of Psychology* 24 (2021).
26 Martha C Monroe et al., "Identifying Effective Climate Change Education Strategies: A Systematic Review of the Research," *Environmental Education Research* 25, no. 6 (2019); Doug Lombardi et al., "Categorising Students' Evaluations of Evidence and Explanations about Climate Change," *International Journal of Global Warming* 12, no. 3-4 (2017); Stephen J Aguilar, Morgan S Polikoff, and Gale M Sinatra, "Refutation Texts: A New Approach to Changing Public Misconceptions about Education Policy," *Educational Researcher* 48, no. 5 (2019).
27 From Anne Marthe Van der Bles et al., "Communicating Uncertainty about Facts, Numbers and Science," *Royal Society Open Science* 6, no. 5 (2019).
28 Lauren C Howe et al., "Acknowledging Uncertainty Impacts Public Acceptance of Climate Scientists' Predictions," *Nature Climate Change* 9, no. 11 (2019).
29 Jordan Harold et al., "Cognitive and Psychological Science Insights to Improve Climate Change Data Visualization," *Nature Climate Change* 6, no. 12 (2016).
30 Ibid.

31 Juliette N Rooney-Varga et al., "The Climate Action Simulation," *Simulation & Gaming* 51, no. 2 (2020).
32 Leela Velautham, Michael Andrew Ranney, and Quinlan S Brow, "Communicating Climate Change Oceanically: Sea Level Rise Information Increases Mitigation, Inundation, and Global Warming Acceptance," *Frontiers in Communication* 4 (2019).
33 Toby Bolsen, Risa Palm, and Justin T Kingsland, "Counteracting Climate Science Politicization with effective Frames and Imagery," *Science Communication* 41, no. 2 (2019).
34 See https://virtualplanet.tech if you have an Oculus headset, also see www.npr.org/2019/11/24/779136094/climate-planners-turn-to-virtual-reality-and-hope-seeing-is-believing.
35 Liselotte J Roosen, Christian A Klöckner, and Janet K Swim, "Visual Art as a Way to Communicate Climate Change: A Psychological Perspective on Climate Change–Related Art," *World Art* 8, no. 1 (2018).
36 Christian Andreas Klöckner and Laura K Sommer, "Visual Art Inspired by Climate Change: An Analysis of Audience Reactions to 37 Artworks Presented during 21st UN Climate Summit in Paris," *PLoS ONE* 16, no. 2 (2021).

Chapter 7

1 https://climatecommunication.yale.edu/visualizations-data/ycom-us/; https://climatecommunication.yale.edu/about/projects/global-warmings-six-americas/.
2 Christian Martin and Sandor Czellar, "Where Do Biospheric Values Come from? A Connectedness to Nature Perspective," *Journal of Environmental Psychology* 52 (2017); Sander Van der Linden, "The Social-Psychological Determinants of Climate Change Risk Perceptions: Towards a Comprehensive Model." *Journal of Environmental Psychology* 41 (2015).
3 Although it is difficult to get population-level numbers on the valuation of nature, a 2021 Gallup poll of US adults found that only 50 percent agreed that the environment should be given priority over economic growth. https://news.gallup.com/poll/1615/environment.aspx. A 2017 survey in the found that 47 percent of respondents were unaware or unconcerned about biodiversity loss, with a further 42 percent showing only "some engagement." https://data.jncc.gov.uk/data/1d064484-d758–4494-84ec-efa09e16e999/UKBI-2019.pdf.
4 Some other accounts are from David W Orr, "Love It or Lose It: The Coming Biophilia Revolution," in R Kellert and EO Wilson (eds.), *The Biophilia Hypothesis* (Island Press, 1993) and Michael Perelman, *The Invention of Capitalism* (Duke University Press, 2000).
5 In *The Ecology of Freedom*, Bookchin argues that these alienated views of nature go much further back to the breakdown of ancient tribal societies in Europe and the Middle East. See also the Institute for Social Ecology at https://social-ecology.org/wp/ and a short summary of social ecology here: https://roarmag.org/magazine/communalism-climate-chaos/.
6 Gerald Matthews, Ian J Deary, and Martha C Whiteman, *Personality Traits* (Cambridge University Press, 2003).
7 These traits are closely related to the more well-known Big Five Model of Personality, which includes extroversion, agreeableness, openness, conscientiousness, and neuroticism. The HEXACO model has been shown to have validity in that

how one scores on the six factors has been shown to correlate closely with subsequent life outcomes and behaviors; Avshalom Caspi, Brent W Roberts, and Rebecca L Shiner, "Personality Development: Stability and Change," *Annual Review of Psychology* 56 (2005).

8 Each participant filled out the emissions reduction survey, a scale called the New Ecological Paradigm (discussed in Chapter 5), a Connectedness to Nature scale, and the HEXACO scale: Cameron Brick and Gary J Lewis, "Unearthing the 'Green' Personality: Core Traits Predict Environmentally Friendly Behavior," *Environment and Behavior* 48, no. 5 (2016).

9 Kathrin Rothermich et al., "The Influence of Personality Traits on Attitudes towards Climate Change: An Exploratory Study," *Personality and Individual Differences* 168 (2021).

10 Wiebke Bleidorn et al., "The Policy Relevance of Personality Traits," *American Psychologist* 74, no. 9 (2019); Brent W Roberts et al., "A Systematic Review of Personality Trait Change through Intervention," *Psychological Bulletin* 143, no. 2 (2017); www.greenhousetherapycenter.com/storage/app/media/file/Psychotherapy_change_meta-analysis_2017.pdf.

11 However, Hudson et al. found that openness to experience was the least changeable trait. Nathan W Hudson et al., "You Have to Follow through: Attaining Behavioral Change Goals Predicts Volitional Personality Change," *Journal of Personality and Social Psychology* 117, no. 4 (2019).

12 Thijs Bouman, Linda Steg, and Henk AL Kiers, "Measuring Values in Environmental Research: A Test of an Environmental Portrait Value Questionnaire," *Frontiers in Psychology* 9 (2018).

13 Mary Douglas and Aaron Wildavsky, *Risk and Culture* (University of California Press, 1983).

14 Hulme, Why We Disagree about Climate Change.

15 Hulme adopted the schematic representations of these "four myths of nature" from Buzz Holling, "The Resilience" and the representative quotations from Ereaut and Segnit. Gill Ereaut and Nat Segnit, *Warm Words: How We Are Telling the Climate Story and Can We Tell It Better*? IPPR, 2006, www.ippr.org/files/images/media/files/publication/2011/05/warm_words_1529.pdf; Crawford S Holling, "The Resilience of Terrestrial Ecosystems: Local Surprise and Global Change," *Sustainable Development of the Biosphere* 14 (1986).

16 Anthony Leiserowitz, "Climate Change Risk Perception and Policy Preferences: The Role of Affect, Imagery, and Values," *Climatic Change* 77, no. 1 (2006). Two other studies finding a relationship between egalitarianism and climate risk were Dan M Kahan et al., "The Tragedy of the Risk-Perception Commons: Culture Conflict, Rationality Conflict, and Climate Change," Temple University Legal Studies Research Paper, no. 2011-26 (2011) and Dan M Kahan et al., "Culture and Identity-Protective Cognition: Explaining the White-Male Effect in Risk Perception," *Journal of Empirical Legal Studies* 4, no. 3 (2007).

17 Wen Xue et al., "Cultural Worldviews and Environmental Risk Perceptions: A Meta-analysis," *Journal of Environmental Psychology* 40 (2014).

18 Helen Kopnina, "Environmental Justice and Biospheric Egalitarianism: Reflecting on a Normative-Philosophical View of Human–Nature Relationship," *Earth Perspectives* 1, no. 1 (2014).

19 Jennifer Sheehy-Skeffington and Lotte Thomsen, "Egalitarianism: Psychological and Socio-Ecological Foundations," *Current Opinion in Psychology* 32 (2020).

20 Wendy A Horwitz, "Developmental Origins of Environmental Ethics: The Life Experiences of Activists," *Ethics & Behavior* 6, no. 1 (1996).

21 Scott R Fisher, "Life Trajectories of Youth Committing to Climate Activism," *Environmental Education Research* 22, no. 2 (2016).
22 MR Sisco, V Bosetti, and EU Weber, "When Do Extreme Weather Events Generate Attention to Climate Change?" *Climatic Change* 143 (2017).
23 Hilary Boudet et al., "Event Attribution and Partisanship Shape Local Discussion of Climate Change after Extreme Weather," *Nature Climate Change* 10, no. 1 (2020).
24 Peter D Howe et al., "How Will Climate Change Shape Climate Opinion?" *Environmental Research Letters* 14, no. 11 (2019). A problem was that nearly half of the studies were restricted to examining the association between participants' *self-reported* subjective experience with extreme weather events and their opinions about climate. While there was evidence for the relationship, those people who thought they experienced extreme weather also tended to believe that global warming was happening, without objective measures such as meteorological data, it could not be established if people were being accurate in what they said. Fortunately, about half the studies could examine the association between *objectively measured* weather variables and climate opinions. But the results were inconsistent. Other studies examining droughts and precipitation also found mixed results in relation to climate opinions.
25 Boudet et al., "Event Attribution."
26 Marshall, *Don't Even Think about It.* The gambler's fallacy is the incorrect belief that if a particular event occurred more frequently than normal during the past, it is less likely to happen in the future. In reality, the probability of such events, such as new extreme weather, does not depend on what has happened in the past.
27 Masashi Soga and Kevin J Gaston, "Shifting Baseline Syndrome: Causes, Consequences, and Implications," *Frontiers in Ecology and the Environment* 16, no. 4 (2018).
28 www.vox.com/energy-and-environment/2020/7/7/21311027/covid-19-climate-change-global-warming-shifting-baselines.
29 www.columbia.edu/~jeh1/mailings/2020/20200706_ShiftingBellCurvesUpdated.pdf.
30 Frances C Moore et al., "Rapidly Declining Remarkability of Temperature Anomalies May Obscure Public Perception of Climate Change," *Proceedings of the National Academy of Sciences* 116, no. 11 (2019).
31 Shifting baselines syndrome can also happen at the personal level, something Roberts refers to as personal amnesia. It is well established that we sometimes adapt to circumstances that are dreadful; think of the London Blitz, when bombs fell for months on end and people got on with life, often cheerily. Research studies of how people think about then react to divorce, failing to get tenure, or other kinds of rejection, show that people hugely overestimate how much these will affect their happiness: often they *can* adjust. And Roberts argues they do this not only emotionally but also cognitively; they forget about their earlier experience and what happened recently is the most important thing in defining the baseline. Such observations have led to suggestions that we have a kind of psychological immune system – although this does not apply to everyone to be sure, see Daniel T Gilbert et al., "Immune Neglect: A Source of Durability Bias in Affective Forecasting," *Journal of Personality and Social Psychology* 75, no. 3 (1998).
32 https://today.yougov.com/topics/politics/articles-reports/2021/08/02/many-americans-feel-recent-effects-climate-change.
33 www.baselinefilm.com.
34 Van der Linden, "The Social-Psychological Determinants of Climate Change Risk Perceptions."

35 It is important to note a caveat to this study, which, like many we have seen in this book, was based on an online survey of self-report with no measurement of whether those reported attitudes mattered to the individual enough to change behavior. Consistent with this, while the risk perceptions of 65 percent of US adults reached the "worry level" in 2018, that same year an Associated Press poll found that only 28 percent of US adults said there were prepared to pay even an extra $10 per month to combat climate change (https://apnorc.org/projects/is-the-public-willing-to-pay-to-help-fix-climate-change/). To put this in context, the average citizen pays $200 in taxes per month to fund the US military against mostly imagined threats to the country (Lieven, "Climate Change and the Nation State").
36 Wolfgang Knorr argues the climate crisis requires new kinds of thinking for some physical scientists: www.resilience.org/stories/2020-08-04/the-climate-crisis-demands-new-ways-of-thinking-from-climate-scientists/. Meanwhile, others feel the problem very acutely: www.motherjones.com/environment/2019/07/weight-of-the-world-climate-change-scientist-grief/.
37 Elke U Weber, "Experience-Based and Description-Based Perceptions of Long-Term Risk: Why Global Warming Does Not Scare Us (Yet)," *Climatic Change* 77, no. 1 (2006).
38 https://coast.noaa.gov/digitalcoast/tools/slr.html.
39 www.newyorker.com/magazine/2022/01/31/can-science-fiction-wake-us-up-to-our-climate-reality-kim-stanley-robinson?utm_source=onsite-share&utm_medium=email&utm_campaign=onsite-share&utm_brand=the-new-yorker.
40 K Stanley-Robinson, *The Ministry of the Future* (Orbit, 2020).
41 Michael E Mann, The New Climate War: The Fight to Take Back Our Planet (Hachette UK, 2021).
42 https://time.com/5735388/climate-change-eco-anxiety/ and
Pihkala Panu, "Anxiety and the Ecological Crisis: An Analysis of Eco-Anxiety and Climate Anxiety," *Sustainability* 12, no. 19 (2020).
43 Caroline Hickman et al., "Young People's Voices on Climate Anxiety, Government Betrayal and Moral Injury: A Global Phenomenon," *Lancet* (2022).
44 www.nytimes.com/2022/02/06/health/climate-anxiety-therapy.html.
45 Graham Lawton, "I Have Eco-Anxiety but That's Normal," *New Scientist* 244, no. 3251 (2019).
46 Catriona McKinnon, "Climate Change: Against Despair," *Ethics & the Environment* 19, no. 1 (2014): 42. Andreas Malm, *How to Blow up a Pipeline* (Verso, 2021). Malm goes on to say that "[people who act] have always been told that … we're doomed… from the slave barracks onwards, every revolt has been discouraged by the elders of defeatism."
47 Emma F Thomas, Craig McGarty, and Kenneth I Mavor, "Transforming 'Apathy into Movement': The Role of Prosocial Emotions in Motivating Action for Social Change," *Personality and Social Psychology Review* 13, no. 4 (2009).
48 Matthew J Hornsey and Kelly S Fielding, "A Cautionary Note about Messages of Hope: Focusing on Progress in Reducing Carbon Emissions Weakens Mitigation Motivation," *Global Environmental Change* 39 (2016).
49 Brandi S Morris et al., "Optimistic vs. Pessimistic Endings in Climate Change Appeals," *Humanities and Social Sciences Communications* 7, no. 1 (2020).
50 Daniel A Chapman, Brian Lickel, and Ezra M Markowitz, "Reassessing Emotion in Climate Change Communication," *Nature Climate Change* 7, no. 12 (2017).
51 Ereanut and Segnit, "Warm Words," www.ippr.org/files/images/media/files/publication/2011/05/warm_words_1529.pdf.

52 Jonathan Lear, Radical Hope: Ethics in the Face of Cultural Devastation (Harvard University Press, 2006).
53 Paul Hoggert, "Embracing 'Radical Hope' in Our Fight to Save the Earth," https://theecologist.org/2015/apr/05/embracing-radical-hope-our-fight-save-earth.
54 John Drury and Steve Reicher, "Collective Psychological Empowerment as a Model of Social Change: Researching Crowds and Power," *Journal of Social Issues* 65, no. 4 (2009).

Chapter 8

1 Note that the phrases "climate justice" and "environmental justice" are sometimes interchangeable and sometimes not. *Climate justice* refers to climate impacts, whereas *environmental justice* can encompass those as well as local effects, such as the poisoning of water or air pollution from nearby freeways, that reflect the injustice experienced by frontline communities. Indeed, the core founding issue of the environmental justice movement was the proximity to and siting of toxic facilities, including active and abandoned waste dumps.
2 Tokar, Towards Climate Justice.
3 The movement issued a declaration that read: "We cannot trust the market with our future, nor put our faith in unsafe, unproven, and unsustainable technologies ... Contrary to those who put their faith in 'green capitalism', we know that it is impossible to have infinite growth on a finite planet."
4 https://jacobinmag.com/2021/11/cop26-deal-betrayal-climate-justice-crisis-western-governments.
5 For example, in 2009, wealthy nations had promised $100 billion per year in climate finance to help vulnerable countries adapt, but twelve years later they had still not fulfilled that promise, which was a major focus of justice concerns at COP 26: www.nature.com/articles/d41586-021-02846-3.
6 Nadja Popovich and Brad Plumer, "Who Has the Most Historical Responsibility for Climate Change," *New York Times*, November 12, 2021.
7 https://ourworldindata.org/co2-emissions.
8 https://climateactiontracker.org/countries/usa/.
9 www.npr.org/2021/07/13/1015581092/biden-promised-to-end-new-drilling-on-federal-land-but-approvals-are-up.
10 https://foe.org/wp-content/uploads/2021/04/USA_Fair_Shares_NDC.pdf.
11 http://piketty.pse.ens.fr/files/ChancelPiketty2015.pdf.
12 www.forbes.com/sites/scottsnowden/2020/09/24/carbon-emissions-of-richest-1-more-than-double-than-those-of-the-poorest-50/?sh=1b56fd8d3c98; https://oxfamilibrary.openrepository.com/bitstream/handle/10546/621052/mb-confronting-carbon-inequality-210920-en.pdf.
13 Overpopulation is, however, a major problem in relation to the ecological crisis.
14 Benedikt Bruckner et al., "Impacts of Poverty Alleviation on National and Global Carbon Emissions," *Nature Sustainability* (2022).
15 https://pdfs.semanticscholar.org/c590/7d86b02c6e15fdbba4CO2369d8bc3c9788ae.pdf.
16 www.peoplesworld.org/article/polluter-elite-lifestyles-of-the-rich-are-incompatible-with-climate-justice/.
17 www.bbc.com/news/uk-england-hampshire-57858103.

18 Norihiko Yamano and Joaquim Guilhoto, "CO_2 Emissions Embodied in International Trade and Domestic Final Demand: Methodology and Results Using the OECD Inter-country Input–Output Database," OECD Science, Technology and Industry Working Papers, No. 2020/11, OECD Publishing, Paris (2020); www.carbontrust.com/resources/international-carbon-flows.
19 Dieter Helm, "Climate-Change Policy: Why Has So Little Been Achieved?" *Oxford Review of Economic Policy* 24, no. 2 (2008).
20 Lucy Baker, "Of Embodied Emissions and Inequality: Rethinking Energy Consumption," *Energy Research & Social Science* 36 (2018): 52.
21 Haakon Lindstad, Bjørn E Asbjørnslett, and Anders H Strømman, "The Importance of Economies of Scale for Reductions in Greenhouse Gas Emissions from Shipping," *Energy Policy* 46 (2012).
22 www.reuters.com/article/us-shipping-environment-imo/shippings-share-of-global-carbon-emissions-increases-idUSKCN2502AY.
23 Hans Sanderson, "Who Is Responsible for Embodied CO_2?" *Climate* 9, no. 3 (2021).
24 Baker, "Of Embodied Emissions"; https://theconversation.com/after-paris-uks-latest-carbon-budget-just-isnt-ambitious-enough-61925; personal communication with climate justice advocate Brian Tokar, November 21, 2021.
25 https://ec.europa.eu/commission/presscorner/detail/en/qanda_21_3661; https://ec.europa.eu/taxation_customs/green-taxation-0/carbon-border-adjustment-mechanism_en.
26 https://thehill.com/homenews/campaign/480518-sanders-vows-to-renegotiate-disastrous-north-american-trade-deal.
27 https://climatestrategies.org/wp-content/uploads/2018/07/CS-Report-_Trade-WP4.pdf.
28 Mehdi Azadi et al., "Transparency on Greenhouse Gas Emissions from Mining to Enable Climate Change Mitigation," *Nature Geoscience* 13, no. 2 (2020); https://waronwant.org/sites/default/files/Post-Extractivist_Transition_WEB_0.pdf.
29 www.oecd-ilibrary.org/environment/global-material-resources-outlook-to-2060_9789264307452-en; Azadi et al., "Transparency on Greenhouse Gas Emissions."
30 The problem of extractivism is discussed in "A Just(ice) Transition is a Post-extractive Transition," https://waronwant.org/sites/default/files/Post-Extractivist_Transition_WEB_0.pdf and "What Is Extractivism?" www.columbancenter.org/what-extractivism.
31 Hannah Appel, The Licit Life of Capitalism: US Oil in Equatorial Guinea (Duke University Press, 2019).
32 This quote was displayed by Hannah Appel in a public presentation at UC San Diego in 2019.
33 www.foei.org/news/oil-spills-ogoniland-nigeria-shell www.theguardian.com/commentisfree/2022/feb/08/chevron-amazon-ecuador-steven-donziger-erin-brockovich.
34 https://waronwant.org/sites/default/files/Post-Extractivist_Transition_WEB_0.pdf.
35 https://ejatlas.org.
36 For example, while lithium has been the go-to option for electric cars, already new options are materializing such as iron-nickel batteries (iron is one of the most common and easily accessible metals in the earth). https://techcrunch.com/2021/07/28/what-teslas-bet-on-iron-based-batteries-means-for-manufacturers/. Further, there are many kinds of batteries in development, with some stationary ones relying on a cousin of common table salt. https://scitechdaily.com/a-cousin-of-table-salt-could-make-rechargeable-batteries-faster-and-safer/.

37 See www.dol.gov/agencies/ilab/resources/reports/child-labor/congo-democratic-republic-drc; www.newyorker.com/magazine/2021/05/31/the-dark-side-of-congos-cobalt-rush.
38 www.designnews.com/content/talk-lithium-expert/82024188259151.
39 www.washingtonpost.com/graphics/business/batteries/tossed-aside-in-the-lithium-rush/.
40 https://grist.org/energy/the-world-needs-lithium-can-bolivias-new-president-deliver-it/; www.irinsider.org/environment-1/2020/10/29/bolivias-lithium-challenges-its-democracy-and-environment.
41 https://magazine.scienceforthepeople.org/vol22–1/agua-es-vida-solidarity-science-against-false-climate-change-solutions/#easy-footnote-bottom-29-3175.
42 https://en.wikipedia.org/wiki/Lithium_mining_in_Australia.
43 www.theguardian.com/us-news/2021/sep/27/salton-sea-california-lithium-mining.
44 https://insideclimatenews.org/news/13012022/inside-clean-energy-battery-recycling/.
45 www.nrdc.org/stories/millions-leaky-and-abandoned-oil-and-gas-wells-are-threatening-lives-and-climate; Mark Z Jacobson, *100% Clean, Renewable Energy and Storage for Everything* (Cambridge University Press, 2020).
46 Vohra et al., "Global Mortality."
47 Emilio F Moran et al., "Sustainable Hydropower in the 21st Century," *Proceedings of the National Academy of Sciences* 115, no. 47 (2018).
48 www.forbes.com/sites/jeffopperman/2018/08/10/the-unexpectedly-large-impacts-of-small-hydropower/?sh=6b93d1de7b9d; Jacobson, "100% Clean."
49 www.prweb.com/releases/the-nature-conservancy/dams/prweb4103384.htm.
50 www.nytimes.com/2016/11/11/world/canada/clean-energy-dirty-water-canadas-hydroelectric-dams-have-a-mercury-problem.html.
51 Alexis Lathem, "Grassroots Megadam Resistance at Muskrat Falls, Labrador," in Brian Tokar and Tamra Gilbertson (eds.), *Climate Justice and Community Renewal: Resistance and Grassroots Solutions* (Routledge, 2020).
52 In a 2016 paper, the contribution was estimated at 0.8 Gt of CO_2 equivalents per year, which is about 2 percent of the total. Bridget R Deemer et al., "Greenhouse Gas Emissions from Reservoir Water Surfaces: A New Global Synthesis," *BioScience* 66, no. 11 (2016). A different study of the same issue also reached similar conclusions. Laura Scherer and Stephan Pfister, "Hydropower's Biogenic Carbon Footprint," *PLoS ONE* 11, no. 9 (2016); www.globalcarbonproject.org/carbonbudget/19/highlights.htm\.
53 Ilissa B Ocko and Steven P Hamburg, "Climate Impacts of Hydropower: Enormous Differences among Facilities and over Time," *Environmental Science & Technology* 53, no. 23 (2019).
54 Deemer et al., "Greenhouse Gas Emissions from Reservoir Water Surfaces: A New Global Synthesis."
55 www.theguardian.com/sustainable-business/2016/nov/06/hydropower-hydroelectricity-methane-clean-climate-change-study.
56 www.theguardian.com/environment/2015/jan/29/biofuels-are-not-the-green-alternative-to-fossil-fuels-they-are-sold-as.
57 Okbazghi Yohannes, *The Biofuels Deception: Going Hungry on the Green Carbon Diet* (NYU Press, 2018). See a precis here: https://monthlyreview.org/product/the-biofuels-deception-going-hungry-on-the-green-carbon-diet/.
58 www.vox.com/energy-and-environment/2020/2/14/21131109/california-natural-gas-renewable-socalgas.

59 www.theguardian.com/environment/2015/dec/03/nuclear-power-paves-the-only-viable-path-forward-on-climate-change; www.theguardian.com/commentisfree/2011/mar/21/pro-nuclear-japan-fukushima.
60 www.nationalgeographic.co.uk/environment/2019/05/chernobyl-disaster-what-happened-and-long-term-impact.
61 Lynas, *Our Final Warning*; www.world-nuclear.org/information-library/country-profiles/countries-a-f/france.aspx.
62 www.sortirdunucleaire.org/Nuclear-power-a-false-solution-to-climate-change-44206; www.gao.gov/key_issues/disposal_of_highlevel_nuclear_waste/issue_summary#t=0.
63 These decommissioning costs may pale by comparison to the massive construction costs. And the fact that new nuclear construction (even in Finland, where a project was supervised by the much-praised French nuclear operator) typically costs 2–3 times as much as initially projected.
64 www.solarreviews.com/blog/what-is-a-solar-farm-do-i-need-one.
65 www.renewableenergyworld.com/solar/142-gw-of-solar-capacity-will-be-added-to-the-global-market-in-2020-says-ihs/#gref; www.pv-magazine.com/2021/09/28/renewables-vs-nuclear-256-0/.
66 Jacobson, "100% Clean," 115; www.sortirdunucleaire.org/Nuclear-power-a-false-solution-to-climate-change-44206.
67 According to Tokar, "[n]uclear proponents have been promising affordable smaller reactors since the 1980s. Despite new infusions of capital from Bill Gates, I don't think these promises are any more realistic now than they were forty years ago. Former nuclear engineer Arjun Makhijani has some of the most thorough critiques of those proposals."
68 www.sciencemag.org/news/2020/05/us-department-energy-rushes-build-advanced-new-nuclear-reactors.
69 www.bloomberg.com/news/articles/2020-09-07/china-gives-operating-license-to-first-homegrown-nuclear-reactor.
70 www.theguardian.com/commentisfree/2011/mar/21/pro-nuclear-japan-fukushima.
71 https://cleantechnica.com/2019/04/16/fukushimas-final-costs-will-approach-one-trillion-dollars-just-for-nuclear-disaster/.
72 https://monthlyreview.org/2021/07/01/from-sandstorm-and-smog-to-sustainability-and-justice-chinas-challenges/.
73 Daniel J Madigan, Zofia Baumann, and Nicholas S Fisher, "Pacific Bluefin Tuna Transport Fukushima-Derived Radionuclides from Japan to California," *Proceedings of the National Academy of Sciences* 109, no. 24 (2012).
74 https://monthlyreview.org/2021/07/01/from-sandstorm-and-smog-to-sustainability-and-justice-chinas-challenges/.
75 www.cleanenergywire.org/factsheets/coal-germany.
76 www.cleanenergywire.org/factsheets/very-brief-timeline-germanys-energiewende.
77 https://monthlyreview.org/2018/09/01/making-war-on-the planet/; www.science.org/content/article/us-geoengineering-research-gets-lift-4-million-congress.
78 Raymond Pierrehumbert, "There Is No Plan B for Dealing with the Climate Crisis," *Bulletin of the Atomic Scientists* 75, no. 5 (2019); Alan Robock, "20 Reasons Why Geoengineering May Be a Bad Idea," *Bulletin of the Atomic Scientists* 64, no. 2 (2008); ibid.
79 www.greenpeace.to/publications/iron_fertilisation_critique.pdf.
80 www.sciencemag.org/news/2018/02/vast-bioenergy-plantations-could-stave-climate-change-and-radically-reshape-planet; www.geoengineeringmonitor.org/2021/04/bio-energy-with-carbon-capture-and-storage-beccs/; www.imperial.ac.uk/media/imperial-college/grantham-institute/public/publications/briefing-papers/BECCS-deployment—a-reality-check.pdf.

81 www.forbes.com/sites/scottcarpenter/2020/08/04/bps-new-renewables-push-redolent-of-abandoned-beyond-petroleum-rebrand/?sh=ea3ae2a1ceb3.
82 www.nrdc.org/stories/fracking-101.
83 www.vox.com/energy-and-environment/2019/5/30/18643819/climate-change-natural-gas-middle-ground.
84 https://electrifyuc.org/carbon-neutrality-is-business-as-usual/.
85 https://theconversation.com/climate-scientists-concept-of-net-zero-is-a-dangerous-trap-157368.
86 www.imperial.ac.uk/media/imperial-college/grantham-institute/public/publications/briefing-papers/BECCS-deployment—a-reality-check.pdf.
87 www.theguardian.com/world/2021/jan/14/carbon-neutrality-is-a-fairy-tale-how-the-race-for-renewables-is-burning-europes-forests; https://elfond.ee/biomassreport.
88 www.theguardian.com/environment/2021/jul/20/a-shocking-failure-chevron-criticised-for-missing-carbon-capture-target-at-wa-gas-project.
89 Ahmed Abdulla et al., "Explaining Successful and Failed Investments in US Carbon Capture and Storage Using Empirical and Expert Assessments," *Environmental Research Letters* 16, no. 1 (2020).
90 Hansen, "Young People's Burden."
91 www.theguardian.com/environment/2021/sep/09/worlds-biggest-plant-to-turn-carbon-dioxide-into-rock-opens-in-iceland-orca.
92 Hansen et al., "Young People's Burden."
93 Wallace Wells, "Uninhabitable Earth."
94 www.nytimes.com/2008/12/11/business/worldbusiness/11carbon.html.
95 https://ww2.arb.ca.gov/sites/default/files/classic/cc/ghg_inventory_trends_00–19.pdf.
96 www.propublica.org/article/cap-and-trade-is-supposed-to-solve-climate-change-but-oil-and-gas-company-emissions-are-up.
97 Danny Cullenward, Mason Inman, and Michael D Mastrandrea, "Tracking Banking in the Western Climate Initiative Cap-and-Trade Program," *Environmental Research Letters* 14, no. 12 (2019).
98 www.propublica.org/article/cap-and-trade-is-supposed-to-solve-climate-change-but-oil-and-gas-company-emissions-are-up.
99 Ibid.
100 https://newpol.org/open-letter-climate-justice-movement/.
101 www.brookings.edu/research/reforming-global-fossil-fuel-subsidies-how-the-united-states-can-restart-international-cooperation/; www.iisd.org/system/files/2020-11/g20-scorecard-report.pdf.
102 www.imf.org/en/Publications/WP/Issues/2021/09/23/Still-Not-Getting-Energy-Prices-Right-A-Global-and-Country-Update-of-Fossil-Fuel-Subsidies-466004.
103 David Coady et al., "How Large Are Global Fossil Fuel Subsidies?" *World Development* 91 (2017).
104 www.ucop.edu/energy-services/carbon-offsets/uc-initiated-offsets/ucs-pilot-uc-initiated-offset-projects-selected-for-cni-awards.html.
105 https://ec.europa.eu/clima/sites/clima/files/ets/docs/clean_dev_mechanism_en.pdf.
106 www.technologyreview.com/2019/08/26/133261/whoops-californias-carbon-offsets-program-could-extend-the-life-of-coal-mines/.
107 David W Keith et al., "A Process for Capturing CO_2 from the Atmosphere," *Joule* 2, no. 8 (2018); www.smithsonianmag.com/smart-news/worlds-largest-carbon-capture-plant-opens-iceland-180978620/.

108 This point made in AR Aron et al., "How Can Neuroscientists Respond to the Climate Emergency?" *Neuron* 106, no. 1 (2020).
109 Jessica L DeShazo, Chandra Lal Pandey, and Zachary A Smith, *Why REDD Will Fail* (Routledge, 2016).
110 https://features.propublica.org/brazil-carbon-offsets/inconvenient-truth-carbon-credits-dont-work-deforestation-redd-acre-cambodia/. In one Bolivian project, generally agreed to be the first REDD-type project, leakage was estimated to be as high as 60 percent. DeShazo et al., "Why REDD Will Fail."
111 https://features.propublica.org/brazil-carbon-offsets/inconvenient-truth-carbon-credits-dont-work-deforestation-redd-acre-cambodia/.
112 Matto Mildenberger, Carbon Captured: How Business and Labor Control Climate Politics (MIT Press, 2020).
113 www.nytimes.com/2018/10/08/climate/carbon-tax-united-nations-report-nordhaus.html.
114 Julius J Andersson, "Carbon Taxes and CO_2 Emissions: Sweden as a Case Study," *American Economic Journal: Economic Policy* 11, no. 4 (2019); Boqiang Lin and Xuehui Li, "The Effect of Carbon Tax on Per Capita CO_2 Emissions," *Energy Policy* 39, no. 9 (2011).
115 https://jacobinmag.com/2019/09/carbon-pricing-green-new-deal-fossil-fuel-environment.
116 Ibid.
117 www.nytimes.com/2018/12/06/world/europe/france-fuel-carbon-tax.html.
118 https://institute.smartprosperity.ca/content/rise-and-fall-australian-carbon-tax.
119 Ibid.
120 www.theguardian.com/commentisfree/2021/jan/05/simple-way-green-economy-cash-prizes-carbon-dividend; https://citizensclimatelobby.org/carbon-fee-and-dividend/.
121 James K Boyce, "Carbon Pricing: Effectiveness and Equity," *Ecological Economics* 150 (2018).
122 www.imf.org/en/Publications/WP/Issues/2019/05/02/Global-Fossil-Fuel-Subsidies-Remain-Large-An-Update-Based-on-Country-Level-Estimates-46509; Coady et al., "How Large Are Global."
123 Senator Sheldon Whitehouse: https://youtu.be/Nny7SnBkbTM.
124 www.nationalaffairs.com/publications/detail/the-conservative-roots-of-carbon-pricing.

Chapter 9

1 www.eia.gov/tools/faqs/faq.php?id=427&t=3
2 www.carbonbrief.org/in-depth-qa-does-the-world-need-hydrogen-to-solve-climate-change; IEA Net Zero report: www.iea.org/reports/net-zero-by-2050
3 Jacobson, "100% Clean, Renewable Energy."
4 "How Electrification Is Changing Mining," September 22, 2021, www.newscientist.com/article/2290944-how-electrification-is-changing-mining/
5 www.eia.gov/tools/faqs/faq.php?id=427&t=3; "Reducing Agricultural Greenhouse Gases," www2.gov.bc.ca. More radically, one could close down cattle feedlots and convert as much as 100 million acres of farmland currently in monocultures into deep-rooted pasture and prairie and later into diverse, perennial food-grain/forage systems. www.resilience.org/stories/2021-11-23/to-keep-fossil-carbon-out-of-the-air-just-stop-pulling-it-out-of-the-earth/

6 Jacobson, "100% Clean, Renewable Energy."
7 Practically, for power generation and transportation, it is much more feasible to aim for, say, 95 percent renewables and retain 5 percent of natural gas. More generally, the "last 20 percent" of emissions from shipping, aviation, cement manufacture and industrial production and mining will be very challenging.
Marc Perez et al., "Overbuilding and Curtailment: The Cost-Effective Enablers of Firm PV Generation," *Solar Energy* 180 (2019).
8 The net-zero America project from Princeton University has a near total reliance on renewables option, https://acee.princeton.edu/rapidswitch/projects/net-zero-america-project/; also see Saul Griffith, *Electrify: An Optimist's Playbook for Our Clean Energy Future* (MIT Press, 2021). Also see the 2022 IPCC Working Group 3 report.
9 Jacobson et al., "100% Clean, Renewable Energy and Storage." See also https://web.stanford.edu/group/efmh/jacobson/Articles/I/WWS-50-USState-plans.html; Mark Z Jacobson et al., "100% Clean and Renewable Wind, Water, and Sunlight (WWS) All-Sector Energy Roadmaps for the 50 United States," *Energy & Environmental Science* 8, no. 7 (2015).
10 Jacobson et al., "100% Clean and Renewable Wind."
11 Mark Z Jacobson et al., "Impacts of Green New Deal Energy Plans on Grid Stability, Costs, Jobs, Health, and Climate in 143 Countries," *One Earth* 1, no. 4 (2019).
12 One such criticism identified what it called "technical flaws," including unrealistic assumptions about future energy efficiency and inadequate means of maintaining power grid stability in the face of the intermittency of wind and solar (Heard et al.). Another took issue with the high reliance on hydrogen as an energy carrier, including the idea that the aviation and steel industries could be converted to use it; criticized the idea of underground thermal energy storage in almost every building, and argued that the installation of wind and solar would need to happen at fifteen times the annual rate of new electricity installation from all sources over the last fifty years (Clack et al.). Tellingly, however, it called instead for the use of other "commercially available" technologies such as new nuclear and BECCS. Jacobson et al. responded, also see Brown et al. Benjamin P Heard et al., "Burden of Proof: A Comprehensive Review of the Feasibility of 100% Renewable-Electricity Systems," *Renewable and Sustainable Energy Reviews* 76 (2017); Christopher TM Clack et al., "Evaluation of a Proposal for Reliable Low-Cost Grid Power with 100% Win"; Mark Z Jacobson et al., "The United States Can Keep the Grid Stable at Low Cost with 100% Clean, Renewable Energy in All Sectors Despite Inaccurate Claims," *Proceedings of the National Academy of Sciences* 114, no. 26 (2017); Tom W Brown et al., "Response to 'Burden of Proof: A Comprehensive Review of the Feasibility of 100% Renewable-Electricity Systems'," *Renewable and Sustainable Energy Reviews* 92 (2018).
13 www.iea.org/reports/net-zero-by-2050; www.ipcc.ch/site/assets/uploads/2018/04/ipcc_srren_leaflet.pdf.
14 www.ipcc.ch/report/ar6/wg3/.
15 www.wsj.com/articles/batteries-challenge-natural-gas-elecric-power-generation-11620236583.
16 In California, for example, electric utility companies such as SDG&E are owned by fossil fuel companies that have enormous interests in fracked natural gas (methane) and do all they can to maintain high rates for consumers, block rooftop solar, and defeat legislation around building electrification (which would cancel gas lines to new houses and apartments). Stokes, "Short Circuiting Policy."

17 www.lrb.co.uk/the-paper/v43/n23/david-wallace-wells/ten-million-a-year.
18 Miguel San Sebastián et al., "Exposures and Cancer Incidence near Oil Fields in the Amazon Basin of Ecuador," *Occupational and Environmental Medicine* 58, no. 8 (2001).
19 Mark Diesendorf and Ben Elliston, "The Feasibility of 100% Renewable Electricity Systems: A Response to Critics," *Renewable and Sustainable Energy Reviews* 93 (2018).
20 Ibid.
21 At each time-step of one hour or less, the simulation compares actual electricity demand with actual or synthetic data on renewable supply, taking into account detailed information about weather patterns. Jacobson et al., "Low-Cost Solution." Also see estimates from a different research group, led by Steven Davis at University of California Irvine, which estimated that wind-heavy electricity systems could meet the demands of forty-two countries 72–91 percent of the time even without storage. Dan Tong et al., "Geophysical Constraints on the Reliability of Solar and Wind Power Worldwide," *Nature Communications* 12, no. 1 (2021); Mark Z Jacobson et al., "Low-Cost Solution to the Grid Reliability Problem with 100% Penetration of Intermittent Wind, Water, and Solar for All Purposes," *Proceedings of the National Academy of Sciences* 112, no. 49 (2015).
22 The failure of the electric grid in the Texas "deep freeze" of early 2021 demonstrated the importance of integrating the long-distance grid for other reasons.
23 www.scientificamerican.com/article/how-solar-heavy-europe-avoided-a-blackout-during-total-eclipse/.
24 Jacobson et al., "Impacts of Green New Deal Energy Plans on Grid Stability, Costs, Jobs, Health, and Climate in 143 Countries."
25 www.nytimes.com/2021/04/22/climate/climate-change-economy.html www.swissre.com/media/press-release/nr-20210422-economics-of-climate-change-risks.html
26 R Pollin, "An Industrial Policy Framework to Advance a Global Green New Deal," in A Oqubay, et al. (ed.), *The Oxford Handbook of Industrial Policy* (Oxford University Press, 2020); ibid.
27 www.carbonbrief.org/solar-is-now-cheapest-electricity-in-history-confirms-iea.
28 https://carbontracker.org/reports/the-trillion-dollar-energy-windfall/.
29 www.eia.gov/todayinenergy/detail.php?id=44636; www.wri.org/blog/2019/07/natural-gas-beat-coal-us-will-renewables-and-storage-soon-beat-natural-gas.
30 https://ieefa.org/ieefa-u-s-surging-generation-from-solar-wind-on-track-to-push-renewable-market-share-to-30-percent-by-2026/. Meanwhile, in 2021, another 100 gigawatt installation of wind and solar was installed in China, in line with the ambition of 1,200 GW by 2030. (Recall that one MW is enough to supply about 164 US homes per year, so 100 GW would supply about 16.4 million US homes per year). www.pv-tech.org/china-signals-construction-start-of-100gw-first-phase-of-desert-renewables-rollout/.
31 https://ieefa.org/articles/renewable-energy-accounted-234-us-electricity-february.
32 www.solarpowerworldonline.com/2022/04/california-grid-set-record-with-97-percent-renewable-power-april-3/.
33 Jacobson et al., "100% Clean and Renewable Wind, Water, and Sunlight (WWS) All-Sector Energy Roadmaps for the 50 United States." Renewable energy engineer Saul Griffith estimated that the shift to solar would require 1 percent of US land, assuming a power need of 1,500–1,800 GW, three times as much as the total electricity now generated. Griffith, *Electrify*.

34 The Bloomberg estimate that 1 percent of the world's surface is needed for solar and wind is calculated as follows. The world in 2018 used 52,000 km^2 to produce 2 PWh (peta is 10 to the power of 15) of renewable electricity from solar and wind. Thus, solar and wind require 26,000 km^2 per PWh. Currently there is about 68 PWh of total electricity demand. At 26,000 km^2 per PWh. this would require 1.75 million km^2. The global land mass is 168 million km^2, meaning that the total land use requirement for solar and wind is about 1 percent. Moreover, future energy supply will be more efficient, and there will be the opportunity to use offshore wind. Further, 1 percent of the total land is the amount needed in 50 years: in the meantime, the levels of land demand will be much lower. Other estimates vary depending how much of the electricity mix is solar vs. wind. Dirk-Jan Van de Ven et al., "The Potential Land Requirements and Related Land Use Change Emissions of Solar Energy," *Scientific Reports* 11, no. 1 (2021).
35 Jacobson, "100% Clean and Renewable Wind."
36 Diesendorf and Elliston, "Feasibility of 100%."
37 Alicia Valero et al., "Material Bottlenecks in the Future Development of Green Technologies," *Renewable and Sustainable Energy Reviews* 93 (2018); www.earthworks.org/cms/assets/uploads/2019/04/MCEC_UTS_Report_lowres-1.pdf.
38 https://insideclimatenews.org/news/13012022/inside-clean-energy-battery-recycling/; www.sciencedaily.com/releases/2019/04/190402081606.htm; Simon M Jowitt et al., "Recycling of the Rare Earth Elements," *Current Opinion in Green and Sustainable Chemistry* 13 (2018).
39 For more details on overcoming the deleterious effects of current extractivist practices, see report by the global solidarity group War on Want, https://waronwant.org/sites/default/files/Post-Extractivist_Transition_WEB_0.pdf, and https://communemag.com/between-the-devil-and-the-green-new-deal/.
40 www.world-nuclear.org/information-library/energy-and-the-environment/carbon-dioxide-emissions-from-electricity.aspx.
41 www.newscientist.com/lastword/mg24332461-400-what-is-the-carbon-payback-period-for-a-wind-turbine/.
42 www.iea.org/reports/net-zero-by-2050.
43 See www.eia.gov/outlooks/steo/; https://download.dnvgl.com/eto-2020-download. York and Bell, "Energy Transitions or Additions," also expand on why renewable are additional to fossil fuels instead of replacing them. Richard York and Shannon Elizabeth Bell, "Energy Transitions or Additions? Why a Transition from Fossil Fuels Requires More than the Growth of Renewable Energy," *Energy Research & Social Science* 51 (2019).
44 www.nytimes.com/2021/12/07/business/airports-solar-farms.html; www.statista.com/topics/2086/tesla/#dossierKeyfigures; www.axios.com/climate-private-equity-renewable-investment-4d0504d1-bc4e-4e0a-b397-71371cc30735.html.
45 Governments are also needed because the social and environmental costs and benefits of energy (the externalities referred to in earlier chapters) are not captured in the price of energy. https://sitn.hms.harvard.edu/flash/2012/energy-finance/.
46 Kai Graylee, "Beyond the Debate: The Role of Government in Renewable Energy Finance," *Science in the News Boston*, Harvard University, December 15, 2012.
47 Griffith, *Electrify*.
48 John Bellamy Foster, "On Fire This Time," https://monthlyreview.org/2019/11/01/on-fire-this-time/.
49 Georgia Piggot et al., "Curbing Fossil Fuel Supply to Achieve Climate Goals," *Climate Policy* 20, no. 8 (2020).

50 Kate Aronoff et al., A Planet to Win: Why We Need a Green New Deal (Verso, 2019); www.vox.com/energy-and-environment/2017/10/6/16428458/us-energy-coal-oil-subsidies; www.brookings.edu/research/reforming-global-fossil-fuel-subsidies-how-the-united-states-can-restart-international-cooperation/. Note the International Monetary Fund's estimate that the overall subsidies are over $5 trillion per year: www.imf.org/en/Publications/WP/Issues/2021/09/23/Still-Not-Getting-Energy-Prices-Right-A-Global-and-Country-Update-of-Fossil-Fuel-Subsidies-466004.
51 Piggot et al., "Curbing Fossil Fuel"; www.spglobal.com/platts/en/market-insights/latest-news/electric-power/051421-spain-passes-climate-bill-banning-new-oil-gas-exploration; www.bloomberg.com/news/articles/2021-09-21/l-a-county-drilling-ban-is-an-environmental-justice-feat; www.nationalobserver.com/2021/10/19/news/environmental-groups-applaud-quebecs-ban-fossil-fuel-exploration.
52 Griffith, *Electrify*. As noted earlier in the chapter, the International Energy Agency and DNV-GL predict that if matters are left to the market and business as usual, the fossil fuel companies will remain substantial players for decades. Companies such as Exxon-Mobil have massive interests at stake in preventing reductions in fossil fuel consumption and wield enormous political power. Griffith estimates that the remaining fossil fuels owned by these industries could be bought out for perhaps 10 percent of the total possible value, or as much as $10 trillion.
53 https://thesolutionsjournal.com/2020/09/01/cap-and-adapt-failsafe-policy-for-the-climate-emergency/.
54 Governments would issue permits each year to the companies that extract the fuels, with the permits being in quantities (barrels of oil) and not in dollars. No company could pull any fuel out of the ground without handing over the permits. The policy would also cover embodied emissions: those relating to the greenhouse gases released during production and delivery.
55 Because nationwide energy shortages would occur during the transition, price controls and rationing would be needed to make sure there is equitable access to electricity.
56 The average salaries in the fossil fuel sector were estimated to be $130,000 compared to $85,000 in the clean energy sector; the proposed transition package was $470 million per year, or 0.02 percent of GDP, to 2030. www.californiaclimatejobsplan.com.
57 www.congress.gov/bill/116th-congress/house-resolution/109.
58 As historians have pointed out, most farm and domestic workers were left out of social security programs, and as Cox discusses in "A Green New Deal," the New Deal exacerbated rather than dissolved institutional racism. For example, the Works Progress Administration allowed payment of locally prevailing wages which hurt people in predominantly Black areas, and the Federal Housing Administration required banks to maintain racial segregation.
59 https://jacobinmag.com/2021/05/green-new-deal-climate-change.
60 www.latimes.com/opinion/story/2021-10-23/oil-gas-jobs-clean-energy-california; https://peri.umass.edu/images/CA-CleanEnergy-6-8-21.pdf.
61 www.nrdc.org/experts/khalil-shahyd/residential-energy-efficiency-largest-source-co2-reduction-potential.
62 As many as 2.5 billion people will be added to urban areas by 2050: https://espas.secure.europarl.europa.eu/orbis/document/global-trends-2030-future-urbanization-and-megacities-0.
63 Cox, Path to a Livable Future.
64 The plan, which aims to cut emissions in California by 50 percent by 2030, rests on the following key assumptions: that about 72 percent of energy in California comes

from combusting oil and fossil gas to generate electricity, and therefore emissions reductions would come from scaling up wind and solar; that nearly two-thirds of total energy available is wasted when generating electricity from primary fossil fuel sources, so switching to renewable energy would cut that waste by 50 percent; and that much energy is lost through inefficiencies in buildings and transportation, and therefore substantial investments must be made to improve those efficiencies. The report assumes that emissions could be cut by 50 percent by 2030 even while growing GDP at 2.5 percent per year, which relies on the prediction that energy consumption will decline by 2030 relative to business-as-usual owing to the above-mentioned efficiency savings in shifting to electric power generation, and the efficiencies from newly insulated buildings, expanded public transportation, new industrial processes, and EVs. However, the report did not apparently consider the rebound effects from people having more money in their pockets from reduced energy costs (see Vivanco et al.) and did not discuss the emissions that would arise from the new infrastructure built. David Font Vivanco, René Kemp, and Ester van der Voet, "How to Deal with the Rebound Effect? A Policy-Oriented Approach," *Energy Policy* 94 (2016).

65 Hickel, Less Is More.
66 https://9to5mac.com/2021/03/01/apple-lawsuit-portugal-planned-obsolescence/.
67 www.repair.org/stand-up/. See European policy: https://ideas.ted.com/how-right-to-repair-legislation-can-reduce-waste/.
68 www.fairphone.com/en/; www.nextpit.com/fairphone-4-review.
69 In one survey cited by Hickel, 51 percent of advertising executives admitted that the products they were pushing were things that people do not actually need. Hickel, *Less Is More*, 211. See also www.ama.org/2021/04/15/key-policy-considerations-for-reducing-public-consumption-of-vice-products/.
70 www.ipsnews.net/2020/08/growing-global-movement-end-outdoor-advertising/.
71 www.sustainalytics.com/esg-research/resource/investors-esg-blog/2020-the-year-of-the-flexitarian; www.greenpeace.org/static/planet4-eu-unit-stateless/2021/04/20210408-Greenpeace-report-Marketing-Meat.pdf.
72 www.worldwildlife.org/stories/fight-climate-change-by-preventing-food-waste.
73 C Alan Rotz et al., "Environmental Footprints of Beef Cattle Production in the United States," *Agricultural Systems* 169 (2019).
74 Key Facts and Findings of the Food and Agricultural Organization of the United Nations: www.fao.org/news/story/en/item/197623/icode/.
75 https://worldpopulationreview.com/country-rankings/countries-with-universal-basic-income; Andreas Siemoneit, "An Offer You Can't Refuse: Enhancing Personal Productivity through 'Efficiency Consumption'," *Technology in Society* 59 (2019).
76 David Rosnick and Mark Weisbrot, "Are Shorter Work Hours Good for the Environment? A Comparison of US and European Energy Consumption," *International Journal of Health Services* 37, no. 3 (2007).
77 Some of these are referred to as engaging with the "commons"; see David Bollier and Silke Helfrich, *Free, Fair, and Alive: The Insurgent Power of the Commons* (New Society Publishers, 2019).
78 Felix Creutzig et al., "Towards Demand-Side Solutions for Mitigating Climate Change," *Nature Climate Change* 8, no. 4 (2018).
79 www.amsterdam.nl/en/policy/sustainability/circular-economy/.
80 www.barcelona.cat/infobarcelona/en/tema/enterprise/the-green-and-circular-economy-a-priority-for-municipal-public-policies_756590.html; https://cooperationjackson.org.

Workers' cooperatives in many countries share some of these concepts, and in western Europe Transition Towns aim to increase self-sufficiency and reduce fossil fuel use. www.democracyatwork.info; https://transitionnetwork.org. In the US some of these have been part of the social ecology movement: https://roarmag.org/magazine/communalism-climate-chaos/.

81 www.ipcc.ch/report/sixth-assessment-report-working-group-3/.

82 Mariana Mazzucato, "Financing the Green New Deal," *Nature Sustainability* (2021).

83 www.npr.org/2022/05/18/1099937734/biden-invokes-defense-production-act-for-baby-formula-shortage.

Chapter 10

1 https://grist.org/politics/la-county-votes-ban-oil-gas-drilling/; www.nationalobserver.com/2021/10/19/news/environmental-groups-applaud-quebecs-ban-fossil-fuel-exploration.

2 For example, at the University of California San Diego, over a two-year period, a vigorous climate justice group made considerable headway in its campaigns for fossil fuel–free energy and fossil fuel–free finance on campus, yet the numbers turning out for its main rallies were about 0.5 percent of the overall campus population, and the core members doing the vital organizing work constituted less than one-tenth of that. And whereas the largest international set of climate-related rallies in late 2019 had numbers in the range of 7.6 million, that was a miniscule fraction of overall populations; and, again, that was the number rallying on a day, not the number organizing in a sustained way. Further, the dramatic interventions of Extinction Rebellion in the streets of London numbered in perhaps the ten to twenty thousand range, but out of a wider London population of 12 million that is only around 0.1 percent, and many participants came from further afield. www.vox.com/energy-and-environment/2019/9/20/20876143/climate-strike-2019-september-20-crowd-estimate.

3 M Engler and P Engler. *This Is an Uprising* (Bold Type Books, 2017).

4 Saul David Alinsky, Rules for Radicals: A Practical Primer for Realistic Radicals (Vintage, 1989).

5 Engler and Engler, *This Is an Uprising*, 37.

6 Frances Fox Piven and Richard Cloward, *Poor People's Movements: Why They Succeed, How They Fail* (Vintage, 2012).

7 The UC system currently emits more than a million tons per year of CO_2 from burning fossil gas on campus for power generation and heating/cooling: https://electrifyuc.org.

8 Ibid.

9 D McAdam, "Social Movement Theory and the Prospects for Climate Change Activism in the United States," *Annual Review of Political Science* (2017).

10 Suzanne Staggenborg, *Social Movements* (Oxford University Press, 2016).

11 McAdam, "Social Movement Theory," 200.

12 www.carbonbrief.org/in-depth-qa-what-is-climate-justice.

13 www.lrb.co.uk/the-paper/v43/n22/james-butler/a-coal-mine-for-every-wildfire.

14 Matthew T Huber, Climate Change as Class War: Building Socialism on a Warming Planet (Verso, 2022).

15 Robert J Brulle, "Institutionalizing Delay: Foundation Funding and the Creation of US Climate Change Counter-Movement Organizations," *Climatic Change* 122,

no. 4 (2014); www.influencewatch.org/non-profit/environmental-defense-fund/; www.activistfacts.com/organizations/194-sierra-club/.

16 www.youtube.com/watch?v=wczb8CJ-cO0.

17 Mildenberger, "Carbon Captured."

18 For example, the fight for a $15 minimum wage started locally in Seattle and then spread to many states, reaching the point in 2021 when all Federal contractors had to offer it. www.theguardian.com/us-news/2016/apr/09/fight-for-15-meet-the-enfant-terrible-behind-the-battle-for-us-minimum-wage www.nytimes.com/2021/11/22/business/economy/minimum-wage-federal-contractors.html.

19 Drawn from an interview with Brian Tokar: www.counterpunch.org/2021/06/03/climate-justice-and-movement-building-an-interview-with-brian-tokar/.

20 S Bamberg, J Rees, and M Schulte, "Environmental Protection through Societal Change: What Psychology Knows about Collective Climate Action – and What It Needs to Find Out," *Psychology and Climate Change* (2018). Although this chapter uses the language of psychological variables, as this might be better understood by most readers, what these really are is psychological constructs, or hypothesized causes for a specific behavior. The term *construct* refers to the fact that it is a mental construction. Scientifically, one observes phenomena in the world (e.g., people fidgeting or marking survey scales in certain ways), infers the common features of those observations, and constructs a label for the hypothesized underlying cause.

21 Robyn Gulliver et al., *The Psychology of Effective Activism* (Cambridge University Press, 2021).

22 Martijn Van Zomeren, Tom Postmes, and Russell Spears, "Toward an Integrative Social Identity Model of Collective Action: A Quantitative Research Synthesis of Three Socio-Psychological Perspectives," *Psychological Bulletin* 134, no. 4 (2008).

23 Henri Tajfel, "Social Identity and Intergroup Behaviour," *Social Science Information* 13, no. 2 (1974).69; Jacquelien Van Stekelenburg, "Collective Identity," *The Wiley-Blackwell Encyclopedia of Social and Political Movements* (Wiley-Blackwell, 2013).

24 Bernd Simon and Bert Klandermans, "Politicized Collective Identity: A Social Psychological Analysis," *American Psychologist* 56, no. 4 (2001).

25 Albert Bandura, "Exercise of Human Agency through Collective Efficacy," *Current Directions in Psychological Science* 9, no. 3 (2000).

26 Bandura, "Exercise of Human Agency."

27 S Bamberg, J Rees, and S Seebauer, "Collective Climate Action: Determinants of Participation Intention in Community-Based Pro-environmental Initiatives," *Journal of Environmental Psychology* 43 (2015).

28 Two examples are Hannah Bührle and Joachim Kimmerle, "Psychological Determinants of Collective Action for Climate Justice: Insights from Semi-structured Interviews and Content Analysis," *Frontiers in Psychology* 12 (2021); Connie Roser-Renouf et al., "The Genesis of Climate Change Activism: From Key Beliefs to Political Action," *Climatic Change* 125, no. 2 (2014).

29 https://psyarxiv.com/jqz29/; Anna Castiglione, Cameron Brick, Stefanie Holden, Ella Miles-Urdan, and Adam Aron. "Discovering the Psychological Building Blocks underlying Climate Action: A Longitudinal Study of Real-World Activism." *Royal Society Open Science* 9, no. 6 (2022).

30 https://350.org.

31 https://arxiv.org/abs/0804.1126.

32 www.thefifthestate.com.au/articles/the-350org-global-day-of-climate-action-"how-the-most-widespread-day-of-political-action-in-the-planet's-history"-came-off/.

33 https://math.350.org.
34 https://gofossilfree.org/divestment/what-is-fossil-fuel-divestment/.
35 A list of universities is here: www.divest101.com/list, not the late date of 2050 for many; and a larger list of claims of divestment is here: https://divestmentdatabase.org.
36 www.youtube.com/watch?v=IC7IfbaLXls&t=4016s.
37 https://newrepublic.com/article/121848/does-divestment-work; https://papers.ssrn.com/sol3/papers.cfm?abstract_id=3376183; https://theconversation.com/fossil-fuel-divestment-will-increase-carbon-emissions-not-lower-them-heres-why-126392.
38 https://climatesafepensions.org/wp-content/uploads/2021/12/CSPN-The-Quiet-Culprit.pdf. A bill to divest main public sector pensions was brought to the California legislature in February 2022: https://leginfo.legislature.ca.gov/faces/billNavClient.xhtml?bill_id=202120220SB1173.
39 www.bankingonclimatechaos.org.
40 https://middleburycampus.com/48399/news/mckibben-talks-arrest-upcoming-mass-action/; https://sandiego350.org/events/.
41 https://medium.com/insurge-intelligence/the-flawed-science-behind-extinction-rebellions-change-strategy-af077b9abb4d; Rupert J Read, *Extinction Rebellion: Insights from the Inside* (Simplicity Institute, 2020).
42 Extinction Rebellion co-founder Roger Hallam on Gene Sharp, http://m.koreatimes.co.kr/pages/article.asp?newsIdx=267234.
43 www.standard.co.uk/news/london/extinction-rebellion-protests-activists-hold-faux-funeral-procession-as-arrests-top-total-from-april-a4260126.html.
44 Karen Bell and Gnisha Bevan, "Beyond Inclusion? Perceptions of the Extent to Which Extinction Rebellion Speaks to, and for, Black, Asian and Minority Ethnic (BAME) and Working-Class Communities," *Local Environment* (2021).
45 https://medium.com/insurge-intelligence/the-flawed-science-behind-extinction-rebellions-change-strategy-af077b9abb4d.
46 www.bbc.com/news/uk-england-54038591.
47 https://time.com/5864702/extinction-rebellion-climate-activism/; www.independent.co.uk/climate-change/news/extinction-rebellion-farnborough-airport-climate-protest-b1931117.html.
48 www.insulatebritain.com.
49 https://juststopoil.org.
50 www.theguardian.com/environment/2022/may/10/criminalising-our-right-to-protest-green-groups-anger-over-public-order-bill-queens-speech.
51 www.vox.com/the-highlight/2019/9/10/20847401/sunrise-movement-climate-change-activist-millennials-global-warming.
52 These candidates included Alexandria Ocasio-Cortez, Rashida Tliab, and Ilhan Omar, who later came to be known as Justice Democrats. https://en.wikipedia.org/wiki/Sunrise_Movement.
53 www.vox.com/the-highlight/2019/9/10/20847401/sunrise-movement-climate-change-activist-millennials-global-warming.
54 The Sunrise-produced video that went viral and was carried by mainstream media shows Feinstein rebuking these climate activists because they were too young to vote for her. www.youtube.com/watch?v=Sb4ddeF91-I.
55 https://newrepublic.com/article/153037/story-behind-green-new-deals-meteoric-rise.
56 Ibid. A video made by Congresswoman Alexandra Ocasio-Cortez and the artist Molly Crabapple is here: www.youtube.com/watch?v=d9uTH0iprVQ

57 www.jacobinmag.com/2021/08/sunrise-movement-green-new-deal-left-politics-local-organizing.
58 https://jacobinmag.com/2021/05/green-new-deal-climate-change.
59 www.greennewdealsd.org.
60 www.iea.org/commentaries/global-suv-sales-set-another-record-in-2021-setting-back-efforts-to-reduce-emissions.
61 Malm, How to Blow up.
62 Some such actions have already taken place through the small valve-turner's movement in the USA, where a few individuals temporarily shut down oil pipelines, and went to prison for their efforts. www.nytimes.com/2018/02/13/magazine/afraid-climate-change-prison-valve-turners-global-warming.html.
63 www.thenation.com/article/environment/environmental-sabotage-infrastructure/.
64 www.transportenvironment.org/wp-content/uploads/2021/08/2021_05_05_Electric_vehicle_price_parity_and_adoption_in_Europe_Final.pdf.
65 www.bankingonclimatechaos.org. Chase alone has financed fossil fuel extraction by over $300 billion over the last four years.
66 Cited in Cox, *Green New Deal*, 92.
67 www.timeshighereducation.com/blog/how-i-quit-neuroscience-focus-preventing-climate-breakdown.
68 https://twitter.com/climatehuman/status/1208086460978454528?lang=en.
69 https://gwec.net/wp-content/uploads/2012/06/Wind-climate-fact-sheet-low-res.pdf.

Conclusion

1 www.earthsblackbox.com.
2 British naturalist and television presenter Chris Packham labeled that corporate interference as evil, since, as he argued, those people know what they are doing and what the stakes are for planet Earth yet are bent on extracting short-term wealth at the expense of the biosphere.
www.youtube.com/watch?v=wczb8CJ-cO0.
3 www.theguardian.com/environment/2022/may/18/eu-plans-massive-increase-in-green-energy-to-rid-itself-of-reliance-on-russia.
4 www.straight.com/news/un-secretary-general-antonio-guterres-accuses-worlds-biggest-polluters-of-committing-arson-on.
5 www.ipcc.ch/report/ar6/wg3/resources/spm-headline-statements/.

Bibliography

Abdulla, Ahmed, Ryan Hanna, Kristen R Schell, Oytun Babacan, and David G Victor. "Explaining Successful and Failed Investments in US Carbon Capture and Storage Using Empirical and Expert Assessments." *Environmental Research Letters* 16, no. 1 (2020): 014036.

Aguilar, Stephen J, Morgan S Polikoff, and Gale M Sinatra. "Refutation Texts: A New Approach to Changing Public Misconceptions about Education Policy." *Educational Researcher* 48, no. 5 (2019): 263–72.

Albert, Simon, Javier X Leon, Alistair R Grinham, John A Church, Badin R Gibbes, and Colin D Woodroffe. "Interactions between Sea-Level Rise and Wave Exposure on Reef Island Dynamics in the Solomon Islands." *Environmental Research Letters* 11, no. 5 (2016): 054011.

Alinsky, Saul David. *Rules for Radicals: A Practical Primer for Realistic Radicals.* Vintage, 1989.

Andersson, Julius J. "Carbon Taxes and CO_2 Emissions: Sweden as a Case Study." *American Economic Journal: Economic Policy* 11, no. 4 (2019): 1–30.

Appel, Hannah. *The Licit Life of Capitalism: US Oil in Equatorial Guinea.* Duke University Press, 2019.

Arneth, Almut, Stephen Sitch, Julia Pongratz, BD Stocker, Philippe Ciais, Benjamin Poulter, Anita D Bayer, et al. "Historical Carbon Dioxide Emissions Caused by Land-Use Changes Are Possibly Larger than Assumed." *Nature Geoscience* 10, no. 2 (2017): 79–84.

Aron, AR, RB Ivry, KJ Jeffery, RA Poldrack, R Schmidt, C Summerfield, and AE Urai. "How Can Neuroscientists Respond to the Climate Emergency?" *Neuron* 106, no. 1 (2020): 17–20.

Aronoff, Kate, Alyssa Battistoni, Daniel Aldana Cohen, and Thea Riofrancos. *A Planet to Win: Why We Need a Green New Deal.* Verso, 2019.

Azadi, Mehdi, Stephen A Northey, Saleem H Ali, and Mansour Edraki. "Transparency on Greenhouse Gas Emissions from Mining to Enable Climate Change Mitigation." *Nature Geoscience* 13, no. 2 (2020): 100–4.

Bain, Paul G, Taciano L Milfont, Yoshihisa Kashima, Michał Bilewicz, Guy Doron, Ragna B Garðarsdóttir, Valdiney V Gouveia, et al. "Co-benefits of Addressing

Climate Change Can Motivate Action around the World." *Nature Climate Change* 6, no. 2 (2016): 154–7.

Baker, Lucy. "Of Embodied Emissions and Inequality: Rethinking Energy Consumption." *Energy Research & Social Science* 36 (2018): 52–60.

Bamberg, S, J Rees, and M Schulte. "Environmental Protection through Societal Change: What Psychology Knows about Collective Climate Action – and What It Needs to Find Out." In S Clayton and C Manning (eds.), *Psychology and Climate Change Human Perceptions, Impacts, and Responses*. Academic Press, 2018.

Bamberg, S, J Rees, and S Seebauer. "Collective Climate Action: Determinants of Participation Intention in Community-Based Pro-environmental Initiatives." *Journal of Environmental Psychology* 43 (2015): 155–65.

Bandura, Albert. "Exercise of Human Agency through Collective Efficacy." *Current Directions in Psychological Science* 9, no. 3 (2000): 75–8.

Bell, Karen, and Gnisha Bevan. "Beyond Inclusion? Perceptions of the Extent to Which Extinction Rebellion Speaks to, and for, Black, Asian and Minority Ethnic (BAME) and Working-Class Communities." *Local Environment* 26, no. 10 (2021): 1–16.

Bendell, Jem, and Rupert Read. *Deep Adaptation: Navigating the Realities of Climate Chaos*. John Wiley & Sons, 2021.

Benelli, Giovanni, and Heinz Mehlhorn. *Mosquito-Borne Diseases*. Springer, 2018.

Benelli, Giovanni, Marco Pombi, and Domenico Otranto. "Malaria in Italy: Migrants Are Not the Cause." *Trends in Parasitology* 34, no. 5 (2018): 351–4.

Bennett, J. *The Global Warming Primer*. Big Kid Science, 2016.

Bhatia, Kieran T, Gabriel A Vecchi, Thomas R Knutson, Hiroyuki Murakami, James Kossin, Keith W Dixon, and Carolyn E Whitlock. "Recent Increases in Tropical Cyclone Intensification Rates." *Nature Communications* 10, no. 1 (2019): 1–9.

Bleidorn, Wiebke, Patrick L Hill, Mitja D Back, Jaap JA Denissen, Marie Hennecke, Christopher J Hopwood, Markus Jokela, et al. "The Policy Relevance of Personality Traits." *American Psychologist* 74, no. 9 (2019): 1056.

Bollier, David, and Silke Helfrich. *Free, Fair, and Alive: The Insurgent Power of the Commons*. New Society Publishers, 2019.

Bolsen, Toby, Risa Palm, and Justin T Kingsland. "Counteracting Climate Science Politicization with effective Frames and Imagery." *Science Communication* 41, no. 2 (2019): 147–71.

Bookchin, Murray. *The Ecology of Freedom*. New Dimensions Foundation, 1982.

Boudet, Hilary, Leanne Giordono, Chad Zanocco, Hannah Satein, and Hannah Whitley. "Event Attribution and Partisanship Shape Local Discussion of Climate Change after Extreme Weather." *Nature Climate Change* 10, no. 1 (2020): 69–76.

Bouman, Thijs, Linda Steg, and Henk AL Kiers. "Measuring Values in Environmental Research: A Test of an Environmental Portrait Value Questionnaire." *Frontiers in Psychology* 9 (2018): 564.

Boyce, James K. "Carbon Pricing: Effectiveness and Equity." *Ecological Economics* 150 (2018): 52–61.

Boykoff, Maxwell T, and Jules M Boykoff. "Balance as Bias: Global Warming and the US Prestige Press." *Global Environmental Change* 14, no. 2 (2004): 125–36.

Brick, Cameron, and Gary J Lewis. "Unearthing the 'Green' Personality: Core Traits Predict Environmentally Friendly Behavior." *Environment and Behavior* 48, no. 5 (2016): 635–58.

Brockway, Paul E, Steve Sorrell, Gregor Semieniuk, Matthew Kuperus Heun, and Victor Court. "Energy Efficiency and Economy-Wide Rebound Effects: A Review of the Evidence and Its Implications." *Renewable and Sustainable Energy Reviews* (2021): 110781.

Brouillette, Monique. "How Microbes in Permafrost Could Trigger a Massive Carbon Bomb." *Nature* 591, no. 7850 (2021): 360–2.

Brown, Tom W, Tobias Bischof-Niemz, Kornelis Blok, Christian Breyer, Henrik Lund, and Brian Vad Mathiesen. "Response to 'Burden of Proof: A Comprehensive Review of the Feasibility of 100% Renewable-Electricity Systems'." *Renewable and Sustainable Energy Reviews* 92 (2018): 834–47.

Bruckner, Benedikt, Klaus Hubacek, Yuli Shan, Honglin Zhong, and Kuishuang Feng. "Impacts of Poverty Alleviation on National and Global Carbon Emissions." *Nature Sustainability* (2022): 1–10.

Brulle, Robert J. "Institutionalizing Delay: Foundation Funding and the Creation of US Climate Change Counter-Movement Organizations." *Climatic Change* 122, no. 4 (2014): 681–94.

Brysse, Keynyn, Naomi Oreskes, Jessica O'Reilly, and Michael Oppenheimer. "Climate Change Prediction: Erring on the Side of Least Drama?" *Global Environmental Change* 23, no. 1 (2013): 327–37.

Buck, Holly Jean. *Ending Fossil Fuels: Why Net Zero Is Not Enough*. Verso, 2021.

Bührle, Hannah, and Joachim Kimmerle. "Psychological Determinants of Collective Action for Climate Justice: Insights from Semi-structured Interviews and Content Analysis." *Frontiers in Psychology* 12 (2021): 3217.

Burke, Kevin D, John W Williams, Mark A Chandler, Alan M Haywood, Daniel J Lunt, and Bette L Otto-Bliesner. "Pliocene and Eocene Provide Best Analogs for near-Future Climates." *Proceedings of the National Academy of Sciences* 115, no. 52 (2018): 13288–93.

Burrows, Kate, and Patrick L Kinney. "Exploring the Climate Change, Migration and Conflict Nexus." *International Journal of Environmental Research and Public Health* 13, no. 4 (2016): 443.

Caesar, Levke, GD McCarthy, DJR Thornalley, N Cahill, and Stefan Rahmstorf. "Current Atlantic Meridional Overturning Circulation Weakest in Last Millennium." *Nature Geoscience* 14, no. 3 (2021): 118–20.

Capstick, Stuart Bryce, and Nicholas Frank Pidgeon. "What Is Climate Change Scepticism? Examination of the Concept Using a Mixed Methods Study of the UK Public." *Global Environmental Change* 24 (2014): 389–401.

Carlson, Colin J, Gregory F Albery, Cory Merow, Christopher H Trisos, Casey M Zipfel, Evan A Eskew, Kevin J Olival, Noam Ross, and Shweta Bansal. "Climate Change Increases Cross-Species Viral Transmission Risk." *Nature* (2022). https://doi.org/10.1038/s41586-022-04788-w.

Caspi, Avshalom, Brent W Roberts, and Rebecca L Shiner. "Personality Development: Stability and Change." *Annual Review of Psychology* 56 (2005): 453–84.

Castiglione, A, C Brick, S Holden, E Miles-Urdan, and AR Aron. "Discovering the Psychological Building Blocks Underlying Climate Action: A Longitudinal Study of Real-World Activism." *Royal Society Open Science* 9 (2022): 210006. https://doi.org/10.1098/rsos.210006.

Ceballos, Gerardo, Paul R Ehrlich, and Rodolfo Dirzo. "Biological Annihilation via the Ongoing Sixth Mass Extinction Signaled by Vertebrate Population Losses and Declines." *Proceedings of the National Academy of Sciences* 114, no. 30 (2017): E6089–E6096.

Chadburn, SE, EJ Burke, PM Cox, P Friedlingstein, G Hugelius, and S Westermann. "An Observation-Based Constraint on Permafrost Loss as a Function of Global Warming." *Nature Climate Change* 7, no. 5 (2017): 340–4.

Chapman, Daniel A, Brian Lickel, and Ezra M Markowitz. "Reassessing Emotion in Climate Change Communication." *Nature Climate Change* 7, no. 12 (2017): 850–2.

Cheng, Lijing, John Abraham, Jiang Zhu, Kevin E Trenberth, John Fasullo, Tim Boyer, Ricardo Locarnini, et al. "Record-Setting Ocean Warmth Continued in 2019." *Advances in Atmospheric Sciences* 37 (2020): 137–42.

Chomsky, Noam, Robert Pollin, and CJ Polychroniou. *Climate Crisis and the Global Green New Deal: The Political Economy of Saving the Planet.* Verso, 2020.

Coady, David, Ian Parry, Louis Sears, and Baoping Shang. "How Large Are Global Fossil Fuel Subsidies?" *World Development* 91 (2017): 11–27.

Collaborators, GBD Risk Factors. "Global, Regional, and National Comparative Risk Assessment of 79 Behavioural, Environmental and Occupational, and Metabolic Risks or Clusters of Risks in 188 Countries, 1990–2013: A Systematic Analysis for the Global Burden of Disease Study 2013." *Lancet* 386, no. 10010 (2015): 2287–323.

Conca, Ken. "The WTO and the Undermining of Global Environmental Governance." *Review of International Political Economy* 7, no. 3 (2000): 484–94.

Cook, Benjamin I, Kevin J Anchukaitis, Ramzi Touchan, David M Meko, and Edward R Cook. "Spatiotemporal Drought Variability in the Mediterranean over the Last 900 Years." *Journal of Geophysical Research: Atmospheres* 121, no. 5 (2016): 2060–74.

Cook, John, Naomi Oreskes, Peter T Doran, William RL Anderegg, Bart Verheggen, Ed W Maibach, J Stuart Carlton, et al. "Consensus on Consensus: A Synthesis of Consensus Estimates on Human-Caused Global Warming." *Environmental Research Letters* 11, no. 4 (2016): 048002.

Cox, Stan. *The Green New Deal and beyond: Ending the Climate Emergency While We Still Can.* City Lights Books, 2020.

Cox, Stan. *The Path to a Livable Future: A New Politics to Fight Climate Change, Racism, and the next Pandemic.* City Lights Books, 2021.

Creutzig, Felix, Joyashree Roy, William F Lamb, Inês ML Azevedo, Wändi Bruine De Bruin, Holger Dalkmann, Oreane Y Edelenbosch, et al. "Towards Demand-Side Solutions for Mitigating Climate Change." *Nature Climate Change* 8, no. 4 (2018): 260–3.

Cullenward, Danny, Mason Inman, and Michael D Mastrandrea. "Tracking Banking in the Western Climate Initiative Cap-and-Trade Program." *Environmental Research Letters* 14, no. 12 (2019): 124037.

D'Aspremont, Alexandre, Thomas Lauvaux, Clément Giron, Matthieu Mazzolini, Riley Duren, Dan Cusworth, Drew Shindell, and Philippe Ciais. "Global Assessment of Oil and Gas Methane Ultra-Emitters." *Science* 375, no. 6580 (2021): 557–61.

Daly, HE. "The Perils of Free Trade." *Scientific American* 269, no. 5 (1993): 50–5.

Deemer, Bridget R, John A Harrison, Siyue Li, Jake J Beaulieu, Tonya DelSontro, Nathan Barros, José F Bezerra-Neto, et al. "Greenhouse Gas Emissions from Reservoir Water Surfaces: A New Global Synthesis." *BioScience* 66, no. 11 (2016): 949–64.

DeShazo, Jessica L, Chandra Lal Pandey, and Zachary A Smith. *Why REDD Will Fail*. Routledge, 2016.

Dessler, Andrew. *Introduction to Modern Climate Change*. Cambridge University Press, 2015.

Díaz, Sandra, Unai Pascual, Marie Stenseke, Berta Martín-López, Robert T Watson, Zsolt Molnár, Rosemary Hill, et al. "Assessing Nature's Contributions to People." *Science* 359, no. 6373 (2018): 270–2.

Diesendorf, Mark, and Ben Elliston. "The Feasibility of 100% Renewable Electricity Systems: A Response to Critics." *Renewable and Sustainable Energy Reviews* 93 (2018): 318–30.

Diffenbaugh, Noah S, Daniel L Swain, and Danielle Touma. "Anthropogenic Warming Has Increased Drought Risk in California." *Proceedings of the National Academy of Sciences* 112, no. 13 (2015): 3931–36.

Douglas, Karen M, and Robbie M Sutton. "Climate Change: Why the Conspiracy Theories Are Dangerous." *Bulletin of the Atomic Scientists* 71, no. 2 (2015): 98–106.

Douglas, Mary, and Aaron Wildavsky. *Risk and Culture*. University of California Press, 1983.

Draxl, Caroline, Andrew Clifton, Bri-Mathias Hodge, and Jim McCaa. "The Wind Integration National Dataset (Wind) Toolkit." *Applied Energy* 151 (2015): 355–66.

Drury, John, and Steve Reicher. "Collective Psychological Empowerment as a Model of Social Change: Researching Crowds and Power." *Journal of Social Issues* 65, no. 4 (2009): 707–25.

Dumitru, Oana A, Jacqueline Austermann, Victor J Polyak, Joan J Fornós, Yemane Asmerom, Joaquín Ginés, Angel Ginés, and Bogdan P Onac. "Constraints on Global Mean Sea Level during Pliocene Warmth." *Nature* 574, no. 7777 (2019): 233–6.

Dunlap, REVL, K Val Liere, Angela Mertig, and Robert Emmet Jones. "Measuring Endorsement of the New Ecological Paradigm: A Revised NEP Scale." *Journal of Social Issues* 56, no. 3 (2000): 425–42.

Dunlap, Riley E, and Robert J Brulle. "Sources and Amplifiers of Climate Change Denial." In David C Holmes and Lucy M Richardson (eds.), *Research Handbook on Communicating Climate Change*. Edward Elgar, 2020.

Eaton, Emily M, and Nick A Day. "Petro-Pedagogy: Fossil Fuel Interests and the Obstruction of Climate Justice in Public Education." *Environmental Education Research* 26, no. 4 (2020): 457–73.

Ecker, Ullrich KH, Stephan Lewandowsky, John Cook, Philipp Schmid, Lisa K Fazio, Nadia Brashier, Panayiota Kendeou, Emily K Vraga, and Michelle A Amazeen. "The Psychological Drivers of Misinformation Belief and Its Resistance to Correction." *Nature Reviews Psychology* 1, no. 1 (2022): 13–29.

Engler, M, and P Engler. *This Is an Uprising*. Bold Type Books, 2017.

Ereaut, Gill, and Nat Segnit. *Warm Words: How We Are Telling the Climate Story and Can We Tell It Better*. IPPR, 2006.

Erickson, Timothy B, Julia Brooks, Eric J Nilles, Phuong N Pham, and Patrick Vinck. "Environmental Health Effects Attributed to Toxic and Infectious Agents following Hurricanes, Cyclones, Flash Floods and Major Hydrometeorological Events." *Journal of Toxicology and Environmental Health, Part B* 22, no. 5–6 (2019): 157–71.

Esposito, Luigi, and Fernando Pérez. "The Global Addiction and Human Rights: Insatiable Consumerism, Neoliberalism, and Harm Reduction." *Perspectives on Global Development and Technology* 9, no. 1–2 (2010): 84–100.

Feng, Huihui, and Mingyang Zhang. "Global Land Moisture Trends: Drier in Dry and Wetter in Wet over Land." *Scientific Reports* 5, no. 1 (2015): 1–6.

Feygina, Irina, John T Jost, and Rachel E Goldsmith. "System Justification, the Denial of Global Warming, and the Possibility of 'System-Sanctioned Change'." *Personality and Social Psychology Bulletin* 36, no. 3 (2010): 326–38.

Fielding, Kelly S, Matthew J Hornsey, Ha Anh Thai, and Li Li Toh. "Using Ingroup Messengers and Ingroup Values to Promote Climate Change Policy." *Climatic Change* 158, no. 2 (2020): 181–99.

Fischer, Erich M, and Reto Knutti. "Anthropogenic Contribution to Global Occurrence of Heavy-Precipitation and High-Temperature Extremes." *Nature Climate Change* 5, no. 6 (2015): 560–4.

Fisher, Scott R. "Life Trajectories of Youth Committing to Climate Activism." *Environmental Education Research* 22, no. 2 (2016): 229–47.

Franta, Benjamin. "Weaponizing Economics: Big Oil, Economic Consultants, and Climate Policy Delay." *Environmental Politics* (2021): 1–21.

Gerth, Karl. *Unending Capitalism: How Consumerism Negated China's Communist Revolution*. Cambridge University Press, 2020.

Gilbert, Daniel T, Elizabeth C Pinel, Timothy D Wilson, Stephen J Blumberg, and Thalia P Wheatley. "Immune Neglect: A Source of Durability Bias in Affective Forecasting." *Journal of Personality and Social Psychology* 75, no. 3 (1998): 617.

Gleick, Peter H. "Water, Drought, Climate Change, and Conflict in Syria." *Weather, Climate, and Society* 6, no. 3 (2014): 331–40.

Goldberg, Matthew H, Sander van der Linden, Matthew T Ballew, Seth A Rosenthal, Abel Gustafson, and Anthony Leiserowitz. "The Experience of Consensus: Video as an Effective Medium to Communicate Scientific Agreement on Climate Change." *Science Communication* 41, no. 5 (2019): 659–73.

Griffith, Saul. *Electrify: An Optimist's Playbook for Our Clean Energy Future*. MIT Press, 2021.

Gulliver, Robyn, Susilo Wibisono, Kelly S Fielding, and Winnifred R Louis. *The Psychology of Effective Activism*. Cambridge University Press, 2021.

Haberl, Helmut Dominik Wiedenhofer, Doris Virág, Gerald Kalt, Barbara Plank, Paul Brockway, Tomer Fishman, et al. "A Systematic Review of the Evidence on Decoupling of GDP, Resource Use and GHG Emissions, Part II: Synthesizing the Insights." *Environmental Research Letters* 15, no. 6 (2020): 065003.

Hansen, James, Makiko Sato, Pushker Kharecha, Karina Von Schuckmann, David J Beerling, Junji Cao, Shaun Marcott, et al. "Young People's Burden: Requirement of Negative CO_2 Emissions." *Earth System Dynamics* 8, no. 3 (2017): 577–616.

Hansen, James, Makiko Sato, Gary Russell, and Pushker Kharecha. "Climate Sensitivity, Sea Level and Atmospheric Carbon Dioxide." *Philosophical Transactions of the Royal Society A: Mathematical, Physical and Engineering Sciences* 371, no. 2001 (2013): 20120294.

Harold, Jordan, Irene Lorenzoni, Thomas F Shipley, and Kenny R Coventry. "Cognitive and Psychological Science Insights to Improve Climate Change Data Visualization." *Nature Climate Change* 6, no. 12 (2016): 1080–89.

Harvey, D. *A Brief History of Neoliberalism*. Oxford University Press, 2005.

Heard, Benjamin P, Barry W Brook, Tom ML Wigley, and Corey JA Bradshaw. "Burden of Proof: A Comprehensive Review of the Feasibility of 100% Renewable-Electricity Systems." *Renewable and Sustainable Energy Reviews* 76 (2017): 1122–33.

Helm, Dieter. "Climate-Change Policy: Why Has So Little Been Achieved?" *Oxford Review of Economic Policy* 24, no. 2 (2008): 211–38.

Herman, Edward S, and Noam Chomsky. *Manufacturing Consent: The Political Economy of the Mass Media*. Random House, 2010.

Herrando-Pérez, Salvador, Corey JA Bradshaw, Stephan Lewandowsky, and David R Vieites. "Statistical Language Backs Conservatism in Climate-Change Assessments." *BioScience* 69, no. 3 (2019): 209–19.

Hickel, Jason. *Less Is More: How Degrowth Will Save the World*. Random House, 2020.

Hickel, Jason. "What Does Degrowth Mean? A Few Points of Clarification." *Globalizations* (2020): 1–7.

Hickel, Jason, Paul Brockway, Giorgos Kallis, Lorenz Keyßer, Manfred Lenzen, Aljoša Slameršak, Julia Steinberger, and Diana Ürge-Vorsatz. "Urgent Need for Post-growth Climate Mitigation Scenarios." *Nature Energy* 6, no. 8 (2021): 766–8.

Hickman, Caroline, Elizabeth Marks, Panu Pihkala, Susan Clayton, Eric R Lewandowski, Elouise E Mayall, Britt Wray, Catriona Mellor, and Lise van Susteren. "Young People's Voices on Climate Anxiety, Government Betrayal and Moral Injury: A Global Phenomenon." *Lancet* 5, no. 12 (2022): e863–e873.

Holling, Crawford S. "The Resilience of Terrestrial Ecosystems: Local Surprise and Global Change." *Sustainable Development of the Biosphere* 14 (1986): 292–317.

Hönisch, Bärbel, N Gary Hemming, David Archer, Mark Siddall, and Jerry F McManus. "Atmospheric Carbon Dioxide Concentration across the Mid-Pleistocene Transition." *Science* 324, no. 5934 (2009): 1551–54.

Hornsey, Matthew J, and Kelly S Fielding. "Attitude Roots and Jiu Jitsu Persuasion: Understanding and Overcoming the Motivated Rejection of Science." *American Psychologist* 72, no. 5 (2017): 459–73.

Hornsey, Matthew J, and Kelly S Fielding. "A Cautionary Note about Messages of Hope: Focusing on Progress in Reducing Carbon Emissions Weakens Mitigation Motivation." *Global Environmental Change* 39 (2016): 26–34.

Hornsey, Matthew J, and Kelly S Fielding. "Understanding (and Reducing) Inaction on Climate Change." *Social Issues and Policy Review* 14, no. 1 (2020): 3–35.

Hornsey, Matthew J, Emily A Harris, Paul G Bain, and Kelly S Fielding. "Meta-analyses of the Determinants and Outcomes of Belief in Climate Change." *Nature Climate Change* 6, no. 6 (2016): 622–6.

Hornsey, Matthew J, Emily A Harris, and Kelly S Fielding. "Relationships among Conspiratorial Beliefs, Conservatism and Climate Scepticism across Nations." *Nature Climate Change* 8, no. 7 (2018): 614–20.

Horwitz, Wendy A. "Developmental Origins of Environmental Ethics: The Life Experiences of Activists." *Ethics & Behavior* 6, no. 1 (1996): 29–53.

Howe, Lauren C, Bo MacInnis, Jon A Krosnick, Ezra M Markowitz, and Robert Socolow. "Acknowledging Uncertainty Impacts Public Acceptance of Climate Scientists' Predictions." *Nature Climate Change* 9, no. 11 (2019): 863–7.

Howe, Peter D, Jennifer R Marlon, Matto Mildenberger, and Brittany S Shield. "How Will Climate Change Shape Climate Opinion?" *Environmental Research Letters* 14, no. 11 (2019): 113001.

Howe, Peter D, Matto Mildenberger, Jennifer R Marlon, and Anthony Leiserowitz. "Geographic Variation in Opinions on Climate Change at State and Local Scales in the USA." *Nature Climate Change* 5, no. 6 (2015): 596–603.

Huber, Matthew T. "The Case for Socialist Modernism." *Political Geography* (2021): 102352.

Huber, Matthew T. *Climate Change as Class War: Building Socialism on a Warming Planet*. Verso, 2022.

Hudson, Nathan W, Daniel A Briley, William J Chopik, and Jaime Derringer. "You Have to Follow through: Attaining Behavioral Change Goals Predicts Volitional Personality Change." *Journal of Personality and Social Psychology* 117, no. 4 (2019): 839.

Hulme, Mike. *Why We Disagree about Climate Change: Understanding Controversy, Inaction and Opportunity*. Cambridge University Press, 2009.

Immerzeel, Walter W, AF Lutz, M Andrade, A Bahl, H Biemans, Tobias Bolch, S Hyde, et al. "Importance and Vulnerability of the World's Water Towers." *Nature* 577, no. 7790 (2020): 364–9.

Jackson, Robert B, Marielle Saunois, Philippe Bousquet, Josep G Canadell, Benjamin Poulter, Ann R Stavert, Peter Bergamaschi, et al. "Increasing Anthropogenic Methane Emissions Arise Equally from Agricultural and Fossil Fuel Sources." *Environmental Research Letters* 15, no. 7 (2020): 071002.

Jackson, Roland. "Eunice Foote, John Tyndall and a Question of Priority." *Notes and Records* 74, no. 1 (2020): 105–18.

Jacobson, Mark Z. *100% Clean, Renewable Energy and Storage for Everything*. Cambridge University Press, 2020.

Jacobson, Mark Z, Mark A Delucchi, Mary A Cameron, Stephen J Coughlin, Catherine A Hay, Indu Priya Manogaran, Yanbo Shu, and Anna-Katharina von Krauland. "Impacts of Green New Deal Energy Plans on Grid Stability, Costs, Jobs, Health, and Climate in 143 Countries." *One Earth* 1, no. 4 (2019): 449–63.

Jacobson, Mark Z, Mark A Delucchi, Mary A Cameron, and Bethany A Frew. "Low-Cost Solution to the Grid Reliability Problem with 100% Penetration of Intermittent Wind, Water, and Solar for All Purposes." *Proceedings of the National Academy of Sciences* 112, no. 49 (2015): 15060–5.

Jacobson, Mark Z, Mark A Delucchi, Mary A Cameron, and Bethany A Frew. "The United States Can Keep the Grid Stable at Low Cost with 100% Clean, Renewable Energy in All Sectors Despite Inaccurate Claims." *Proceedings of the National Academy of Sciences* 114, no. 26 (2017): E5021–E5023.

Jacobson, Mark Z, Mark A Delucchi, Guillaume Bazouin, Zack AF Bauer, Christa C Heavey, Emma Fisher, Sean B Morris, et al. "100% Clean and Renewable Wind, Water, and Sunlight (WWS) All-Sector Energy Roadmaps for the 50 United States." *Energy & Environmental Science* 8, no. 7 (2015): 2093–117.

Jacobson, Mark Z, Mark A Delucchi, Zack AF Bauer, Savannah C Goodman, William E Chapman, Mary A Cameron, Cedric Bozonnat, et al. "100% Clean and Renewable Wind, Water, and Sunlight All-Sector Energy Roadmaps for 139 Countries of the World." *Joule* 1, no. 1 (2017): 108–21.

Joslyn, Susan, and Raoni Demnitz. "Communicating Climate Change: Probabilistic Expressions and Concrete Events." *Weather, Climate, and Society* 11, no. 3 (2019): 651–64.

Jowitt, Simon M, Timothy T Werner, Zhehan Weng, and Gavin M Mudd. "Recycling of the Rare Earth Elements." *Current Opinion in Green and Sustainable Chemistry* 13 (2018): 1–7.

Kahan, Dan M, Donald Braman, John Gastil, Paul Slovic, and CK Mertz. "Culture and Identity-Protective Cognition: Explaining the White-Male Effect in Risk Perception." *Journal of Empirical Legal Studies* 4, no. 3 (2007): 465–505.

Kahan, Dan M, and Katherine Carpenter. "Out of the Lab and into the Field." *Nature Climate Change* 7, no. 5 (2017): 309–11.

Kahan, Dan M, Hank Jenkins-Smith, and Donald Braman. "Cultural Cognition of Scientific Consensus." *Journal of Risk Research* 14, no. 2 (2011): 147–74.

Kahan, Dan M, Maggie Wittlin, Ellen Peters, Paul Slovic, Lisa Larrimore Ouellette, Donald Braman, and Gregory N Mandel. "The Tragedy of the Risk-Perception Commons: Culture Conflict, Rationality Conflict, and Climate Change." Temple University Legal Studies Research Paper, no. 2011–26 (2011).

Kalmus, Peter. *Fly Less to Convey Urgency*. American Association for the Advancement of Science, 2019.

Karl, Thomas R, Jerry M Melillo, Thomas C Peterson, and Susan J Hassol. *Global Climate Change Impacts in the United States*. Cambridge University Press, 2009.

Keeling, Charles D, Stephen C Piper, Robert B Bacastow, Martin Wahlen, Timothy P Whorf, Martin Heimann, and Harro A Meijer. "Exchanges of Atmospheric CO_2 and $13CO_2$ with the Terrestrial Biosphere and Oceans from 1978 to 2000. I. Global Aspects." In IT Baldwin, MM Caldwell, G Heldmaier, RB Jackson, OL Lange, HA Mooney, E-D Schulze, et al. (eds.), *A History of Atmospheric CO_2 and Its Effects on Plants, Animals, and Ecosystems. Ecological Studies*, vol. 177. Springer, 2001.

Keen, Steve. "The Appallingly Bad Neoclassical Economics of Climate Change." *Globalizations* (2020): 1–29.

Keith, David W, Geoffrey Holmes, David St Angelo, and Kenton Heidel. "A Process for Capturing CO_2 from the Atmosphere." *Joule* 2, no. 8 (2018): 1573–94.

Kelley, Colin P, Shahrzad Mohtadi, Mark A Cane, Richard Seager, and Yochanan Kushnir. "Climate Change in the Fertile Crescent and Implications of the Recent Syrian Drought." *Proceedings of the National Academy of Sciences* 112, no. 11 (2015): 3241–6.

Klein, N. *This Changes Everything: Capitalism vs. the Climate*. Simon and Schuster, 2014.

Klöckner, Christian Andreas, and Laura K Sommer. "Visual Art Inspired by Climate Change: An Analysis of Audience Reactions to 37 Artworks Presented during 21st UN Climate Summit in Paris." *PloS One* 16, no. 2 (2021): e0247331.

Kopnina, Helen. "Environmental Justice and Biospheric Egalitarianism: Reflecting on a Normative-Philosophical View of Human–Nature Relationship." *Earth Perspectives* 1, no. 1 (2014): 1–11.

Kossin, James P, Kenneth R Knapp, Timothy L Olander, and Christopher S Velden. "Global Increase in Major Tropical Cyclone Exceedance Probability over the Past Four Decades." *Proceedings of the National Academy of Sciences* 117, no. 22 (2020): 11975–80.

Kukal, Meetpal S, and Suat Irmak. "Us Agro-Climate in 20th Century: Growing Degree Days, First and Last Frost, Growing Season Length, and Impacts on Crop Yields." *Scientific Reports* 8, no. 1 (2018): 1–14.

Ladd, Anthony E. "Priming the Well: 'Frackademia' and the Corporate Pipeline of Oil and Gas Funding into Higher Education." *Humanity & Society* 44, no. 2 (2020): 151–77.

Lambert, Julie L, and Robert E Bleicher. "Argumentation as a Strategy for Increasing Preservice Teachers' Understanding of Climate Change, a Key Global Socioscientific Issue." *International Journal of Education in Mathematics, Science and Technology* 5, no. 2 (2017): 101–12.

Landrigan, Philip J, Richard Fuller, Nereus JR Acosta, Olusoji Adeyi, Robert Arnold, Abdoulaye Bibi Baldé, Roberto Bertollini, et al. "The Lancet Commission on Pollution and Health." *The Lancet* 391, no. 10119 (2018): 462–512.

Lathem, Alexis. "Grassroots Megadam Resistance at Muskrat Falls, Labrador." In Brian Tokar and Tamra Gilbertson (eds.), *Climate Justice and Community Renewal: Resistance and Grassroots Solutions*. Routledge, (2020): 26–38.

Lawton, Graham. "I Have Eco-Anxiety but That's Normal." *New Scientist* 244, no. 3251 (2019): 22.

Le Quéré, Corinne, Jan Ivar Korsbakken, Charlie Wilson, Jale Tosun, Robbie Andrew, Robert J Andres, Josep G Canadell, et al. "Drivers of Declining CO_2 Emissions in 18 Developed Economies." *Nature Climate Change* 9, no. 3 (2019): 213–17.

Lear, Jonathan. *Radical Hope: Ethics in the Face of Cultural Devastation*. Harvard University Press, 2006.

Leiserowitz, Anthony. "Climate Change Risk Perception and Policy Preferences: The Role of Affect, Imagery, and Values." *Climatic Change* 77, no. 1 (2006): 45–72.

Lenton, TM, J Rockstrom, O Gaffney, S Rahmstorf, K Richardson, W Steffen, and HJ Schellnhuber. "Climate Tipping Points: Too Risky to Bet against." *Nature* 575, no. 7784 (2019): 592–5.

Lewandowsky, S, K Oberauer, and GE Gignac. "NASA Faked the Moon Landing – Therefore (Climate) Science Is a Hoax: An Anatomy of the Motivated Rejection of Science." *Psychological Science* 24, no. 5 (May 2013): 622–33.

Lewandowsky, S. "Climate Change Disinformation and How to Combat It." *Annual Review of Public Health* 42 (2021): 1–21.

Lieven, Anatol. *Climate Change and the Nation State: The Case for Nationalism in a Warming World*. Oxford University Press, 2020.

Lin, Boqiang, and Xuehui Li. "The Effect of Carbon Tax on Per Capita CO_2 Emissions." *Energy Policy* 39, no. 9 (2011): 5137–46.

Lindstad, Haakon, Bjørn E Asbjørnslett, and Anders H Strømman. "The Importance of Economies of Scale for Reductions in Greenhouse Gas Emissions from Shipping." *Energy Policy* 46 (2012): 386–98.

Loeb, Norman G, Gregory C Johnson, Tyler J Thorsen, John M Lyman, Fred G Rose, and Seiji Kato. "Satellite and Ocean Data Reveal Marked Increase in Earth's Heating Rate." *Geophysical Research Letters* 48, no. 13 (2021): e2021GL093047.

Lombardi, Doug, Elliot S Bickel, Carol B Brandt, and Colin Burg. "Categorising Students' Evaluations of Evidence and Explanations about Climate Change." *International Journal of Global Warming* 12, no. 3–4 (2017): 313–30.

Lord, Charles G, Lee Ross, and Mark R Lepper. "Biased Assimilation and Attitude Polarization: The Effects of Prior Theories on Subsequently Considered Evidence." *Journal of Personality and Social Psychology* 37, no. 11 (1979): 2098.

Lovejoy, Thomas E, and Carlos Nobre. "*Amazon Tipping Point: Last Chance for Action*." American Association for the Advancement of Science, 2019.

Lynas, M. *Our Final Warning: Six Degrees of Climate Emergency*. 4th Estate, 2020.

MacDougall, Andrew H, Thomas L Frölicher, Chris D Jones, Joeri Rogelj, H Damon Matthews, Kirsten Zickfeld, Vivek K Arora, et al. "Is there Warming in the Pipeline? A Multi-model Analysis of the Zero Emissions Commitment from CO_2." *Biogeosciences* 17, no. 11 (2020): 2987–3016.

Macfarling Meure, Cecelia, David Etheridge, Cathy Trudinger, P Steele, R Langenfelds, T Van Ommen, A Smith, and J Elkins. "Law Dome CO2, CH4 and N2O Ice Core Records Extended to 2000 Years BP." *Geophysical Research Letters* 33, no. 14 (2006).

Madigan, Daniel J, Zofia Baumann, and Nicholas S Fisher. "Pacific Bluefin Tuna Transport Fukushima-Derived Radionuclides from Japan to California." *Proceedings of the National Academy of Sciences* 109, no. 24 (2012): 9483–6.

Malm, Andreas. *Fossil Capital: The Rise of Steam Power and the Roots of Global Warming*. Verso Books, 2016.

Malm, Andreas. *How to Blow up a Pipeline*. Verso Books, 2021.

Mann, Michael E. *The New Climate War: The Fight to Take Back Our Planet*. Hachette UK, 2021.

Marshall, George. *Don't Even Think about It: Why Our Brains Are Wired to Ignore Climate Change*. Bloomsbury Publishing USA, 2015.

Martin, Christian, and Sandor Czellar. "Where Do Biospheric Values Come From? A Connectedness to Nature Perspective." *Journal of Environmental Psychology* 52 (2017): 56–68.

Mata, André, Steven J Sherman, Mário B Ferreira, and Cristina Mendonça. "Strategic Numeracy: Self-Serving Reasoning about Health Statistics." *Basic and Applied Social Psychology* 37, no. 3 (2015): 165–73.

Matthews, Gerald, Ian J Deary, and Martha C Whiteman. *Personality Traits*. Cambridge University Press, 2003.

Maurer, Josh M, JM Schaefer, S Rupper, and A Corley. "Acceleration of Ice Loss across the Himalayas over the Past 40 Years." *Science Advances* 5, no. 6 (2019): eaav7266.

Mauritsen, Thorsten, and Robert Pincus. "Committed Warming Inferred from Observations." *Nature Climate Change* 7, no. 9 (2017): 652–5.

Mazzucato, Mariana. "Financing the Green New Deal." *Nature Sustainability* (2021): 1–2.

McAdam, D. "Social Movement Theory and the Prospects for Climate Change Activism in the United States." *Annual Review of Political Science* (2017).

McAllister, Lucy, Meaghan Daly, Patrick Chandler, Marisa McNatt, Andrew Benham, and Maxwell Boykoff. "Balance as Bias, Resolute on the Retreat? Updates and Analyses of Newspaper Coverage in the United States, United Kingdom, New Zealand, Australia and Canada over the Past 15 Years." *Environmental Research Letters* 16, no. 9 (2021): 094008.

McKinnon, Catriona. "Climate Change: Against Despair." *Ethics & the Environment* 19, no. 1 (2014): 31–48.

Mihaylov, Nikolay L, and Douglas D Perkins. "Local Environmental Grassroots Activism: Contributions from Environmental Psychology, Sociology and Politics." *Behavioral Sciences* 5, no. 1 (2015): 121–53.

Mildenberger, Matto. *Carbon Captured: How Business and Labor Control Climate Politics*. MIT Press, 2020.

Millward-Hopkins, Joel, Julia K Steinberger, Narasimha D Rao, and Yannick Oswald. "Providing Decent Living with Minimum Energy: A Global Scenario." *Global Environmental Change* 65 (2020): 102168.

Miralles, Diego G, Adriaan J Teuling, Chiel C Van Heerwaarden, and Jordi Vila-Guerau De Arellano. "Mega-heatwave Temperatures Due to Combined Soil Desiccation and Atmospheric Heat Accumulation." *Nature Geoscience* 7, no. 5 (2014): 345–9.

Mitchell, Timothy. *Carbon Democracy*. Verso, 2013.

Monroe, Martha C, Richard R Plate, Annie Oxarart, Alison Bowers, and Willandia A Chaves. "Identifying Effective Climate Change Education Strategies: A Systematic Review of the Research." *Environmental Education Research* 25, no. 6 (2019): 791–812.

Moore, Frances C, Nick Obradovich, Flavio Lehner, and Patrick Baylis. "Rapidly Declining Remarkability of Temperature Anomalies May Obscure Public Perception of Climate Change." *Proceedings of the National Academy of Sciences* 116, no. 11 (2019): 4905–10.

Moran, Emilio F, Maria Claudia Lopez, Nathan Moore, Norbert Müller, and David W Hyndman. "Sustainable Hydropower in the 21st Century." *Proceedings of the National Academy of Sciences* 115, no. 47 (2018): 11891–8.

Morris, Brandi S, Polymeros Chrysochou, Simon T Karg, and Panagiotis Mitkidis. "Optimistic vs. Pessimistic Endings in Climate Change Appeals." *Humanities and Social Sciences Communications* 7, no. 1 (2020): 1–8.

Mueller, Brigitte, Mathias Hauser, Carley Iles, Ruksana Haque Rimi, Francis W Zwiers, and Hui Wan. "Lengthening of the Growing Season in Wheat and Maize Producing Regions." *Weather and Climate Extremes* 9 (2015): 47–56.

Niang, Isabelle, Oliver C Ruppel, Mohamed A Abdrabo, Ama Essel, Christopher Lennard, Jonathan Padgham, and Penny Urquhart. "Africa." In VR Barros, CB Field, DJ Dokken, MD Mastrandrea, KJ Mach, TE Bilir, M Chatterjee, et al. (eds.), *Climate Change 2014: Impacts, Adaptation, and Vulnerability. Part B: Regional Aspects. Contribution of Working Group II to the Fifth Assessment Report of the Intergovernmental Panel on Climate Change*. Cambridge University Press, 2014.

Ocko, Ilissa B, and Steven P Hamburg. "Climate Impacts of Hydropower: Enormous Differences among Facilities and over Time." *Environmental Science & Technology* 53, no. 23 (2019): 14070–82.

Oreskes, Naomi, and Erik M Conway. *Merchants of Doubt: How a Handful of Scientists Obscured the Truth on Issues from Tobacco Smoke to Global Warming*. Bloomsbury Publishing USA, 2011.

Orr, David W. "Love It or Lose It: The Coming Biophilia Revolution." In R Kellert and EO Wilson (eds.), *The Biophilia Hypothesis*. Island Press, 1993.

Panu, Pihkala. "Anxiety and the Ecological Crisis: An Analysis of Eco-Anxiety and Climate Anxiety." *Sustainability* 12, no. 19 (2020): 7836.

Pattyn, Frank, and Mathieu Morlighem. "The Uncertain Future of the Antarctic Ice Sheet." *Science* 367, no. 6484 (2020): 1331–5.

Penn, Justin L, and Curtis Deutsch. "Avoiding Ocean Mass Extinction from Climate Warming." *Science* 376, no. 6592 (2022): 524–6.

Perelman, Michael. *The Invention of Capitalism*. Duke University Press, 2000.

Perez, Marc, Richard Perez, Karl R Rábago, and Morgan Putnam. "Overbuilding and Curtailment: The Cost-Effective Enablers of Firm PV Generation." *Solar Energy* 180 (2019): 412–22.

Piao, Shilong, Xuhui Wang, Taejin Park, Chi Chen, XU Lian, Yue He, Jarle W Bjerke, et al. "Characteristics, Drivers and Feedbacks of Global Greening." *Nature Reviews Earth & Environment* 1, no. 1 (2020): 14–27.

Pierrehumbert, Raymond. "There Is No Plan B for Dealing with the Climate Crisis." *Bulletin of the Atomic Scientists* 75, no. 5 (2019): 215–21.

Piggot, Georgia, Cleo Verkuijl, Harro van Asselt, and Michael Lazarus. "Curbing Fossil Fuel Supply to Achieve Climate Goals." *Climate Policy* 20, no. 8 (2020): 881–7.

Pimm, Stuart L, Clinton N Jenkins, Robin Abell, Thomas M Brooks, John L Gittleman, Lucas N Joppa, Peter H Raven, Callum M Roberts, and Joseph O Sexton. "The Biodiversity of Species and Their Rates of Extinction, Distribution, and Protection." *Science* 344, no. 6187 (2014). https://doi.org/10.1126/science.1246752.

Piven, Frances Fox, and Richard Cloward. *Poor People's Movements: Why They Succeed, How They Fail*. Vintage, 2012.

Plutzer, Eric, Mark McCaffrey, A Lee Hannah, Joshua Rosenau, Minda Berbeco, and Ann H Reid. "Climate Confusion among US Teachers." *Science* 351, no. 6274 (2016): 664–5.

Pollin, R. "De-growth vs. a Green New Deal." *New Left Review* 112, July–August (2018).

Pollin, R. "An Industrial Policy Framework to Advance a Global Green New Deal." In A Oqubay, C Cramer, H-J Chang, and R Kozul-Wright (eds.), *The Oxford Handbook of Industrial Policy*. Oxford: Oxford University Press, 394–428.

Pun, Iam-Fei, I-I Lin, and Min-Hui Lo. "Recent Increase in High Tropical Cyclone Heat Potential Area in the Western North Pacific Ocean." *Geophysical Research Letters* 40, no. 17 (2013): 4680–4.

Ramanathan, Veerabhadran, Roger Aines, Max Auffhammer, Matt Barth, Jonathan Cole, Fonna Forman, Hahrie Han, et al. *Bending the Curve: Climate Change Solutions*. Regents of the University of California, 2019.

Ranney, Michael Andrew, and Dav Clark. "Climate Change Conceptual Change: Scientific Information Can Transform Attitudes." *Topics in Cognitive Science* 8, no. 1 (2016): 49–75.

Raupach, Michael R, Gregg Marland, Philippe Ciais, Corinne Le Quéré, Josep G Canadell, Gernot Klepper, and Christopher B Field. "Global and Regional Drivers of Accelerating CO2 Emissions." *Proceedings of the National Academy of Sciences* 104, no. 24 (2007): 10288–93.

Raworth, Kate. *Doughnut Economics: Seven Ways to Think like a 21st-Century Economist*. Chelsea Green Publishing, 2017.

Read, Rupert J. *Extinction Rebellion: Insights from the Inside. Simplicity Institute*, 2020.

Rignot, Eric, Jérémie Mouginot, Bernd Scheuchl, Michiel Van Den Broeke, Melchior J Van Wessem, and Mathieu Morlighem. "Four Decades of Antarctic Ice Sheet Mass Balance from 1979–2017." *Proceedings of the National Academy of Sciences* 116, no. 4 (2019): 1095–103.

Ripple, William, Christopher Wolf, Thomas Newsome, Phoebe Barnard, William Moomaw, and Philippe Grandcolas. "World Scientists' Warning of a Climate Emergency." *BioScience* 71, no. 9 (2021): 894–8.

Roberts, Brent W, Jing Luo, Daniel A Briley, Philip I Chow, Rong Su, and Patrick L Hill. "A Systematic Review of Personality Trait Change through Intervention." *Psychological Bulletin* 143, no. 2 (2017): 117.

Robine, Jean-Marie, Siu Lan K Cheung, Sophie Le Roy, Herman Van Oyen, Clare Griffiths, Jean-Pierre Michel, and François Richard Herrmann. "Death Toll Exceeded 70,000 in Europe during the Summer of 2003." *Comptes Rendus Biologies* 331, no. 2 (2008): 171–8.

Robock, Alan. "20 Reasons Why Geoengineering May Be a Bad Idea." *Bulletin of the Atomic Scientists* 64, no. 2 (2008): 14–18.

Romps, David M, and Jean P Retzinger. "Climate News Articles Lack Basic Climate Science." *Environmental Research Communications* 1, no. 8 (2019): 081002.

Rooney-Varga, Juliette N, Florian Kapmeier, John D Sterman, Andrew P Jones, Michele Putko, and Kenneth Rath. "The Climate Action Simulation." *Simulation & Gaming* 51, no. 2 (2020): 114–40.

Roosen, Liselotte J, Christian A Klöckner, and Janet K Swim. "Visual Art as a Way to Communicate Climate Change: A Psychological Perspective on Climate Change–Related Art." *World Art* 8, no. 1 (2018): 85–110.

Roser-Renouf, Connie, Edward W Maibach, Anthony Leiserowitz, and Xiaoquan Zhao. "The Genesis of Climate Change Activism: From Key Beliefs to Political Action." *Climatic Change* 125, no. 2 (2014): 163–78.

Rosnick, David, and Mark Weisbrot. "Are Shorter Work Hours Good for the Environment? A Comparison of US and European Energy Consumption." *International Journal of Health Services* 37, no. 3 (2007): 405–17.

Rothermich, Kathrin, Erika Katherine Johnson, Rachel Morgan Griffith, and Monica Marie Beingolea. "The Influence of Personality Traits on Attitudes towards Climate Change: An Exploratory Study." *Personality and Individual Differences* 168 (2021): 110304.

Rotz, C Alan, Senorpe Asem-Hiablie, Sara Place, and Greg Thoma. "Environmental Footprints of Beef Cattle Production in the United States." *Agricultural Systems* 169 (2019): 1–13.

Rubino, M, DM Etheridge, CM Trudinger, CE Allison, MO Battle, RL Langenfelds, LP Steele, et al. "A Revised 1000 Year Atmospheric Δ13C–CO2 Record from Law

Dome and South Pole, Antarctica." *Journal of Geophysical Research: Atmospheres* 118, no. 15 (2013): 8482–99.

San Sebastián, Miguel, Ben Armstrong, Juan Antonio Córdoba, and Carolyn Stephens. "Exposures and Cancer Incidence near Oil Fields in the Amazon Basin of Ecuador." *Occupational and Environmental Medicine* 58, no. 8 (2001): 517–22.

Sanderson, Hans. "Who Is Responsible for Embodied CO2?" *Climate* 9, no. 3 (2021): 41.

Saunois, Marielle, Philippe Bousquet, Ben Poulter, Anna Peregon, Philippe Ciais, Josep G Canadell, Edward J Dlugokencky, et al. "The Global Methane Budget 2000–2012." *Earth System Science Data* 8, no. 2 (2016): 697–751.

Scherer, Laura, and Stephan Pfister. "Hydropower's Biogenic Carbon Footprint." *PloS One* 11, no. 9 (2016): e0161947.

Schultz, P Wesley, Jessica M Nolan, Robert B Cialdini, Noah J Goldstein, and Vladas Griskevicius. "The Constructive, Destructive, and Reconstructive Power of Social Norms." *Psychological Science* 18, no. 5 (2007): 429–34.

Schwalm, Christopher R, Spencer Glendon, and Philip B Duffy. "RCP8: 5 Tracks Cumulative CO2 Emissions." *Proceedings of the National Academy of Sciences* 117, no. 33 (2020): 19656–7.

Sengupta, Manajit, Yu Xie, Anthony Lopez, Aron Habte, Galen Maclaurin, and James Shelby. "The National Solar Radiation Data Base (NSRDB)." *Renewable and Sustainable Energy Reviews* 89 (2018): 51–60.

Sheehy-Skeffington, Jennifer, and Lotte Thomsen. "Egalitarianism: Psychological and Socio-Ecological Foundations." *Current Opinion in Psychology* 32 (2020): 146–52.

Shepherd, Andrew, Erik Ivins, Eric Rignot, Ben Smith, Michiel Van Den Broeke, Isabella Velicogna, Pippa Whitehouse, et al. "Mass Balance of the Greenland Ice Sheet from 1992 to 2018." *Nature* 579, no. 7798 (2020): 233–9.

Sherwood, SC, Mark J Webb, James D Annan, KC Armour, Piers M Forster, Julia C Hargreaves, Gabriele Hegerl, et al. "An Assessment of Earth's Climate Sensitivity Using Multiple Lines of Evidence." *Reviews of Geophysics* 58, no. 4 (2020): e2019RG000678.

Siemoneit, Andreas. "An Offer You Can't Refuse: Enhancing Personal Productivity through 'Efficiency Consumption'." *Technology in Society* 59 (2019): 101181.

Simon, Bernd, and Bert Klandermans. "Politicized Collective Identity: A Social Psychological Analysis." *American Psychologist* 56, no. 4 (2001): 319.

Sisco, MR, V Bosetti, and E. U. Weber. "When Do Extreme Weather Events Generate Attention to Climate Change?" *Climatic Change* 143 (2017): 227–41.

Soga, Masashi, and Kevin J Gaston. "Shifting Baseline Syndrome: Causes, Consequences, and Implications." *Frontiers in Ecology and the Environment* 16, no. 4 (2018): 222–30.

Staggenborg, Suzanne. *Social Movements*. Oxford University Press, 2016.

Stanley-Robinson K. *The Ministry of the Future*. Orbit, 2020.

Sterman, John D. "Risk Communication on Climate: Mental Models and Mass Balance." *Science* 322, no. 5901 (2008): 532–3.

Sterman, John D, and Linda Booth Sweeney. "Understanding Public Complacency about Climate Change: Adults' Mental Models of Climate Change Violate Conservation of Matter." *Climatic Change* 80, no. 3 (2007): 213–38.

Stilwell, Frank. *Political Economy: The Contest of Economic Ideas*. Oxford University Press, 2011.

Stokes, Leah Cardamore. *Short Circuiting Policy: Interest Groups and the Battle over Clean Energy and Climate Policy in the American States*. Oxford University Press, 2020.

Swain, Daniel L, Deepti Singh, Danielle Touma, and Noah S Diffenbaugh. "Attributing Extreme Events to Climate Change: A New Frontier in a Warming World." *One Earth* 2, no. 6 (2020): 522–7.

Tajfel, Henri. "Social Identity and Intergroup Behaviour." *Social Science Information* 13, no. 2 (1974): 65–93.

Taube, Oliver, Michael Andrew Ranney, Laura Henn, and Florian G Kaiser. "Increasing People's Acceptance of Anthropogenic Climate Change with Scientific Facts: Is Mechanistic Information More Effective for Environmentalists?" *Journal of Environmental Psychology* 73 (2021): 101549.

Thier, Hadas. *A People's Guide to Capitalism: An Introduction to Marxist Economics*. Haymarket Books, 2018.

Thomas, Emma F, Craig McGarty, and Kenneth I Mavor. "Transforming 'Apathy into Movement': The Role of Prosocial Emotions in Motivating Action for Social Change." *Personality and Social Psychology Review* 13, no. 4 (2009): 310–33.

Tokar, Brian. *Toward Climate Justice: Perspectives on the Climate Crisis and Social Change*. New Compass Press, 2014.

Tol, Richard SJ. "The Economic Impacts of Climate Change." *Review of Environmental Economics and Policy* (2020).

Tollefson, Jeff. "Covid Curbed Carbon Emissions in 2020 – but Not by Much." *Nature* 589, no. 7842 (2021): 343.

Tollefson, Jeff. "How Hot Will Earth Get by 2100?" *Nature* 580, no. 7804 (2020): 443–6.

Tong, Dan, David J Farnham, Lei Duan, Qiang Zhang, Nathan S Lewis, Ken Caldeira, and Steven J Davis. "Geophysical Constraints on the Reliability of Solar and Wind Power Worldwide." *Nature Communications* 12, no. 1 (2021): 1–12.

Tong, Dan, Qiang Zhang, Yixuan Zheng, Ken Caldeira, Christine Shearer, Chaopeng Hong, Yue Qin, and Steven J Davis. "Committed Emissions from Existing Energy Infrastructure Jeopardize 1.5 C Climate Target." *Nature* 572, no. 7769 (2019): 373–7.

Turetsky, Merritt R, Benjamin W Abbott, Miriam C Jones, Katey Walter Anthony, David Olefeldt, Edward AG Schuur, Charles Koven, et al. *Permafrost Collapse Is Accelerating Carbon Release*. Nature Publishing Group, 2019.

Vaks, Anton, Oxana S Gutareva, Sebastian FM Breitenbach, Erdenedalai Avirmed, Andrew J Mason, Alexander L Thomas, Alexander V Osinzev, Alexander M Kononov, and Gideon M Henderson. "Speleothems Reveal 500,000-Year History of Siberian Permafrost." *Science* 340, no. 6129 (2013): 183–6.

Valero, Alicia, Antonio Valero, Guiomar Calvo, and Abel Ortego. "Material Bottlenecks in the Future Development of Green Technologies." *Renewable and Sustainable Energy Reviews* 93 (2018): 178–200.

Van de Ven, Dirk-Jan, Iñigo Capellan-Peréz, Iñaki Arto, Ignacio Cazcarro, Carlos de Castro, Pralit Patel, and Mikel Gonzalez-Eguino. "The Potential Land Requirements and Related Land Use Change Emissions of Solar Energy." *Scientific Reports* 11, no. 1 (2021): 1–12.

Van der Bles, Anne Marthe, Sander Van Der Linden, Alexandra LJ Freeman, James Mitchell, Ana B Galvao, Lisa Zaval, and David J Spiegelhalter. "Communicating Uncertainty about Facts, Numbers and Science." *Royal Society Open Science* 6, no. 5 (2019): 181870.

Van der Linden, Sander. "The Social-Psychological Determinants of Climate Change Risk Perceptions: Towards a Comprehensive Model." *Journal of Environmental Psychology* 41 (2015): 112–24.

Van der Linden, Sander, Anthony Leiserowitz, Geoffrey Feinberg, and Edward Maibach. "The Scientific Consensus on Climate Change as a Gateway Belief: Experimental Evidence." *PloS One* 10, no. 2 (2015): e0118489.

Van der Linden, Sander, Anthony Leiserowitz, and Edward Maibach. "The Gateway Belief Model: A Large-Scale Replication." *Journal of Environmental Psychology* 62 (2019): 49–58.

Van der Linden, Sander, Anthony Leiserowitz, and Edward Maibach. "Perceptions of Scientific Consensus Predict Later Beliefs about the Reality of Climate Change Using Cross-Lagged Panel Analysis: A Response to Kerr and Wilson (2018)." *Journal of Environmental Psychology* 60 (2018): 110–11.

Van der Linden, Sander, Anthony Leiserowitz, Seth Rosenthal, and Edward Maibach. "Inoculating the Public against Misinformation about Climate Change." *Global Challenges* 1, no. 2 (2017): 1600008.

Van Der Linden, Sander, Edward Maibach, John Cook, Anthony Leiserowitz, Michael Ranney, Stephan Lewandowsky, Joseph Árvai, and Elke U Weber. "Culture versus Cognition Is a False Dilemma." *Nature Climate Change* 7, no. 7 (2017): 457–7.

Van der Linden, Sander, Jon Roozenbeek, Rakoen Maertens, Melisa Basol, Ondřej Kácha, Steve Rathje, and Cecilie Steenbuch Traberg. "How Can Psychological Science Help Counter the Spread of Fake News?" *The Spanish Journal of Psychology* 24 (2021).

Van Stekelenburg, Jacquelien. "Collective Identity." In David A Snow, Donatella Della Porta, Bert Klandermans, and Doug McAdam (eds.), *The Wiley-Blackwell Encyclopedia of social and Political Movements* (Wiley-Blackwell, 2013).

Van Zomeren, Martijn, Tom Postmes, and Russell Spears. "Toward an Integrative Social Identity Model of Collective Action: A Quantitative Research Synthesis of Three Socio-Psychological Perspectives." *Psychological Bulletin* 134, no. 4 (2008): 504.

Veblen, Thorstein. *The Theory of the Leisure Class*. Houghton Mifflin Boston, 1973.

Velautham, Leela, Michael Andrew Ranney, and Quinlan S Brow. "Communicating Climate Change Oceanically: Sea Level Rise Information Increases Mitigation, Inundation, and Global Warming Acceptance." *Frontiers in Communication* 4 (2019): 7.

Vivanco, David Font, René Kemp, and Ester van der Voet. "How to Deal with the Rebound Effect? A Policy-Oriented Approach." *Energy Policy* 94 (2016): 114–25.

Vohra, Karn, Alina Vodonos, Joel Schwartz, Eloise A Marais, Melissa P Sulprizio, and Loretta J Mickley. "Global Mortality from Outdoor Fine Particle Pollution Generated by Fossil Fuel Combustion: Results from Geos-Chem." *Environmental Research* 195 (2021): 110754.

Wallace-Wells, David. *The Uninhabitable Earth*. Columbia University Press, 2019.

Wang, Daoping, Dabo Guan, Shupeng Zhu, Michael MacKinnon, Guannan Geng, Qiang Zhang, Heran Zheng, et al. "Economic Footprint of California Wildfires in 2018." *Nature Sustainability* 4, no. 3 (2021): 252–60.

Watts, Nick, Markus Amann, Nigel Arnell, Sonja Ayeb-Karlsson, Kristine Belesova, Maxwell Boykoff, Peter Byass, et al. "The 2019 Report of the Lancet Countdown on Health and Climate Change: Ensuring that the Health of a Child Born Today Is Not Defined by a Changing Climate." *The Lancet* 394, no. 10211 (2019): 1836–78.

Weber, Elke U. "Experience-Based and Description-Based Perceptions of Long-Term Risk: Why Global Warming Does Not Scare Us (Yet)." *Climatic Change* 77, no. 1 (2006): 103–20.

Weintrobe, Sally. *Psychological Roots of the Climate Crisis: Neoliberal Exceptionalism and the Culture of Uncare.* Bloomsbury Publishing USA, 2021.

Weisman, Kara, and Ellen M Markman. "Theory-Based Explanation as Intervention." *Psychonomic Bulletin & Review* 24, no. 5 (2017): 1555–62.

Wentz, Frank J, Lucrezia Ricciardulli, Kyle Hilburn, and Carl Mears. "How Much More Rain Will Global Warming Bring?" *Science* 317, no. 5835 (2007): 233–5.

Wiedmann, Thomas, Manfred Lenzen, Lorenz T Keyßer, and Julia K Steinberger. "Scientists' Warning on Affluence." *Nature Communications* 11, no. 1 (2020): 1–10.

Wolske, Kimberly S, Kenneth T Gillingham, and P Wesley Schultz. "Peer Influence on Household Energy Behaviours." *Nature Energy* 5, no. 3 (2020): 202–12.

Wuebbles, Donald J, David W Fahey, and Kathy A Hibbard. *Climate Science Special Report: Fourth National Climate Assessment*, vol. 1. US Global Change Research Program, 2017.

Xifra, Jordi. "Climate Change Deniers and Advocacy: A Situational Theory of Publics Approach." *American Behavioral Scientist* 60, no. 3 (2016): 276–87.

Xu, Chi, Timothy A Kohler, Timothy M Lenton, Jens-Christian Svenning, and Marten Scheffer. "Future of the Human Climate Niche." *Proceedings of the National Academy of Sciences* 117, no. 21 (2020): 11350–5.

Xue, Wen, Donald W Hine, Natasha M Loi, Einar B Thorsteinsson, and Wendy J Phillips. "Cultural Worldviews and Environmental Risk Perceptions: A Meta-analysis." *Journal of Environmental Psychology* 40 (2014): 249–58.

Yamano, Norihiko, and Joaquim Guilhoto. "CO2 Emissions Embodied in International Trade and Domestic Final Demand: Methodology and Results Using the OECD Inter-country Input–Output Database." OECD Science, Technology and Industry Working Papers, No. 2020/11, OECD Publishing, Paris (2020).

Yohannes, Okbazghi. *The Biofuels Deception: Going Hungry on the Green Carbon Diet.* NYU Press, 2018.

York, Richard, and Shannon Elizabeth Bell. "Energy Transitions or Additions? Why a Transition from Fossil Fuels Requires More than the Growth of Renewable Energy." *Energy Research & Social Science* 51 (2019): 40–3.

Yu, Hao, Bing Wang, Yue-Jun Zhang, Shouyang Wang, and Yi-Ming Wei. "Public Perception of Climate Change in China: Results from the Questionnaire Survey." *Natural Hazards* 69, no. 1 (2013): 459–72.

Zhou, Chen, Mark D Zelinka, Andrew E Dessler, and Minghuai Wang. "Greater Committed Warming after Accounting for the Pattern Effect." *Nature Climate Change* 11, no. 2 (2021): 132–6.

Index

Note: Page numbers in **bold** indicate the entries that also appeared in the Glossary.

activism, 167–8, 264, 281, 288
  climate, 167, 194, 272
activist, 2, 254, 260, 262, 270, 275–6, 284, 293
  climate, 21, 117, 176–7, 179, 185, 200, 258, 264, 274
  environmental, 167, 175
  environmental justice, 28, 190, 217
  killing of, 198
advertising, 99, 101, 249–50
affluence, 103, 194
affluent. *See* affluence
affordable housing, 247, 252
Africa, 25, 95, 197, 281
  East Africa, 211
  offsets in, 213
  South, 67, 256
  Southern, 2, 77
agriculture: industrial, 7, 60, 71, 203
  regenerative, 223, 242–3, 284
  working outdoors, 60
air pollution, 70–1, 137, 157, 181, 199–200, 205, 209, 211, 223–4, 228, 235, 259
alarmist, 55, 120, 178
Alinsky, Saul, 254, 271
Amazon Jungle, 28, 73, 80, 201
American Legislative Exchange Council, 122
American Petroleum Institute, 14–15
Amnesty International, 198
amplifier of climate change, **35**
Antarctica, 13, 36, 47, 61, 69, 72, 78
Apple Corporation, 24, 198
Arctic, 28, 47, 50, 61, 68–9, 75
  permafrost, 55
Arrhenius, Svante, 12
artificial scarcity, 101, 248
Asia, 25, 50, 77, 95
astroturfing, **122**
Atlantic Meriodonal Overturning Current, 80
attitude roots, 129, **129**, **131–3**, **135**
Australia, 27–8, 49, 77, 180, 182–3, 192, 209, 260, 289, 292
  agriculture, 71
  carbon tax, 215
  emissions, 191
  hottest day, 67
  mining, 199
  politics, 180
  prime minister of, 50
  skepticism, 111, 113, 118–19, 123
  wildfire, 67
aviation: electrification, 223
  emissions, 20, 194, 212, 272
  flying less movement, 283
  offsets, 212
  social norms, 132

banks, 87
  bail outs, 103
  divestment, 268, 283, 285, 293
  financing of fossil extraction, 2, 269
  targeted by activists, 272
banning: chloroflourocarbons, 16
  fossil extraction, 102, 245, 254
basic income, 250
batteries, 199, 219, 226, 228, 233–4, 236–7, 287, 291
BECCS, **52**, **77**, **202**, **206–9**, **287**
beef, 73, 250

Berry, Wendell, 179
Biden Administration, 28, 242–3, 273
Big Green, 281, 288
bioenergy, **22**, **30**, **200**, **202–3**
bioenergy with carbon capture and sequestration. *See* BECCS
biofuel, **189**, **202**
biomass, **10**, **202**, **206**, **209**
biospheric values, **161**, **164**, **167**, **171**, **173**, **262–4**
Bloomberg, 50, 128, 231
Bolivia, 25, 98, 198
bomb cyclones, 62
Brazil, 77, 118, 213
Britain, 10, 17, 106
British Petroleum, 120, 208, 238
building electrification, 102
Bush, George H.W, 16, 124, 218
Bush, George W., 20, 28, 122, 290

California, 69, 230, 236–7, 255, 274
  air resources board, 71
  Cap and Trade, 23, 211
  climate jobs plan, 240, 243, 246–8
  Diane Feinstein, 273
  drought, 63
  emissions, 211
  energy use, 137
  farmworkers, 28
  Fukushima pollution, 205
  GDP, 65
  geothermal, 199
  La Jolla, 14
  nuclear, 204
  pensions, 268
  renewable electricity, 238, 240
  sea level rise, 157
  solar power, 204
  wildfires, 28, 50, 65, 67, 71
Canada, 17, 24, 27, 41, 191, 201, 226, 260, 278, 292
  dams, 201
  increasing fossil extraction, 32
  NAFTA opposition, 96
  permafrost, 78
  skepticism, 118
  tar sands, 290
cancer, 15, 65, 122, 208, 228, 284
Cap-And-Adapt, 245
cap-and-trade, **20**, **23**, **210–11**, 211
capitalism, 10, 51, 92, 98–9, 101, 108, 161, 248, 286, 292
  curbing, 131
  distinctive ideology, 84, 99
  expansionary compulsion, 84, 88
  features of, 84, 88
  neoclassical model of, 82, 214
  role in emissions, 3, 84
  social democracy, 90
  state capitalism, 98
carbon budget, **8**, **57**, **268**
carbon capture, 123, 128, 200, 209, 238, 287
  and storage, 30, 209
carbon colonialism, **195**
carbon cycle, **40**, **52**
carbon leakage, **195**, **213**
carbon neutrality, **207–8**, **287**
carbon offsets, **20**, **23**, **208**, **210–11**, **287**
  additionality, 23, 212–13, 287
  permanence, 213
  uncertainty, 212
carbon sink, **60**, **80**, **142**, **201–2**
carbon tax, 50, 136, 214–15, 245, 278
  carbon fee and dividend, **215–16**
carbon trading: clean development mechanism, **20**, **212**
carbon-trading, 20
Carson, Rachel, 12, 167
Carter, Jimmy, 13, 219
Cato Institute, 120
cement, 42
Charney, Jule, 13–14, 31, 46
Chernobyl, 203, 205
Chevron, 120, 122, 197, 209
Chile, 93, 103
China, 18, 25, 77, 95, 191, 199, 263, 285
  Conference of Parties, 25
  dams, 225
  emissions, 29, 98, 183, 194
  manufacturing in, 20, 95, 192
  nuclear, 205
  pollution, 95
  rainfall, 68
  refugees from, 50
  skepticism, 114
chloroflourocarbons, 15, 43
Chomsky, Noam, 102
Citizens United, 27, 120, 260
civil rights movement. *See* social movements
Clean Air Act, 13, 218, 262
clean electricity standards, 119

climate change communication, 135, 177
climate fatalism, 177
climate finance, 190, 192
climate forcing, **38**, **45**, **52**, **58**
climate justice, **28**, **51**, **193**, **213**, **217**, **219**, **259**, **267**, **269**
advocates, 207, 210, 213
false solutions, 22, 30–1, 200, 210, 253
frame, 259
movement, 22, 30, 189, 196, 218
organizations, 190
responsibility of nations, 192
Climate Leadership Council, 136
climate opinions, 168
Climate Psychology Alliance, 176, 179
Climate Reality Project, 120
climate risk perceptions, 3, 161, 163, 167–8, 171, 174, 176, 179, 184
climate sensitivity, **13**, **46**
Climategate, 131
Clinton, Bill, 20, 97
CNN, 128, 268
coal, 10–11, 29, 45, 50, 52, 55, 57, 94–5, 102, 120, 154, 180, 208, 212, 226–7, 245, 268
Australia, 180
Britain, 10
burning of, 9, 31
clean, 22
conversion from, 209
demise, 259
Germany, 205, 290
increased demand, 55
methane capture, 212
mining, 269
mountaintop removal, 243
new, 7, 22, 239
phase out, 119, 230
states, 247
stranded asset, 281
co-benefits, **137**, 179, **292**
air pollution, 292
energy security, 292
national security, 137
cognitive biases, **128**, **133**, **135**
Cold War, 12, 98
collective action, 2–3, 84, 183, 221, 253–4, 256, 262–3, 284, 288, 292
psychology of, 3
collective identity, **258**, **262**
colonialism, 10, 162, 196–7, 256, 258
committed warming, **58**
common but differentiated responsibilities, **18**, **20**, **30**, **191**
Community Choice Aggregation, 236
concrete. *See* cement
Conference of the Parties, 189, 260
confirmation bias, **133**
Congress, 15, 17, 102, 119, 214, 242–3, 245, 267, 273
consensus gap, **144**, **146**
consensus of scientists, 1–2, 8, 139
conspiracy-induced misinformation, 131, 146
denial of, 144
feasibility of shift to renewables, 224
global heating, 9, 49, 122
greenhouse effect, 15
human cause, 127, 146
IPCC, 17, 27, 57, 125
messaging about, 146–7
on climate change, 117, 140, 145
conservative, 13, 26, 31, 57–8, 111, 113, 117, 123, 136, 142, 144, 180, 260, 275, 278
conspiracy thinking, **129**, **133**
consumerism, 101
consumption, 73, 84, 88, 100–1, 103–4, 108, 128, 212, 245, 249
growth, 15
limits on, 101
material intensive, 53
reduction, 53, 103–4, 106, 138, 198, 218, 221, 233, 240, 248, 250, 252, 285–6, 291
regulation of, 251
contrarian, 122–3, 146
Copenhagen Accord, 25
Copenhagen summit, 22, 30, 61, 190, 268
coral reefs,, 72–3
COVID-19 pandemic, 9, 32, 103, 142, 200, 252, 272, 274
Cox, Stan, 102–3, 221, 245, 248
crop yields, 49–50, 68, 71
culture of uncare, **99–101**
cyclones, 68

damages, 51, 58, 65, 68–9, 80, 82, 173, 190–1, 210, 220
dark money, 27, 120, 260
data visualization, 154, 292
debunk, **148–9**
decision-making, 165
behavioral study of, 133
behavioral study of, 149
behavioral study of, 161
by class, 209

decouple. *See* emissions:decoupling
deforestation, 31, 42, 73, 80, 201, 213, 250
degrowth. *See* post growth policies
Demand-Side, 251
Democratic party. *See* Democrats
Democrats, 17, 112, 133, 136, 261
demographic, 114, 117, 171
denial, 119
  denial machine, 120, 122–3
  of climate change, 15, 41, 113, 120, 149, 176
  of responsibilty to act, 160
Denmark, 25, 49, 228, 289
derisking, **268**
Di'Aping, Lumumba, 25, 61
direct air capture, **52**, **209**, **212**, **238**, **287**
discourses of climate delay, 284
Don't Look Up, 176
doomism, 174, 176, 286
drought, 49, 62–3, 67, 71, 168–9

earth scientists, 173
earth system models, **39**, **59**
Earth's Black Box, 289
eco-anxiety, **176**, **180**, **182**
ecological crisis, 3, 72–3, 87, 181, 220, 251, 275
economic growth, 12, 16–17, 53, 73, 90, 95, 102–4, 154, 199
economics, 49, 51, 80–3, 95, 100, 134
  economic model, 51, 130
  neoclassical tradition, 82
economists, 3, 14, 23, 80, 82–3, 93, 99, 106, 214, 237
efficacy beliefs, 177–8, 182, 262
  collective efficacy, **262**
  participatory, 263
  self-efficacy, 262
egalitarian, 130, 165–6
electric grid, 228, 234, 238
  transmission lines, 201, 228
electric heat pumps, 225–6
electric utilities, 93, 120, 123, 224, 228, 246, 260, 280
electric vehicles, 1, 138, 198, 219, 223, 225–6, 235, 240, 283, 285, 291
embedded liberalism, **90**, **92**, **100**
embodied emissions, **194–5**
  border carbon adjustments, 195
emissions, 2, 13
  cap, 20, 210
  carbon offsets, 212–13
  consumption-based, 191, 194
  cuts, 2–3, 8
  cuts via consumption reduction, 249, 288
  decoupling, 104, 248
  equity concerns, 215
  from animal agriculture, 250, 284
  from food waste, 249
  from tiny elite, 259, 272
  high pathway, 8, 157, 251
  historically benefits from, 220
  IPCC goal for cuts, 104, 224, 239
  major sources of, 221–3
  medium pathway, 8, 207
  net-zero, 246
  opposition to cuts, 17, 210
  pathways, 1
  per capita, 18
  proposals to cut, 1, 17–18, 210, 221–3, 239, 250, 253
  rise, 1, 7, 23, 210
  short term increase with renewable transition, 237
  shortcomings of emergency declarations, 271
  social cost of, 213
  temporary dip, 9
  trading, 218
emissions scenarios, **52**
  representative concentration pathways, *See* RCPs; **shared socioeconomic pathways**, **52**
emotion, 119, 134, 156, 158, 171, 173–5, 177–9, 184–5
Engler, Paul and Mark, 254, 258
EN-ROADS simulator, 154
Environmental Defense Fund, 96
environmental justice, **69**, **189**, **217**, **254**, **259**, **269**, **293**
Environmental Protection Agency, 15, 95
Eocene, 47
epistemic skepticism, **112**, **114**, **117**, **129**, **132**, **140**, **159–60**, **184**, **292**
Europe, 22–3, 25, 31, 77, 80, 92, 97, 108, 182, 190, 195, 209, 218, 240
  electric grid, 229
  heatwave, 66
  meat sales, 249
  right wing think tanks, 120
  skepticism, 114
European Union, 26, 191
  Emissions Trading System, 210
exchange value, **88**
externalized costs, **95**
extinction, 51, 72, 81, 158, 173–4, 271

Extinction Rebellion, 194, 265, 267, 270, 272, 280
extractivism, **3**, **98**, **162**, **197–8**, **220**, **233**, **285–7**
extreme weather, 1, 3, 7, 12, 28, 41, 50, 61–3, 65, 70, 117, 126, 128, 133, 144, 168–9, 171, 173–4, 183
  extreme event attribution, **62–3**
Exxon, 13, 15, 206
ExxonMobil, 120, 122, 238

Facebook, 133, 149
Fairphone, 249
faith in institutions, **263**
false beliefs, 134
false equivalence, **125**
FDR. *See* Franklin Delano Roosevelt
federal government, 102, 179, 242
feedback loops, 46, 78, 101
fertilizers, 42, 84, 284
flooding, 3, 7, 63, 68–70, 74, 83, 157, 168–9, 174, 183, 201
food production, 70–1, 83, 209
fossil fuel divestment, 1, 259, 267–9, 272
  Stop the Money Pipeline, 268–9, 283
fossil fuel industry, 43, 119–20, 122, 125, 128, 138, 148, 216, 224, 236, 240, 243, 252, 268, 273, 278, 281, 283, 285, 289, 291
fossil fuel subsidies, 119, 196, 211, 215, 245, 291
Fox News, 120, 272
France, 70, 115, 192, 203, 229
  Yellow Vest protests, 215
Franta, Benjamin, 82, 124
free market, 92, 99
  beliefs, 117, 130, 216
  economics, 15, 31
Friedman, Milton, 99
Friends of the Earth, 28, 31, 97, 189
front groups, **120**, **122–3**
frontline communities, 198, 243
Fukushima, 134, 203, 205

Gateway Belief Model, **145**
GDP, 50–1, 65, 81–2, 103–4, 106, 215, 235
geoengineering, 30, 200, 205, 209, 238, 287
geothermal, 225, 228
Germany, 23, 28, 32, 115, 192, 205, 210, 229, 260, 290
glacier, 61, 69, 71, 171
Glasgow, 190, 260, 280, 289
Global North, 30–1, 57, 98, 105, 189, 196, 253, 265, 271, 285
Global South, **22**, **61**, **98**, **177**, **190**, **196**, **198**, **202–3**, **214**, **237**, **249**
Global Warming Petition Project, 147–8
global warming potential, **43**
globalization, 53, 95, 97, 258
Goldilocks zone, **33**, **35**
Google, 24, 273
Gore, Al, 20, 23, 96, 175
Great Acceleration, **84**
Great Depression, 240, 255
green hydrogen. *See* Renewable Energy
Green New Deal, 102, 134, 221, 224, 240, 243, 246, 267, 273–4
greenhouse effect, 3, 12, 15–18, 31, 34–5, 60–1, 127, 141, 144, 158
  accelerated, 7
Greenland, 69, 72, 78
Greenpeace, 97
greenwashing, **148**, **207–8**, **238**, **280**, **285**
Gross Domestic Product. *See* GDP
growth imperative, **88**, **98**, **103**, **128**, **202**
Guardian, 125–6, 202

Hansen, James, 7, 16, 18, 27, 33, 53, 58, 126, 203
Harvard University, 16, 268
health effects, 12, 70
Heartland Institute, 120
heatwaves, 3, 28, 62, 65, 67, 70–1, 74, 78, 83, 168
Heritage Foundation, 120
HEXACO, 163
hierarchical, 117, 130, 177
Himalayas, 60, 72
Hiroshima, 7, 203
homo economicus, **100**
homo socialis, **101**
hopelessness, 173, 184, 293
Huber, Matt, 247
Hulme, David, 56, 165
human nature, 100
human rights, 198, 233, 258–9, 270
hurricanes, 3, 7, 68, 83, 175
  Hurricane Harvey, 68–9
hydroelectric. *See* hydropower
hydroflourocarbons. *See* chloroflourocarbons
hydrogen: made from methane, 123
hydropower, 72, 200, 222, 226, 228
hyperbolic discounting, **133**
hyper-consumption, 101, 292

ice cores, 36–7, 46–7
ice-albedo effect, **78**

imagery, 178, 273
India, 28–9, 63, 74, 96, 174, 191, 194, 225, 250, 256
  drought, 67
Indigenous, 11, 22, 29, 97, 217, 259, 261
  communities, 126, 190, 199, 201, 220, 290, 293
  peoples, 161, 170, 189, 247, 260
  societies, 163
individualistic, 117, 130, 177
industrial policy, 102, 251
Industrial Revolution, 10, 30, 36, 38, 84, 208
inequality, 53, 95, 103, 215, 272
infrared, 34–5, 141
in-group messaging, **136**, **292**
injustice, 29, 177, 196, 209, 286
insurance, 51, 76, 82, 153, 157, 174, 269
interglacial, 46
International Energy Agency, 55, 230, 239
International Monetary Fund, 211, 215
IPCC, 2, 8, 74, 131, 152–3, 206, 275
  2018 report, 57, 75, 104, 125, 204
  goals, 221, 240
  lead author, 206
  mitigation report 2022, 19, 106, 224, 230, 251, 291
  physical science report 2021, 9, 19, 25, 28, 37, 55, 68, 70, 77, 144
  representative carbon pathways, 52
  shared socioeconomic pathways, 52–4
  social cost carbon, 214
  Special Report on Global Warming of 1.5C, 27
IPCC's: physical science report 2021, 80
irradiance, 39, 149
isotope, 37, 47
Italy, 69–70, 115
  climate education, 147

Jacobson, Mark, 224, 229–30, 235
Japan, 17, 25, 68, 134, 191–2
Jevon's Effects. *See* rebound effects
Jevons, William Stanley, 100
jiu jitsu persuasion, **135**
justice: intergenerational, 51, 205

Keeling Curve, 175
Keeling, Charles, 12
Keynesianism, 90
King, Martin Luther, 270
Klein, Naomi, 84, 88, 100
knowledge deficit, 158, 292
Kyoto Protocol, 20, 23, 27, 212, 218

Latin America, 25
liberal, 112, 117, 144, 261
life cycle assessment, 235
Lomborg, Bjørn, 49, 51
Los Angeles, 71, 125, 245, 254

Malm, Andreas, 7, 177, 281
Marshall, George, 133–4, 169
Mauna Loa, 12, 36
Mazzocchi, Tony, 243
McAdam, Doug, 258
McCain, John, 27
McKibben, Bill, 125–6, 268
McKinsey, 49, 74, 82
meat, 249–51
  reducing, 283
mechanistic explanation, **141**
meta-analysis, **104**, **114**, **117**, **166**, **180**, **183**, **262–3**
metals, 194, 197, 232, 237, 249
methane, 44
  biomethane from landfill, 202
  branded as natural gas, 148
  capture from mines, 212
  emitted from hydropower, 201
  in atmosphere, 42
  leaks, 42, 45
  livestock, 42
  permafrost thaw, 55, 78
  release into atmosphere, 7
Midwest, 68, 169
migration, 76, 173–4
Milankovitch cycles, **46**
mining, 82, 96, 181, 191, 194, 196–9, 201, 203, 223, 234–7, 243, 291
  cobalt, 198, 233
  lithium, 198, 226, 237
  substitute materials, 233, 285, 287
misinformation, 3, 119–20, 122, 125, 128, 134, 138, 140, 144, 146–7, 149, 158, 160–1
moral outrage, 167, 275, 279
Morgan Stanley, 50
mosquitoes, 70
motivated cognition, **128**, **132–3**, **158**, **160–1**, **292**
Mozambique, 28–9, 69, 193

NAFTA, 96–7, 196
NASA, 7, 16, 18, 33, 41, 55
National Academy of Sciences, 13, 150
National Oceanic and Atmospheric Administration, 41, 174
Nationally Determined Contributions, 192

natural gas, 7, 34, 43, 208, 223, 230, 240
Natural Resources Defense Council, 96, 205
nature: early experiences with, 164, 167, 262
  extraction from, 88
  harmony with, 98, 167
  human dominion over, 130
  human relationship with, 161, 163, 168
negative emissions, **52**, **77**, **206**, **208**
negative forcing, 38
neocolonialism, 197
neoliberalism, **90**, **92–3**, **99–101**, **103**, **108**
net zero, **31**, **207–8**, **280**
New Deal, 240, 242, 247, 273
New Ecological Paradigm, 117
New York Times, 14, 16, 125, 127–8, 174, 192
news media, 119, 123, 125, 127, 136, 270, 272
Nierenberg, William, 14–15
Nixon, Richard, 13, 262
Nordhaus, William, 82
Nordstream 2, 28, 290
North American Free Trade Act. *See* NAFTA
Norway, 17, 115, 228, 250, 261
nuclear, 128, 204, 238
  accidents, 134, 203, 228
  anti movement, 262, 270
  disarmament, 243
  energy, 131, 203–4, 227, 229
  new, 22, 200, 203–4, 287
  phasing out, 205
  plants, 14, 30, 203, 205, 226, 242
  proliferation, 228
  safety issues, 205

Obama, Barack, 22, 25, 30, 96, 196, 243
Ocasio-Cortez, Alexandria, 243
oceans, 12, 39, 46, 52, 58, 60, 68–9, 72, 83, 237
  temperatures, 69, 289
oil industry, 14, 50, 84, 236
optimistic, 14, 177–8, 224
Oreskes, Naomi, 14, 122
organizer, 2, 243, 274–5
over-population. *See* population growth
ozone hole, 15–16, 206

Paris Accord, 27, 31, 58, 60, 120, 192, 195–6, 269
Patagonia, 249
patriotic, 122, 135
peat, 68
permafrost thaw, 55, 78, 80
personality, **161**, **163**, **262–4**, **275**
  openness, 163
pessimistic, 14, 177–8, 183
petroleum companies, 119
Piven, Fances Fox, 255
planetary boundaries, 92, 106
planned obsolescence, **101**, **291**, *See* artificial scarcity
Pliocene, 49
Polar Vortex, 41, 62
political affiliation, 112, 114, 117, 119
political economy, **83–4**, **100**, **133**
political ideology, 98, 117
Pollin, Robert, 102, 104, 230
population growth, 52–3, 71, 84
positive feedback, 46, 78
positive forcing, 38
post growth policies, **104–6**
  Avoid-Shift-Improve, 251
  circular economy, 250
  ownership to usership, 250
  right to repair, 249
  warranties, 249
Prakash, Varshini, 268, 273
prebunking, **147–8**
privatization, 92–3
psychological barriers, 3, 87, 101, 108
psychologists, 141, 169, 176–7, 179–82
psychotherapy, 99, 135, 176, 179
public transportation, 102, 106, 163, 198, 233, 248
Puerto Rico, 69

r value, **114**, **163**
radical, 92, 177, 179, 243, 245, 247, 251, 278
radical hope, 179
rainforest, 73, 80
Rand, Ayn, 99
rationing, 102, 246
Raworth, Kate, 100, 106
RCP4.5, 52–3, 70
RCP8.5, 52–3, 55, 60, 70, 73, 157
RCPs, 52–3
Reagan, Ronald, 14–16, 92, 99–100
realists, 137
rebound effects, **104**, **215**, **248**, **291**
recycling, 198–9, 233, 237, 250, 285
REDD, 212
Reducing Emissions from Deforestation and Forest Degradation. *See* REDD
remunicipalisation, **102**

renewable energy: green hydrogen, 223–4
intermittency, 228, 291
just transition, 243, 246–7, 252, 259, 274, 284, 292–3
land area, 50, 69, 231
sources of, 10, 23, 104, 200, 221, 224, 239–40, 242, 245–6, 252
supply of, 196
transition to, 3, 93, 120, 196, 198–9, 232, 234, 240, 246–7, 253, 268, 283, 285
Republican party, 118, 123
Republicans, 16–17, 27, 111, 113, 117, 132–3, 136, 171, 261
response skepticism, **112**, **132**, **160**, **177**, **179**, **184**, **292**
Ricardo, David, 94
right wing, 51, 99, 120, 122
Rio Earth Summit, 18–19, 30, 96, 189, 191, 196
risk perception. *See* climate risk perceptions
Roberts, David, 126, 169–70
Robinson, Kim Stanley, 174–5
Roosevelt, Franklin Delano, 240
Russia, 16, 28, 50, 78, 98, 115, 191, 290
heatwave, 71
wildfire, 67

safe space for humanity, 106
same-sex marriage, 135, 256, 262, 277
San Diego, 15, 141, 257, 274, 276
Sanders, Bernie, 242–3
Saudi Arabia, 191, 193
science communication, 112, 140, 146, 158, 173
scientific graphics, 153
Scripps Institution of Oceanography, 12, 14–15
sea level rise, 1, 13, 28, 47–8, 61, 69, 78, 152, 156, 171, 174
Shell, 120, 197, 272
shifting baselines syndrome, **169**
shipping, 20, 223, 249
Siberia, 68
Sierra Club, 97, 281
Sixth Mass Extinction, 72
Smith, Adam, 100
social identity needs, **129**, **132**, **135**
social justice, 167, 197, 217, 259, 274, 279
social media, 120, 123, 127, 133–4, 147, 256, 273, 276
social mobilization. *See* social movements
social movements: Black Lives Matter, 258, 277
civil disobedience, 269, 288
civil rights movement, 271, 273, 281
community-based organizing, **254–5**, **274**
frames, **258**, **269**, **271**, **274**
grassroots, 3, 98, 101, 108, 122, 182, 189, 219, 252–4, 256, 261–4, 268, 270, 274–7, 280–1, 288
mass mobilization, **242**, **254–5**, **271**
momentum-based organizing, **254–5**
radical flank, 280
transactional struggle, **255**, **267**, **269**, **274**
transformational struggle, **256**, **267–9**
social norms, **135**, **137**, **171**, **263**, **276**, **283**
social psychology, 128, 181, 262
socialism: socialist parties, 92
solar, 222
arrays, 197, 204, 225
concentrated solar with storage, 226
plants, 197, 225, 230
rooftop, 138, 197, 227, 231–2, 248, 284
solar radiation, 33, 78, 205, 225
solastalgia, **198**
Southwestern US, 49, 67
Soviet Union, 17, 20
Spain, 70, 102, 104, 106, 115, 226, 229, 245, 250
stagflation, 90
Stanford University, 224, 235, 238
steady-state economy, **88**
Stern, Nicholas, 80–1
stock-and-flow problem, **142**, **158**
storms, 62, 68, 70, 152, 168
stranded assets, **58**, **281**
sunk-cost fallacy, **134**
Sunrise Movement, 265, 267–8, 272
sunspot cycle, 38
Sununu, John, 17
Supreme Court, 27, 256
surface attitude, **129**
surplus value, 88
Swiss Re, 51, 82, 230
Syria, 67
system justification, 130, 135

Tea Party, 27
technofixes, 196, 205, 209, 253
Tesla, 226, 237, 240
Thatcher, Margaret, 17, 92, 99–100
The Inconvenient Truth, 175

theory of change, **263**, **269**, **271**
theory-based explanations, 144
thermal inertia, 13, 58
tipping point, **69**, **78**, **80**
Tokar, Brian, 22, 189, 216
trade, 14, 20, 23, 82, 87–8, 92–3, 95–6, 101, 108, 123, 196, 210, 250
  absolute advantage, **93**
  agreements, 93, 95–6
  comparative advantage, **93**
transmission lines. *See* electric grid
Trans-Pacific Partnership, 96, 196
Trump, Donald, 27–8, 117
Twitter, 133, 149, 168–9
Tyndall, John, 12

UK, 25, 41, 92, 99–100, 102, 104, 131, 171, 176, 178, 192, 194, 209, 267, 270, 272
  2020 election,149
  climate emergency announcement, 119
  research study, 132
Ukraine: war in, 292
Ukraine, war in, 290
uncertainty, 152
  bounded uncertainty, **152**
  future, 152
  irreducible uncertainty, **152**
  overstating, 122
  science communication, 57, 140, 149, 158
  scientific, 60, 149
  sown by denial machine, 123
unemployment, 90, 99, 252
UNFCCC, 18, 27, 96, 194
unions, 95, 247, 255–6, 279–80, 292
United Nations, 16–17, 56, 149
University of California, 1, 15, 24, 208, 256, 268
use-value, **88**

vested interests, **129–30**, **133**, **135**, **158**, **180**
visual arts, 157, 284
volcanic, 38–9, 47
von Hayek, Friederich, 99

Wallace-Wells, David, 126, 160, 174, 209
War Production Board, 102, 246
water vapor, 12, 35, 46, 67–8
weather on steroids, 65
wet-bulb temperature, **74**
wildfires, 16, 28, 67, 70, 83, 168, 171
wildlife, 72, 167
wind turbines, 197, 204, 225, 231–2
working class, 105, 247, 260
World Bank, 214
World Metereological Organization, 17
World Trade Organization. *See* WTO
World War II. *See* WWII
World Wildlife Fund, 97
worldviews, 3, 112, 117, 128, 130, 140, 158, 161, 169, 173, 177, 292
Worshipping at the Church of Technology, **209**
WTO, 93, 96–7, 196, 258
WWII, 12, 14, 84, 90, 92, 102, 245, 251

Yale Climate Change Communication Group, 111–12

zoonotic, 71

Printed in the United States
by Baker & Taylor Publisher Services